LC-MS/MS in Proteomics

Methods and Applications

Edited by

Pedro R. Cutillas

Centre for Cell Signalling, Institute of Cancer, Bart's and the London School of Medicine, Queen Mary, University of London, UK

John F. Timms

Cancer Proteomics Group, EGA Institute for Women's Health, University College London, London, UK

Humana Press

Editors
Pedro R. Cutillas
Institute of Cancer
Centre for Cell Signalling
Bart's and the London School of Medicine
Queen Mary, Universtiy of London
Charterhouse Square
EC1M 6BQ London
3rd Floor John Vane Science Centre
London, UK
p.cutillas@qmul.ac.uk

John F. Timms
Cancer Proteomics Group
EGA Institute for Women's Health
University College London
Gower Street
WC1E 6BT London
Cruciform Building 1.1.09
London, UK
j.timms@wibr.ucl.ac.uk

ISSN 1064-3745 e-ISSN 1940-6029
ISBN 978-1-4939-5744-6 ISBN 978-1-60761-780-8 (eBook)
DOI 10.1007/978-1-60761-780-8
Springer New York Dordrecht Heidelberg London

Printed on acid-free paper

Humana Press is part of Springer Science+Business Media (www.springer.com)

METHODS IN MOLECULAR BIOLOGY™

Series Editor
John M. Walker
School of Life Sciences
University of Hertfordshire
Hatfield, Hertfordshire, AL10 9AB, UK

For other titles published in this series, go to
www.springer.com/series/7651

METHODS IN MOLECULAR BIOLOGY

Series Editor
John M. Walker
School of Life Sciences
University of Hertfordshire
Hatfield, Hertfordshire, AL10 9AB, UK

For further volumes:
http://www.springer.com/series/7651

Preface

The aim of this book is to be a useful resource for experienced proteomics practitioners as well as an aid to newcomers in order to become acquainted with the theory and practice of a wide array of mass spectrometric techniques for proteome research. The role of mass spectrometry (MS) in proteomics has gradually evolved from being a technique used to identify proteins in gels to be the preferred method for protein detection and quantification. Seminal work at the beginning of this century demonstrated that the hyphenation of liquid chromatography with tandem mass spectrometry (LC-MS/MS) allowed the protein biochemist to quantify thousands of proteins within time frames that make these techniques powerful readouts for biological experiments. Consequently, LC-MS/MS has found a niche in many different areas of biological and biomedical research, mainly due to the introduction of user-friendly and high-performance instrumentation and because of the development of new quantitative strategies and of powerful bioinformatics tools to cope with the analysis of the large amounts of data generated in proteomics experiments. These advances are making possible the analysis of proteins on a global scale, meaning that proteomics can now compete with cDNA microarrays for the analysis of whole genomes.

In this volume of *Methods in Molecular Biology*, we provide protocols and up-to-date reviews of the applications of LC-MS/MS, with a particular focus on MS-based methods of protein and peptide quantification and the analysis of post-translational modifications. **Section I** presents overviews of the use of LC-M/MS in protein analysis. Quantifying protein expression changes, sites of modification, enzymatic activities and metabolites can be used in combination as a measure of pathway activity and implies that it may be possible to quantify pathway fluxes at a depth of analysis and scale which has previously not been possible. **Chapter 1** discusses such systems biology approaches, which will undoubtedly provide a better understanding of normal biological processes and insight into the molecular aetiology of disease.

Choosing the most appropriate methods should be based on the question that needs to be addressed and the nature of the samples. The pros and cons of the different quantitative techniques, reviewed in great depth in **Chapter 2**, will assist in the most appropriate use of quantitative LC-MS/MS for biomedical research. **Section I** of this volume also reviews the instrumentation available for LC-MS/MS (**Chapter 3**) and the various bioinformatics tools (**Chapter 4**) used for the analysis of different and often very complex data sets.

LC-MS/MS can also be used to detect and quantify post-translational modifications, the proteomes of cellular organelles, protein–protein interactions and to profile protein abundance in biological fluids – applications that are outside the scope of genomics (these applications are reviewed in **Chapters 1**, **2**, and **5**). However, in contrast with genomics, for which the technology is well developed, off-the-shelf protocols for MS-based proteomics do not yet exist. This volume meets this demand and describes step-by-step protocols for the main applications of LC-MS/MS in protein analysis and some more novel applications.

Section II details protocols for the analysis of post-translational modifications, with particular focus on phosphorylation (**Chapters 6** and 7) and glycosylation (**Chapter 8**). The most popular techniques for quantitative proteomics, including those based on multiple reaction monitoring (**Chapters 9** and **10**), metabolic labelling (**Chapter 11**), chemical tagging (**Chapter 12**) and label-free (**Chapter 13**) approaches are covered in **Section III**. **Section IV** then describes how these quantitative proteomic techniques can be used to investigate cell biochemistry by comparative assessment of membrane proteomes between two or more cell populations (**Chapter 14**), characterizing the proteomes of organelles (**Chapter 15**) and by more accurately and specifically mapping protein–protein interactions (**Chapter 16**). Another popular application of LC-MS/MS is for biomarker discovery in biological fluids and **Section V** gives protocols for such workflows; **Chapter 17** focuses on the analysis of serum proteins, whilst **Chapters 18** and **19** deal with the analysis of proteins and peptides in urine, respectively. Finally, **Section VI** describes relatively novel applications of LC-MS/MS in proteomics; **Chapter 20** shows how LC-MS/MS is not only useful to quantify protein expression, identify (and quantify) their post-translational modifications and map their interactions but also to quantify their enzymatic activity. Although this book focuses on protocols that analyse proteins at the peptide level (the so-called bottom-up approach), the final chapter of the volume, **Chapter 21**, describes a protocol for the analysis of full-length proteins (i.e. for top-down proteomics). This bias is not intentioned, but simply reflects the current relative popularity of the two approaches.

Collectively, these protocols and review chapters illustrate the formidable power and versatility of LC-MS/MS in biological and biomedical research. Although the large number of techniques and applications available means that no single volume can be exhaustive, virtually all the main analytical concepts and applications are discussed throughout the chapters. Indeed, the techniques and concepts learned throughout this volume should allow proteomic practitioners to apply LC-MS/MS to tackle essentially any biological problem, the only limitation perhaps just being our own imagination and creativity.

Pedro R. Cutillas
John F. Timms

Contents

Contributors

MARIA P. ALCOLEA • *Analytical Signalling Group, Centre for Cell Signalling, Institute of Cancer, Bart's and the London School of Medicine, Queen Mary University of London, London, UK*

JENS S. ANDERSEN • *Center for Experimental BioInformatics, Department of Biochemistry and Molecular Biology, University of Southern Denmark, Odense, Denmark*

SABINE BAHN • *Institute of Biotechnology, University of Cambridge, Cambridge, UK*

BLAGOY BLAGOEV • *Center for Experimental BioInformatics, Department of Biochemistry and Molecular Biology, University of Southern Denmark, Odense, Denmark*

BERND BODENMILLER • *Institute for Molecular Systems Biology, Swiss Federal Institute of Technology Zurich, Zurich, Switzerland*

ROBERT CHALKLEY • *Department of Pharmaceutical Chemistry, University of California San Francisco, San Francisco, CA*

EITHNE COSTELLO • *Division of Surgery and Oncology, Royal Liverpool University Hospital, University of Liverpool, Liverpool, UK*

PEDRO R. CUTILLAS • *Analytical Signalling Group, Centre for Cell Signalling, Institute of Cancer, Bart's and the London School of Medicine, Queen Mary University of London, London, UK*

JOERN DENGJEL • *Center for Experimental BioInformatics, Department of Biochemistry and Molecular Biology, University of Southern Denmark, Odense, Denmark; Freiburg Institute for Advanced Studies and Zentrum für Biosystemanalyse, University of Freiburg, Freiburg, Germany*

BRUNO DOMON • *IMSB – ETH Zurich, Zurich, Switzerland*

CLAIRE E. EYERS • *Michael Barber Centre for Mass Spectrometry, School of Chemistry, The University of Manchester, Manchester, UK*

BERTRAN GERRITS • *Functional Genomics Center Zurich, University of Zurich and Swiss Federal Institute of Technology, Zurich, Switzerland*

SARAH R. HART • *Michael Barber Centre for Mass Spectrometry, University of Manchester, Manchester, UK*

RICHARD J. JACOB • *Matrix Science Inc., Boston, MA*

LIS JAKOBSEN • *Center for Experimental BioInformatics, Department of Biochemistry and Molecular Biology, University of Southern Denmark, Odense, Denmark*

ANDREW JAMES • *Samuel Lunenfeld Research Institute, Mount Sinai Hospital, Toronto, ON, Canada*

HANNAH JOHNSON • *Michael Barber Centre for Mass Spectrometry, School of Chemistry, The University of Manchester, Manchester, UK*

CLAUS JORGENSEN • *Samuel Lunenfeld Research Institute, Mount Sinai Hospital, Toronto, ON, Canada*

REIKO KIYONAMI • *ThermoFisher Scientific, San Jose, CA*

IRINA KRATCHMAROVA • *Center for Experimental BioInformatics, Department of Biochemistry and Molecular Biology, University of Southern Denmark, Odense, Denmark*

YISHAI LEVIN • *Institute of Biotechnology, University of Cambridge, Cambridge, UK*

JOHN P. NEOPTOLEMOS • *Division of Surgery and Oncology, Royal Liverpool University Hospital, University of Liverpool, Liverpool, UK*

KRISTOFFER T.G. RIGBOLT • *Center for Experimental BioInformatics, Department of Biochemistry and Molecular Biology, University of Southern Denmark, Odense, Denmark*

ELAINE STEPHENS • *MRC Laboratory of Molecular Biology, Cambridge, UK*

ANDREW J. THOMPSON • *MRC Centre for Neurodegeneration Research, Institute of Psychiatry, King's College London, London, UK*

JOHN F. TIMMS • *Cancer Proteomics Laboratory, EGA Institute for Women's Health, University College London, London, UK*

SARAH TONACK • *Division of Surgery and Oncology, Royal Liverpool University Hospital, University of Liverpool, Liverpool, UK*

THEODORA TRYFONA • *Department of Biochemistry, University of Cambridge, Cambridge, UK*

RICHARD D. UNWIN • *Stem Cell and Leukaemia Proteomics Laboratory, University of Manchester, Manchester, UK*

MARK E. WEEKS • *Molecular Pathogenesis and Genetics Department, Veterinary Laboratories Agency (VLA-Weybridge), Surrey, UK*

RITCHIE WILLIAMSON • *Biomedical Research Institute, Ninewells Medical School, University of Dundee, Dundee, UK*

Part I

Overviews of Techniques and Approaches in LC-MS-Based Proteomics

Chapter 1

Approaches and Applications of Quantitative LC-MS for Proteomics and Activitomics

Pedro R. Cutillas and John F. Timms

Abstract

LC-MS is a powerful technique in biomolecular research. In addition to its uses as a tool for protein and peptide quantization, LC-MS can also be used to quantify the activity of signalling and metabolic pathways in a multiplex and comprehensive manner, i.e. as an 'activitomic' tool. Taking cancer research as an illustrative example of application, this review discusses the concepts of biochemical pathway analysis using LC-MS-based proteomic and activitomic techniques.

Key words: Mass spectrometry, proteomics, metabolomics, lipidomics, activitomics, quantification.

1. Introduction

LC-MS has become a powerful molecular biology tool and multiple strategies for peptide and protein quantitation by LC-MS have been developed and applied to address a wide range of biological questions. It is being used more and more to analyse differences between samples at the protein expression level, in post-translational modifications (PTMs), in the components of protein complexes and in intracellular protein localization, and this on a scale where thousands of proteins can now be compared in a single experiment.

In addition, techniques based on LC-MS are being used to quantify the activities of biochemical pathways at the systems biology level. For example, the phosphoproteome encodes information on the activity of protein kinase/phosphatase reaction pairs, which are in turn a reflection of the activity of key signalling

P.R. Cutillas, J.F. Timms (eds.), *LC-MS/MS in Proteomics*, Methods in Molecular Biology 658,
DOI 10.1007/978-1-60761-780-8_1,

pathways linking activation of surface receptors to biological outputs. Similarly, LC-MS is becoming the method of choice for high-content quantitative analysis of metabolites, whose amounts reflect metabolic pathway activity. Thus, quantitative methods for metabolomics, phosphoproteomics, lipidomics, glycomics, etc., all have in common that these techniques provide a reflection of enzyme (and hence pathway) activity (**Fig. 1.1**). A trend in bioanalytical science is therefore to apply these new techniques, which collectively could be termed '*activitomic*' (**Fig. 1.1**), to enhance our understanding of enzyme and pathway activity which, it may be argued, ultimately determines biological function to a greater extent than protein amounts in cells.

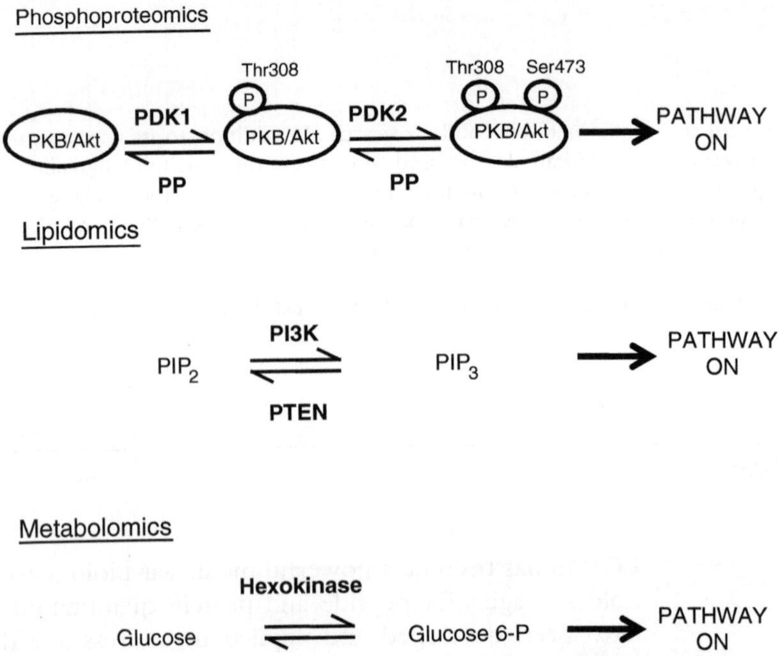

Fig. 1.1. Activitomic techniques provide a measure of enzyme and pathway activation. Examples of biochemical reactions controlled by protein, lipid and carbohydrate kinases are given in *top*, *middle* and *bottom* panels, respectively. In each case, measuring the products of the reactions, being phosphoproteins, phospholipids or glucose phosphates, provides a measure of the equilibrium and the activity of the kinase/phosphatase reaction pairs. Techniques for large-scale and in-depth quantification of phosphoproteins (phosphoproteomics), lipids (lipidomics) and metabolites (metabolomics) have been developed and these in general use LC-MS for identification and quantification. Since all of these techniques provide a reflection of enzyme activity, collectively these can all be considered as methods for activitomic analysis. PDK, Phosphoinositide-3-kinase-dependent kinase; PP, protein kinase; PI3K, phosphoinositide-3-kinase; PTEN, phosphatase and tensin homologue.

Such proteomic and activitomic approaches have applications in understanding the regulation of cellular function through signalling and the response to specific stimuli or stresses; in understanding cellular processes such as proliferation, apoptosis, motility and developmental programmes; and in understanding the protein changes associated with normal versus diseased states. As well as enhancing our basic knowledge of biological processes and systems biology, LC-MS is also contributing to translational and clinical research in terms of drug target identification and disease biomarker discovery and monitoring.

2. Applications

As briefly mentioned above, quantitative LC-MS has many applications, of which the examples given here illustrate the power of this versatile technique for biomolecular research. **Chapter 2** will discuss in more detail the different LC-MS techniques that can be used for performing such analyses.

2.1. Activitomics

A relatively new application of LC-MS is in the quantitative analysis of enzymatic activity. This approach consists of quantifying products of in vitro enzymatic reactions by LC-MS (*see* protocol in **Chapter 20**), a technique that allows multiplexing enzymatic measurements with unrivalled specificity (1). This analytical concept can be extended to the quantification of the activity of many enzymes such as protein kinases that provide a read-out of pathway activity (2). The advantage of measuring enzymatic reactions, rather than the amounts of enzymes (proteins) in cells, is that the modulation of enzymatic activity is a complex process, and protein amounts are often a poor indication of enzymatic and pathway activities. This concept is well understood for metabolic pathways where approaches for metabolic control analysis (MCA) are well developed (3, 4). It is through the application of MCA that biochemists and biotechnologists have realized that measuring protein amounts is uninformative for the prediction of biological output as a result of specific perturbations. Systems biology and modelling thus need to take into account the share of control that enzymes have in their pathway for a full understanding of how biochemical pathways are regulated (3). Advances in systems biology research thus need methods that can be used to quantify enzymatic activity as comprehensively as possible, as this is an essential requirement for the construction of reasonably accurate models of biochemical pathways, the ultimate aim of many biological fields including metabolism and cell signalling research.

2.1.1. Phosphoproteomics

Quantitative analysis of phosphorylation may be regarded as an activitomic approach for comprehensive quantification of pathways controlled by protein kinases. LC-MS allows the quantification of thousands of phosphorylation sites on proteins and these approaches are increasingly used in several biological fields including cell signalling and cancer research (5–13). For example, quantitative LC-MS has been applied to the analysis of growth factor signalling (9). It was found that >2,000 phosphorylation sites are regulated as a result of activating the EGF receptor in HeLa cells, which indicates that kinase pathways downstream of growth factors are more complex than previously thought. These studies used a combination of SCX and TiO_2 to enrich phosphopeptides that were later identified and quantified by LC-MS/MS. Related strategies involve the use of IMAC instead of TiO_2 as the enrichment step (12, 13) or the use of anti-phosphotyrosine antibodies to enrich phosphotyrosine-containing peptides (10, 11). Using these protocols (detailed in **Chapters 6** and 7), it is now possible to 'routinely' identify and quantify thousands of phosphopeptides per experiment.

A question that need to be addressed is what is the value of these analyses (14)? Indeed, these authors believe that the significance of phosphoprotein quantification is sometimes not fully understood. A misconception often found in the literature is that phosphorylation activates the enzymatic activity of proteins. Although this is the case for many enzymes, including protein kinases such as PKB/Akt and Erk1/2, it is not the case for many others such as GSK3β, which is inactivated as a result of being phosphorylated by PKB/Akt. It has also been proposed that much of the phosphorylation occurring in eukaryotes is non-functional (14), with functional events occurring in a background of phosphorylation 'noise' for which there is no selection against. An example of apparent non-functional phosphorylation is that occurring on PKB/Akt, which although phosphorylated at Ser124, Ser126, Ser129, Thr308, Ser473 and Tyr474 (of the human AKT1 isoform), only the phosphorylations at Thr308 and Ser473 are known to modulate the enzymatic activity of PKB/Akt. The other phosphorylation sites do not seem to alter any of the properties of this kinase.

The hypothesis of the existence of non-functional phosphorylation is difficult to prove or disprove, because phosphorylation with no apparent function under determined conditions may have functions under other conditions. It should also be noted that activation of enzymatic activity is only one of the properties that phosphorylation may attribute to a protein. Indeed, not all proteins are enzymes and phosphorylation can also affect protein localization and their ability to interact with other molecules in cells, which may be as important in controlling the biological properties of proteins as the modulation of enzymatic activity.

Regardless of whether or not non-functional phosphorylation exists, in our view, the value of phosphoproteomics lies in its potential ability to serve as a comprehensive read-out of kinase activity in cells. There exist 518 protein kinase genes in cells, and the deregulation of these enzymes is a common occurrence of many diseases including cancer, where kinases have been found to be frequently mutated (15). But because enzymatic activity can be

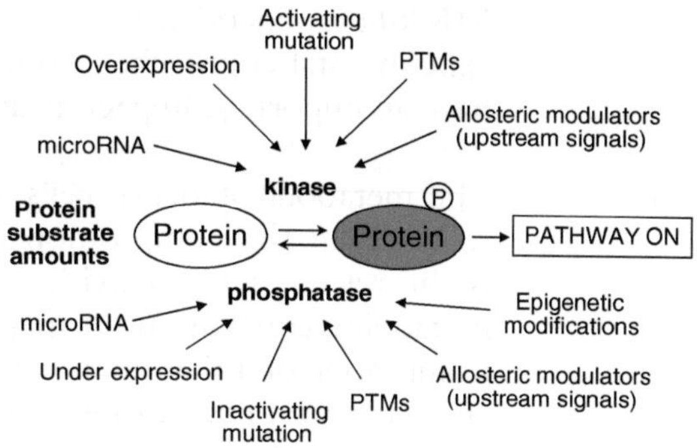

Fig. 1.2. Molecular alterations that can modulate kinase/phosphatase activity in cells. Many different molecular phenomena can affect the activity of kinase/phosphatase equilibrium reactions. Quantifying the extent of phosphorylation is a direct measure of this equilibrium as this considers all the different phenomena that can affect enzymatic activities.

modulated by several molecular phenomena in addition to genetic mutations (**Fig. 1.2**), measuring kinase activity is a more relevant and direct way of assessing activation. By definition, each one of the thousands of phosphorylation sites quantifiable by LC-MS is the result of a kinase reaction (or a kinase/phosphatase reaction pair) (**Fig. 1.1**). Thus measuring phosphorylation amounts in cells provides a measure of the relative activities of kinases and phosphatases acting on these phosphorylation sites. Of course, phosphoprotein amounts are also the result of their expression, in addition to the activity of kinase and phosphatases acting upon them, but since substrate levels control enzyme activity, this view still holds (clearly, when a phosphoprotein substrate is not expressed, the kinase pathway cannot be active regardless of the activity of the kinase). Thus since protein kinases are major players in the control of signalling pathways, in-depth analysis of phosphorylation by LC-MS may be regarded as an activitomic approach that offers the unprecedented prospect for comprehensive quantification of oncogenic signalling at the system level. This is an exciting realization because the ability of quantifying the activity of these pathways in a comprehensive manner could provide a breakthrough in cancer research by identifying the pathways that are specifically altered in individual patients, which could in turn contribute to the advent of personalized therapies for cancer patients. This is recognized to be of utmost importance for translating basic findings in cancer research to clinical applications (16).

The field is now at a stage where thousands of phosphorylation sites can be quantified with relatively good precision, accuracy and throughput (*see* **Chapters 6** and 7 for protocols). The challenge now is to link each of the thousands of phosphorylation sites quantifiable by LC-MS to the kinases responsible for their phosphorylation. This will allow facile activitomic analysis of the

518 kinases encoded in the genome. In addition to brute force experimental approaches, computational techniques are likely to have an important impact in this regard (17–19).

2.1.2. Metabolomics

The metabolic status of cells and whole organisms is ultimately reflected in the composition of metabolites, and this, in turn, is the result of the activities of metabolic enzymes. Measuring metabolite concentrations may therefore be regarded as an activitomic approach that quantifies the activity of metabolic enzymes and pathways. Historically, NMR has been the method of choice to profile metabolites in biological samples. However, the current trend is to use LC-MS-based approaches, which are gradually becoming the methods of choice because of their greater throughput and coverage of the analysed metabolome (20, 21).

Metabolomic techniques based on LC-MS are still in their infancy. Indeed, although it is relatively easy to derive metabolite profiles by MS and LC-MS, obtaining the identity of these metabolites is not straightforward. This is because of the many different chemical structures that form the backbone of these small molecules, which makes deriving general rules for their identification by MS/MS, analogous to those derived for peptides, a very difficult, perhaps impossible, task. The field of metabolomics will ultimately require the construction of comprehensive spectral libraries of metabolite CID MS/MS data, similar to those already available for electron impact and chemical ionization spectra.

Despite these technical limitations, LC-MS is increasingly used for metabolic profiling and to quantify the fluxes of metabolic pathways. In cancer research, metabolomic studies have contributed to the notion that several metabolic pathways may be altered in malignant cells. Cancer cells are characterized by an increased proliferation rate and their ability to invade adjacent tissue in hypoxic environments. Tumour cells therefore have different energy requirements than do normal cells, but as with signalling pathways, different cancer cells, even those of the same pathology, may deregulate different metabolic pathways; in other words, different tumours may obtain their energy via different metabolic routes (22–24). Identifying these different routes and how these are regulated by known oncogenes and tumour suppressors could be exploited in the design of novel therapeutic strategies.

As an example of the use of LC-MS for the investigation of metabolism in cancer, Sreekumar et al. (25) used an LC-MS- and GC-MS-based metabolomic approach to profile 1,126 metabolites in 262 prostate cancer samples. Sarcosine (methylglycine) concentrations were found to be greater in the plasma of prostate cancer patients with metastatic disease relative to patients with localized disease and benign tumours. This elevated sarcosine level probably reflects a greater activity of the enzyme

N-methyltransferase, which catalyses the generation of sarcosine from glycine. The activity of this enzyme was found to be important for cell invasion and metastasis, thus representing a new drug target for the treatment of prostate cancer (25). Importantly, since sarcosine levels can be monitored in urine, it provides a convenient biomarker for the prognosis of the disease and to personalize the treatment of prostate cancer patients based on the metabolic status of tumours.

As mentioned above, different cancers may obtain their energy via different metabolic pathways. Signalling pathways, known to be deregulated in cancer, are also involved in the homeostasis of normal cells by mediating the effects of insulin, growth factors and other hormones that regulate energy levels in cells. These pathways, which include the PI3K/Akt and Ras/MEK/Erk pathways, affect the expression and/or activity of metabolic enzymes. Since different cancers deregulated different signalling pathways, the metabolic pathways by which cancer cells obtain their energy may also be different in different tumours.

LC-MS and GC-MS-based metabolic studies support this hypothesis. For example, certain cancer cells show an elevated glycolytic flux and shut down the citric acid cycle, even in the presence of oxygen (22, 24). This is reflected by an increased expression of glycolytic enzymes in solid cancers (26) and in the presence of abnormally high levels of lactic acid (27), the final product of anaerobic glycolysis. Anaerobic glycolysis in the presence of oxygen, also known as the Warburg effect, seems to be promoted by the activation of signalling pathways, which in normal cells mediate the biological functions of growth factors and insulin; the PI3K/Akt pathway seems to have a prominent role in this respect (22). However, not all cancers may deregulate this pathway (28). Thus in cancers with wild-type PI3K/Akt, mutations on the Myc oncogene promote the utilization of glutamate as the preferred energy source instead of glucose (24). These findings were possible because of the use of an LC-MS-based strategy to quantify the fluxes of glycolytic and glutamate catabolic pathways (24).

The connection between oncogenic signalling and metabolic pathways is also exemplified by a LC-MS-based study (23), which showed that a single mutation on the oncogene *K-Ras* at codon 13 produced markedly different metabolic phenotypes than did mutations on *K-Ras* at codon 12. Thus the mutation on codon 13 made cells increase the pentose phosphate pathway flux, while mutations on codon 12 diverted metabolism to glycolysis (23). Taken together, these studies illustrate the heterogeneous nature of metabolic alterations in cancer cells and the connection between metabolism and signalling. Metabolomics studies based on LC-MS are contributing to the understanding of this complexity, but much more work is required to decipher how the different combinations of mutations on tumour suppressors and oncogenes may affect metabolism in cancer cells. Because of its ability for in-depth analysis of metabolites (which reflect the activities of metabolic enzymes), LC-MS-based activitomic/metabolomic studies should contribute to the advancement of this field.

2.1.3. Lipidomics

LC-MS is increasingly used to multiplex the quantitative analysis of lipids, whose levels, in turn, reflect the activity of enzymes involved in their production, modification and degradation. Lipidomics can therefore be regarded as an activitomic approach to quantify the activity of enzymes with roles in lipid metabolism, including lipid kinases and phosphatases. As with protein kinases and protein phosphatases, lipid kinases and phosphatases have important roles in controlling signalling pathways with roles in fundamental biological processes and in diseases such as cancer. Therefore, quantifying the activity of lipid kinases/phosphatase pairs also has numerous applications in basic and applied research.

Phosphoinositide (PI) research is an example of a lipidomic/activitomic application in which LC-MS could have a major role (29). PI signalling has roles in insulin and antigen signalling and in cancer. Once activated, receptors for growth factors, antigens and certain hormones recruit PI-kinases to plasma membranes and cellular endomembranes, where they phosphorylate PIs to produce PI-phosphates (PIPs). PI 3-kinases (PI3K) are well-studied contributors of the pool of PIP species, many of which are second messengers that activate several intracellular pathways through their ability to interact with proteins containing phospholipid recognition domains (including FYVE and PH domains) (30).

The *PIK3CA* gene, encoding for a catalytic isoform of PI3K termed p110α, is an oncogene found to be frequently activated (via mutations and overexpression) in many different forms of cancer, while the lipid phosphatase opposing the p110α reaction (termed PTEN) is frequently inactivated in several different tumour types. Taken together, the PI signalling pathway is one of the most frequently deregulated pathways in cancer (31). In addition to PI3K and PTEN, several other lipid kinases and phosphatases contribute to the production of PIPs with different phosphate content and with different isometry. Thus class I PI3Ks preferentially phosphorylate PI-4,5-bisphosphate (PI-4,5-P2) to produce PI-3,4,5-trisphosphate (PIP3), while classes II and III PI3K use PI as a substrate to produce PI-3P. PI4K and PI5K lipid kinases also exist that phosphorylated PI at 4 and 5 positions to produce PI4P and PI5P, respectively (32, 33). This complexity is increased by the existence of the lipid phosphatases, SHIP1 and SHIP2, that dephosphorylate PIP3 at different positions than PTEN to form PI-3,5-P2 and PI-3,4-P2 (34).

PI signalling controls fundamental biological processes including insulin, growth factor and antigen signalling, apoptosis, cell cycle progression and proliferation. Understanding the roles of the different PI-modifying enzymes requires methods to quantify their activity status. For PI3K, this is commonly performed by measuring radioactive de-acetylated PIP3 after metabolic labelling of cells with radioactive ^{32}P-ATP. This involves separating the de-acetylated PI pool by ion exchange HPLC and measurement of ^{32}P by scintillation or Cherenkov counting in HPLC fractions.

As an alternative to these methods, several studies have reported the use of LC-MS to quantify PI species. The advantage of using LC-MS for phospholipid analysis is that by using normal-phase LC, one does not need to de-acetylate lipids prior to the analysis (35). Therefore, the analysis can also consider the preference of enzymes for the phosphorylation of lipids with different acyl composition. For example, by using ESI-MS, it has been demonstrated that there are at least eight different PIP3 species that differ in their acyl composition (36). This type of information cannot be obtained by other methods for detecting PI species such as those based on HPLC (briefly outlined above), antibodies or affinity reagents based on lipid-binding domains. Moreover, LC-MS could in principle be used to profile all PI species in a single analysis, thus allowing the quantitative estimation of PI-kinase and phosphatase activities in a single run.

2.2. Proteomics

2.2.1. Protein Expression

One of the first uses of proteomics was as an alternative to cDNA microarrays for the analysis of gene expression (37). The rationale of using proteomics rather than transcriptomics for measuring gene expression is that functional expression of a gene requires not only its transcription to an mRNA molecule but also the translation to protein. Initial studies using 2D gels demonstrated that mRNA levels are often a poor indication of protein levels (38). Later using LC-MS/MS, it was suggested that there was a significant correlation between mRNA and protein expression in 70% of instances, thus implying that 30% of expression is controlled at the level of translation. Therefore, measuring protein expression, rather than mRNA levels, would be a more direct and relevant measure of gene expression as this would account for the contribution of both translation and transcription (*see* **Chapters 11, 12** and **13** for protocols for proteome-wide quantification by LC-MS). Proteomics also has applicability to the analysis of proteins in blood and other biological fluids whose expression cannot be analysed by microarrays and to the analysis of proteins in different cellular compartments (see below).

It is often argued that gene expression microarrays offer a more comprehensive picture of gene expression than do approaches based on protein quantification. This view is based on the fact that probes for whole genomes can now be printed on microarray chips. However, not all genes give a positive signal on the chip, presumably because not all genes are expressed at a given time or tissue, and head to head comparison of LC-MS-based proteomic and microarray data has shown that the two techniques offer a similar depth of analysis (39).

However, although analysis of gene expression at the protein level is conceptually more powerful than is analysis of gene expression at the mRNA level, performing LC-MS-based proteomics is much more technically demanding, time consuming and expensive than is performing microarray analyses. For these reasons, the latter is still the method of choice for the analysis of gene expression for most researchers. This trend may change in the near future as mass spectrometers become less expensive, with greater duty cycle and dynamic range, and novel quantification methods and computer programs for their implementation become available.

2.2.2. Organelle Proteomics

A powerful application of proteomics is in the elucidation of the protein composition of intracellular compartments and organelles (40–44). Assigning proteins to their respective intracellular location aids in annotating the proteome and the genome and may give a hint of the function of novel proteins and genes; conversely, this is also important for understanding the function of the different organelles. For example, the proteome of the centrosome has been extensively studied by quantitative LC-MS. In an influential study, Andersen and colleagues identified 23 novel proteins in this subcellular structure (45). This catalogue of proteins represented a leap in the understanding of the composition of the centrosome since only 60 centrosomal proteins were known prior to this work. These findings have formed the basis for follow-on molecular biology studies aimed at understanding the role of these novel proteins in centrosomal function (46, 47) and structure (48, 49).

A problem encountered with all MS-based organelle proteomic studies is that these experiments require an initial fractionation step to isolate the organelle under investigation; this is normally carried out by differential or sucrose gradient centrifugation. The sensitivity and dynamic range of modern mass spectrometers is such that even minor contaminants are detected in these biochemical preparations of subcellular structures. Therefore, a qualitative analysis by LC-MS/MS does not allow discriminating true organelle components to contaminants. In order to address this problem, researchers developed strategies to quantify the enrichment of proteins in subcellular fractions relative to the enrichment of validated markers of these fractions. This is the basis of protein correlation profiling (PCP) (45) and localization of organelle proteins by isotope tagging (LOPIT) (50). The main difference between these two approaches is that PCP uses label-free LC-MS to quantify the relative enrichment of components in the subcellular fractions, whereas LOPIT relies on the labelling of components before quantification. PCP has been used to investigate the composition and dynamics of nucleolar proteins (45) and for comprehensive characterization of subcellular proteomes (51). LOPIT has been applied to the characterization of membrane proteins (52) and more recently to identify the protein composition of several organelles in *Drosophila* embryos (53). A similar strategy was used to more confidently locate membrane proteins specific for the basolateral and apical membranes of polarized kidney epithelial cells (54). A protocol for PCP is described in **Chapter 15**.

2.2.3. Protein Interactions

Proteins do not work in isolation but instead have to interact with other molecules to function. Thus, protein–protein, protein–lipid and protein–nucleic acid interactions play fundamental roles in biology. LC-MS is routinely used to detect novel protein–protein interactions (55–59). A recent example is the identification of pyruvate kinase M2 as a phosphotyrosine-binding protein (60). This study used LC-MS to quantify the enrichment of binding of proteins to tyrosine-phosphorylated peptides relative to binding of their non-phosphorylated peptide counterparts. In addition to proteins known to bind phosphotyrosines, the screen identified pyruvate kinase M2 as a tyrosine-binding protein. This finding inspired further experiments that linked activation of tyrosine kinase signalling to the activity of pyruvate kinase, an enzyme that controls the fluxes of glycolytic pathways often found to be upregulated in cancer cells (61). The Warburg effect (anaerobic glycolysis in the presence of oxygen, a common occurrence in cancer cells) may in some instances be the result of the activation of oncogenic tyrosine kinase signalling pathways, which in turn affect metabolic pathways through the allosteric activation of metabolic enzymes (61).

As with the case of organelle protein analysis, detecting a protein in a protein complex by LC-MS/MS does not necessarily mean that this protein is part of the complex. Protein complexes are difficult (perhaps impossible) to isolate to homogeneity and the presence of non-specific contaminants is often impossible to avoid. An approach to overcome this problem is to over-express a protein bait with several tags for tandem affinity purification (62). This approach can certainly result in the identification of novel protein binders but because it needs to over-express the protein bait, it may result in the introduction of false positives without biological relevance. An alternative approach is to use quantitative LC-MS to identify dynamic protein complexes that form after cells are stimulated with specific agents. LC-MS can then be used to quantify proteins present in complexes after cell stimulation, thus identifying dynamic members of these complexes that may have a functional role in controlling the molecular biology of the complex under study. In **Chapter 16**, Dengjel et al. exemplify this approach for the identification of dynamic binders of the EGF receptor. As outlined above, this strategy was also used to identify pyruvate kinase M2 as a tyrosine-binding protein (61) and should have wide applicability not only for the identification of protein-binding partners but also for mapping protein–lipid and protein–nucleic acid interactions.

3. Summary and Perspectives

LC-MS has many applications in biomolecular research. In addition to the quantification of gene expression at the protein level, as an alternative to microarray analysis, LC-MS can be used to comprehensively profile the protein composition of subcellular organelles and protein complexes and to provide a picture of the dynamic protein composition of macromolecular structures. Another important and gradually increasing application of LC-MS is for in-depth quantification of enzymatic activities in cells. Enzymes are important mediators of biological functions, but since their levels may not reflect their activity status, standard proteomic approaches cannot be used to infer the activity status of biochemical pathways. Therefore, methods that can be used to quantify enzymatic activities in a comprehensive manner will have many applications in biological research. Here we have exemplified this concept by showing how activitomic approaches, including phosphoproteomics, metabolomics and lipidomics, are being used to provide insights into cancer biology. In the future, related approaches that quantify other protein modifications, such as acetylation or methylation, may also be explored as these should provide equally important biological insights. Quantitative LC-MS is likely to continue to have important roles in these workflows. Several techniques for quantitative LC-MS have been developed over the last decade and these are discussed in detail in **Chapter 2**.

References

1. Greis, K. D. (2007) Mass spectrometry for enzyme assays and inhibitor screening: an emerging application in pharmaceutical research. *Mass Spectrom. Rev.* **26**, 324–339.

2. Cutillas, P. R., Khwaja, A., Graupera, M., Pearce, W., Gharbi, S., Waterfield, M., and Vanhaesebroeck, B. (2006) Ultrasensitive and absolute quantification of the

phosphoinositide 3-kinase/Akt signal transduction pathway by mass spectrometry. *Proc. Natl. Acad. Sci. USA* **103**, 8959–8964.

3. Cascante, M., Boros, L. G., Comin-Anduix, B., de Atauri, P., Centelles, J. J., and Lee, P. W. (2002) Metabolic control analysis in drug discovery and disease. *Nat. Biotechnol.* **20**, 243–249.
4. Comin-Anduix, B., Boren, J., Martinez, S., Moro, C., Centelles, J. J., Trebukhina, R., Petushok, N., Lee, W. N., Boros, L. G., and Cascante, M. (2001) The effect of thiamine supplementation on tumour proliferation. A metabolic control analysis study. *Eur. J. Biochem.* **268**, 4177–4182.
5. Guha, U., Chaerkady, R., Marimuthu, A., Patterson, A. S., Kashyap, M. K., Harsha, H. C., Sato, M., Bader, J. S., Lash, A. E., Minna, J. D., Pandey, A., and Varmus, H. E. (2008) Comparisons of tyrosine phosphorylated proteins in cells expressing lung cancer-specific alleles of EGFR and KRAS. *Proc. Natl. Acad. Sci. USA* **105**, 14112–14117.
6. Kim, J. E., and White, F. M. (2006) Quantitative analysis of phosphotyrosine signaling networks triggered by CD3 and CD28 costimulation in Jurkat cells. *J. Immunol.* **176**, 2833–2843.
7. Kruger, M., Kratchmarova, I., Blagoev, B., Tseng, Y. H., Kahn, C. R., and Mann, M. (2008) Dissection of the insulin signaling pathway via quantitative phosphoproteomics. *Proc. Natl. Acad. Sci. USA* **105**, 2451–2456.
8. Larive, R. M., Urbach, S., Poncet, J., Jouin, P., Mascre, G., Sahuquet, A., Mangeat, P. H., Coopman, P. J., and Bettache, N. (2009) Phosphoproteomic analysis of Syk kinase signaling in human cancer cells reveals its role in cell–cell adhesion. *Oncogene* **28**(24), 2337–2347.
9. Olsen, J. V., Blagoev, B., Gnad, F., Macek, B., Kumar, C., Mortensen, P., and Mann, M. (2006) Global, in vivo, and site-specific phosphorylation dynamics in signaling networks. *Cell* **127**, 635–648.
10. Rikova, K., Guo, A., Zeng, Q., Possemato, A., Yu, J., Haack, H., Nardone, J., Lee, K., Reeves, C., Li, Y., Hu, Y., Tan, Z., Stokes, M., Sullivan, L., Mitchell, J., Wetzel, R., Macneill, J., Ren, J. M., Yuan, J., Bakalarski, C. E., Villen, J., Kornhauser, J. M., Smith, B., Li, D., Zhou, X., Gygi, S. P., Gu, T. L., Polakiewicz, R. D., Rush, J., and Comb, M. J. (2007) Global survey of phosphotyrosine signaling identifies oncogenic kinases in lung cancer. *Cell* **131**, 1190–1203.
11. Rush, J., Moritz, A., Lee, K. A., Guo, A., Goss, V. L., Spek, E. J., Zhang, H., Zha, X. M., Polakiewicz, R. D., and Comb, M. J. (2005) Immunoaffinity profiling of tyrosine phosphorylation in cancer cells. *Nat. Biotechnol.* **23**, 94–101.
12. Trinidad, J. C., Specht, C. G., Thalhammer, A., Schoepfer, R., and Burlingame, A. L. (2006) Comprehensive identification of phosphorylation sites in postsynaptic density preparations. *Mol. Cell. Proteomics* **5**, 914–922.
13. Villen, J., and Gygi, S. P. (2008) The SCX/IMAC enrichment approach for global phosphorylation analysis by mass spectrometry. *Nat. Protoc.* **3**, 1630–1638.
14. Lienhard, G. E. (2008) Non-functional phosphorylations? *Trends Biochem. Sci.* **33**, 351–352.
15. Greenman, C., Stephens, P., Smith, R., Dalgliesh, G. L., Hunter, C., Bignell, G., Davies, H., Teague, J., Butler, A., Stevens, C., Edkins, S., O'Meara, S., Vastrik, I., Schmidt, E. E., Avis, T., Barthorpe, S., Bhamra, G., Buck, G., Choudhury, B., Clements, J., Cole, J., Dicks, E., Forbes, S., Gray, K., Halliday, K., Harrison, R., Hills, K., Hinton, J., Jenkinson, A., Jones, D., Menzies, A., Mironenko, T., Perry, J., Raine, K., Richardson, D., Shepherd, R., Small, A., Tofts, C., Varian, J., Webb, T., West, S., Widaa, S., Yates, A., Cahill, D. P., Louis, D. N., Goldstraw, P., Nicholson, A. G., Brasseur, F., Looijenga, L., Weber, B. L., Chiew, Y. E., DeFazio, A., Greaves, M. F., Green, A. R., Campbell, P., Birney, E., Easton, D. F., Chenevix-Trench, G., Tan, M. H., Khoo, S. K., Teh, B. T., Yuen, S. T., Leung, S. Y., Wooster, R., Futreal, P. A., and Stratton, M. R. (2007) Patterns of somatic mutation in human cancer genomes. *Nature* **446**, 153–158.
16. Sawyers, C. L. (2008) The cancer biomarker problem. *Nature* **452**, 548–552.
17. Miller, M. L., Jensen, L. J., Diella, F., Jorgensen, C., Tinti, M., Li, L., Hsiung, M., Parker, S. A., Bordeaux, J., Sicheritz-Ponten, T., Olhovsky, M., Pasculescu, A., Alexander, J., Knapp, S., Blom, N., Bork, P., Li, S., Cesareni, G., Pawson, T., Turk, B. E., Yaffe, M. B., Brunak, S., and Linding, R. (2008) Linear motif atlas for phosphorylation-dependent signaling. *Sci. Signal.* **1**, ra2.
18. Linding, R., Jensen, L. J., Pasculescu, A., Olhovsky, M., Colwill, K., Bork, P., Yaffe, M. B., and Pawson, T. (2008) NetworKIN: a resource for exploring cellular phosphorylation networks. *Nucleic Acids Res.* **36**, D695–D699.
19. Linding, R., Jensen, L. J., Ostheimer, G. J., van Vugt, M. A., Jorgensen, C., Miron, I. M., Diella, F., Colwill, K., Taylor, L., Elder, K., Metalnikov, P., Nguyen, V., Pasculescu,

A., Jin, J., Park, J. G., Samson, L. D., Woodgett, J. R., Russell, R. B., Bork, P., Yaffe, M. B., and Pawson, T. (2007) Systematic discovery of in vivo phosphorylation networks. *Cell* **129**, 1415–1426.

20. Metz, T. O., Page, J. S., Baker, E. S., Tang, K., Ding, J., Shen, Y., and Smith, R. D. (2008) High resolution separations and improved ion production and transmission in metabolomics. *Trends Anal. Chem.* **27**, 205–214.

21. Metz, T. O., Zhang, Q., Page, J. S., Shen, Y., Callister, S. J., Jacobs, J. M., and Smith, R. D. (2007) The future of liquid chromatography–mass spectrometry (LC-MS) in metabolic profiling and metabolomic studies for biomarker discovery. *Biomark. Med.* **1**, 159–185.

22. Elstrom, R. L., Bauer, D. E., Buzzai, M., Karnauskas, R., Harris, M. H., Plas, D. R., Zhuang, H., Cinalli, R. M., Alavi, A., Rudin, C. M., and Thompson, C. B. (2004) Akt stimulates aerobic glycolysis in cancer cells. *Cancer Res.* **64**, 3892–3899.

23. Vizan, P., Boros, L. G., Figueras, A., Capella, G., Mangues, R., Bassilian, S., Lim, S., Lee, W. N., and Cascante, M. (2005) K-ras codon-specific mutations produce distinctive metabolic phenotypes in NIH3T3 mice [corrected] fibroblasts. *Cancer Res.* **65**, 5512–5515.

24. Wise, D. R., DeBerardinis, R. J., Mancuso, A., Sayed, N., Zhang, X. Y., Pfeiffer, H. K., Nissim, I., Daikhin, E., Yudkoff, M., McMahon, S. B., and Thompson, C. B. (2008) Myc regulates a transcriptional program that stimulates mitochondrial glutaminolysis and leads to glutamine addiction. *Proc. Natl. Acad. Sci. USA* **105**, 18782–18787.

25. Sreekumar, A., Poisson, L. M., Rajendiran, T. M., Khan, A. P., Cao, Q., Yu, J., Laxman, B., Mehra, R., Lonigro, R. J., Li, Y., Nyati, M. K., Ahsan, A., Kalyana-Sundaram, S., Han, B., Cao, X., Byun, J., Omenn, G. S., Ghosh, D., Pennathur, S., Alexander, D. C., Berger, A., Shuster, J. R., Wei, J. T., Varambally, S., Beecher, C., and Chinnaiyan, A. M. (2009) Metabolomic profiles delineate potential role for sarcosine in prostate cancer progression. *Nature* **457**, 910–914.

26. Bi, X., Lin, Q., Foo, T. W., Joshi, S., You, T., Shen, H. M., Ong, C. N., Cheah, P. Y., Eu, K. W., and Hew, C. L. (2006) Proteomic analysis of colorectal cancer reveals alterations in metabolic pathways: mechanism of tumorigenesis. *Mol. Cell. Proteomics* **5**, 1119–1130.

27. Chan, E. C., Koh, P. K., Mal, M., Cheah, P. Y., Eu, K. W., Backshall, A., Cavill, R., Nicholson, J. K., and Keun, H. C. (2009) Metabolic profiling of human colorectal cancer using high-resolution magic angle spinning nuclear magnetic resonance (HR-MAS NMR) spectroscopy and gas chromatography mass spectrometry (GC/MS). *J. Proteome Res.* **8**, 352–361.

28. Parsons, D. W., Wang, T. L., Samuels, Y., Bardelli, A., Cummins, J. M., DeLong, L., Silliman, N., Ptak, J., Szabo, S., Willson, J. K., Markowitz, S., Kinzler, K. W., Vogelstein, B., Lengauer, C., and Velculescu, V. E. (2005) Colorectal cancer: mutations in a signalling pathway. *Nature* **436**, 792.

29. Wakelam, M. J., Pettitt, T. R., and Postle, A. D. (2007) Lipidomic analysis of signaling pathways. *Methods Enzymol.* **432**, 233–246.

30. Engelman, J. A., Luo, J., and Cantley, L. C. (2006) The evolution of phosphatidylinositol 3-kinases as regulators of growth and metabolism. *Nat. Rev. Genet.* **7**, 606–619.

31. Yuan, T. L., and Cantley, L. C. (2008) PI3K pathway alterations in cancer: variations on a theme. *Oncogene* **27**, 5497–5510.

32. Carpenter, C. L., and Cantley, L. C. (1990) Phosphoinositide kinases. *Biochemistry* **29**, 11147–11156.

33. Fruman, D. A., Meyers, R. E., and Cantley, L. C. (1998) Phosphoinositide kinases. *Annu. Rev. Biochem.* **67**, 481–507.

34. Niggli, V. (2005) Regulation of protein activities by phosphoinositide phosphates. *Annu. Rev. Cell. Dev. Biol.* **21**, 57–79.

35. Pettitt, T. R., Dove, S. K., Lubben, A., Calaminus, S. D., and Wakelam, M. J. (2006) Analysis of intact phosphoinositides in biological samples. *J. Lipid Res.* **47**, 1588–1596.

36. Milne, S. B., Ivanova, P. T., DeCamp, D., Hsueh, R. C., and Brown, H. A. (2005) A targeted mass spectrometric analysis of phosphatidylinositol phosphate species. *J. Lipid Res.* **46**, 1796–1802.

37. Mann, M., Hendrickson, R. C., and Pandey, A. (2001) Analysis of proteins and proteomes by mass spectrometry. *Annu. Rev. Biochem.* **70**, 437–473.

38. Gygi, S. P., Rochon, Y., Franza, B. R., and Aebersold, R. (1999) Correlation between protein and mRNA abundance in yeast. *Mol. Cell. Biol.* **19**, 1720–1730.

39. Cox, J., and Mann, M. (2007) Is proteomics the new genomics? *Cell* **130**, 395–398.

40. Brunet, S., Thibault, P., Gagnon, E., Kearney, P., Bergeron, J. J., and Desjardins, M. (2003) Organelle proteomics: looking at less to see more. *Trends Cell. Biol.* **13**, 629–638.

41. Dreger, M. (2003) Subcellular proteomics. *Mass Spectrom. Rev.* **22**, 27–56.

42. Robinson, C. V., Sali, A., and Baumeister, W. (2007) The molecular sociology of the cell. *Nature* **450**, 973–982.
43. Taylor, S. W., Fahy, E., and Ghosh, S. S. (2003) Global organellar proteomics. *Trends Biotechnol.* **21**, 82–88.
44. Yates, J. R., 3rd, Gilchrist, A., Howell, K. E., and Bergeron, J. J. (2005) Proteomics of organelles and large cellular structures. *Nat. Rev. Mol. Cell. Biol.* **6**, 702–714.
45. Andersen, J. S., Wilkinson, C. J., Mayor, T., Mortensen, P., Nigg, E. A., and Mann, M. (2003) Proteomic characterization of the human centrosome by protein correlation profiling. *Nature* **426**, 570–574.
46. Fabbro, M., Zhou, B. B., Takahashi, M., Sarcevic, B., Lal, P., Graham, M. E., Gabrielli, B. G., Robinson, P. J., Nigg, E. A., Ono, Y., and Khanna, K. K. (2005) Cdk1/Erk2- and Plk1-dependent phosphorylation of a centrosome protein, Cep55, is required for its recruitment to midbody and cytokinesis. *Dev. Cell* **9**, 477–488.
47. Guarguaglini, G., Duncan, P. I., Stierhof, Y. D., Holmstrom, T., Duensing, S., and Nigg, E. A. (2005) The forkhead-associated domain protein Cep170 interacts with Polo-like kinase 1 and serves as a marker for mature centrioles. *Mol. Biol. Cell* **16**, 1095–1107.
48. Graser, S., Stierhof, Y. D., and Nigg, E. A. (2007) Cep68 and Cep215 (Cdk5rap2) are required for centrosome cohesion. *J. Cell Sci.* **120**, 4321–4331.
49. Yan, X., Habedanck, R., and Nigg, E. A. (2006) A complex of two centrosomal proteins, CAP350 and FOP, cooperates with EB1 in microtubule anchoring. *Mol. Biol. Cell* **17**, 634–644.
50. Dunkley, T. P., Watson, R., Griffin, J. L., Dupree, P., and Lilley, K. S. (2004) Localization of organelle proteins by isotope tagging (LOPIT). *Mol. Cell. Proteomics* **3**, 1128–1134.
51. Foster, L. J., de Hoog, C. L., Zhang, Y., Xie, X., Mootha, V. K., and Mann, M. (2006) A mammalian organelle map by protein correlation profiling. *Cell* **125**, 187–199.
52. Sadowski, P. G., Dunkley, T. P., Shadforth, I. P., Dupree, P., Bessant, C., Griffin, J. L., and Lilley, K. S. (2006) Quantitative proteomic approach to study subcellular localization of membrane proteins. *Nat. Protoc.* **1**, 1778–1789.
53. Tan, D. J., Dvinge, H., Christoforou, A., Bertone, P., Martinez Arias, A., and Lilley, K. S. (2009) Mapping organelle proteins and protein complexes in *Drosophila melanogaster*. *J Proteome Res.* **8**(6), 2667–2678.
54. Cutillas, P. R., Biber, J., Marks, J., Jacob, R., Stieger, B., Cramer, R., Waterfield, M., Burlingame, A. L., and Unwin, R. J. (2005) Proteomic analysis of plasma membrane vesicles isolated from the rat renal cortex. *Proteomics* **5**, 101–112.
55. Abu-Farha, M., Elisma, F., and Figeys, D. (2008) Identification of protein–protein interactions by mass spectrometry coupled techniques. *Adv. Biochem. Eng. Biotechnol.* **110**, 67–80.
56. Kocher, T., and Superti-Furga, G. (2007) Mass spectrometry-based functional proteomics: from molecular machines to protein networks. *Nat. Methods* **4**, 807–815.
57. Lee, W. C., and Lee, K. H. (2004) Applications of affinity chromatography in proteomics. *Anal. Biochem.* **324**, 1–10.
58. Simpson, R. J., and Dorow, D. S. (2001) Cancer proteomics: from signaling networks to tumor markers. *Trends Biotechnol.* **19**, S40–S48.
59. Yarmush, M. L., and Jayaraman, A. (2002) Advances in proteomic technologies. *Annu. Rev. Biomed. Eng.* **4**, 349–373.
60. Christofk, H. R., Vander Heiden, M. G., Wu, N., Asara, J. M., and Cantley, L. C. (2008) Pyruvate kinase M2 is a phosphotyrosine-binding protein. *Nature* **452**, 181–186.
61. Christofk, H. R., Vander Heiden, M. G., Harris, M. H., Ramanathan, A., Gerszten, R. E., Wei, R., Fleming, M. D., Schreiber, S. L., and Cantley, L. C. (2008) The M2 splice isoform of pyruvate kinase is important for cancer metabolism and tumour growth. *Nature* **452**, 230–233.
62. Rigaut, G., Shevchenko, A., Rutz, B., Wilm, M., Mann, M., and Seraphin, B. (1999) A generic protein purification method for protein complex characterization and proteome exploration. *Nat. Biotechnol.* **17**, 1030–1032.

Chapter 2

Overview of Quantitative LC-MS Techniques for Proteomics and Activitomics

John F. Timms and Pedro R. Cutillas

Abstract

LC-MS is a useful technique for protein and peptide quantification. In addition, as a powerful tool for systems biology research, LC-MS can also be used to quantify post-translational modifications and metabolites that reflect biochemical pathway activity. This review discusses the different analytical techniques that use LC-MS for the quantification of proteins, their modifications and activities in a multiplex manner.

Key words: Mass spectrometry, proteomics, activitomics, quantification, isotope labeling, label free.

1. Introduction

LC-MS has become a powerful molecular biology tool and multiple strategies for peptide, protein and enzyme activity quantitation by LC-MS have been developed and applied to address a wide range of biological questions. **Chapter 1** discusses the concepts of biochemical pathway analysis using LC-MS. In this chapter, we describe how to approach such types of analysis using the different quantitative strategies available to the researcher.

LC-MS protein quantification strategies fall into two broad categories: those that are label free and those which involve protein/peptide labelling (known as differential mass tagging or isotopic labelling). These strategies are mostly used for relative quantification of proteins/peptides between two or more samples of interest. However, these methods have also been adapted for absolute quantitation, where the introduction of a known amount

P.R. Cutillas, J.F. Timms (eds.), *LC-MS/MS in Proteomics*, Methods in Molecular Biology 658,
DOI 10.1007/978-1-60761-780-8_2,

of internal standard (often an isotopically labelled version of the peptide of interest) into the test samples permits the direct quantitation based on the peak intensity (ion current) of the standard versus endogenous peptide. In general terms, label-free quantitation methods involve comparison of peptide ion currents between samples or the comparison of spectral counts for peptide ions from particular proteins across the samples (see below). In terms of quantitative accuracy and coverage, the label-free approaches are more reliant on chromatographic reproducibility and the ability to detect the same ions from LC-MS run to run. In contrast, labelling strategies permit the mixing of samples prior to LC-MS, and in some cases upstream of any fractionation. Thus, multiple specimens can be run simultaneously with the same peptides (or proteins) being identically separated and co-eluted into the mass spectrometer with ion intensities being directly compared in the same MS or MS/MS scans. The labelling strategies thus improve throughput and quantitative accuracy. The following sections describe in more detail both labelling and label-free strategies for quantitative LC-MS, their advantages and disadvantages, and some of their applications.

2. Labelling Methods for Quantitative LC-MS in Proteomics

2.1. General Considerations

Differential labelling approaches for quantitative LC-MS in proteomics fall into two main categories: those which use chemical derivatisation or enzymatic modification of proteins or peptides after sample collection and those which use incorporation of isotope-labelled amino acids in vivo. The chemical labelling approaches make use of tags with the same (isobaric labelling) or different masses (isotopic labelling). Isobaric labelling, exemplified by isobaric tags for relative and absolute quantitation (iTRAQ) (1), consists of peptide tags which generate specific fragment ions by MS/MS. Samples are differentially labelled and then combined and concurrently analysed by LC-MS/MS, with relative quantitation performed by comparison of intensities of the 'reporter' fragments in the MS/MS spectra. In contrast, isotopic labelling methods, such as isotope-coded affinity tags (ICATs) (2) or proteolytic ^{18}O labelling (3), generate pairs (or more) of peptides with a mass difference introduced by the label. The ion intensities of the isotopic forms of the labelled peptides which should have identical LC elution profiles are then compared to give a peptide ratio of the 'heavy'-labelled versus 'light'-labelled peptide. In a similar manner, differential isotopic labelling in vivo allows quantification of peptides following incorporation of 'light' (^{12}C, ^{14}N, ^{1}H) and 'heavy' stable isotope-labelled (^{13}C, ^{15}N, ^{2}H) amino acids and is exemplified by the stable isotope labelling of amino acids in culture (SILAC) strategy (4).

In all of these methods, the ratios of detected 'reporter' fragments or isotopically labelled peptides are computed and integrated into protein ratios which can then be evaluated statistically. Multiple software solutions for the analysis of quantitative information using these labelling strategies are available and have been recently reviewed (5) (*see* also **Chapter 4**). Although many of these solutions are instrument, data or tag dependent, they all work on the same principle whereby isotopically labelled peptide pairs (or reporter ions) are extracted on the basis of their characteristic mass differences and successful MS/MS peptide assignments. Ratios of the extracted isotopic pairs are then computed and statistical evaluation performed. It is important to note that the smaller the mass difference of the tags is, then the more difficult it becomes to interpret the data and perform accurate quantification, since the isotope envelopes of the differentially labelled peptides may overlap.

When using any labelling approach for LC-MS, the labels are best introduced at the earliest point in the workflow to minimise differences introduced into the samples by handling or quantitative differences between LC-MS runs. In this sense, the in vivo labelling strategies outperform the chemical and enzymatic labelling strategies in terms of accuracy of quantitation; however, as will be discussed below, multiplexing using in vivo labelling is presently more limited than chemical labelling, where 12-plex strategies have been reported (6). Another important difference in these tagging strategies is in the analysis of primary tissues and clinical samples such as tissues, body fluids and urine, which are only amenable to the in vitro labelling approaches.

2.2. ICAT Isotope-Coded Affinity Tags and Variations

Although stable isotope labelling for protein quantitation had been previously reported, ICAT was the first robust and universal differential labelling strategy to be developed for quantitative LC-MS and is based on cysteine thiol group modification using iodoacetamide tags. In the first report, ICAT was used to examine the expression profiles of yeast growing on either galactose or ethanol in a single analysis (2). Stable isotopes were incorporated into intact proteins after lysis by selective alkylation of cysteines with either a heavy (deuterium D_8) or a light (deuterium D_0) reagent bearing a biotin tag. Prior to LC-ESI-MS/MS analysis, the protein mixture was digested with trypsin and the ICAT-labelled (cysteine-containing) peptides enriched on monomeric avidin–agarose. This had the advantage of simplifying the peptide mixture for downstream LC-MS/MS-based identification. Subsequent work used the ICAT strategy to successfully identify and determine the ratios of abundance of 491 proteins between the microsomal fractions of naïve and in vitro-differentiated human myeloid leukaemia cells (7), whilst Griffin et al. combined ICAT labelling with LC-MALDI-MS/MS for targeted protein identification for better sample utilisation and reduced MS instrument time (8).

Several technical problems were reported with the original deuterated ICAT reagents. Prominent of these was the differential elution of ICAT-labelled peptide pairs in reversed-phase HPLC (RP-LC) leading to errors in the quantification (9, 10). A further problem was the relatively low efficiencies of labelled peptide collision-induced dissociation (CID), which was speculated to be due to the relatively large size of the ICAT moiety (11). Poor recovery of tagged peptides from avidin may also contribute to reduced proteomic coverage. To improve these issues, a new generation of ICAT reagents were developed where the deuterium atoms were substituted for nine ^{12}C (light) or ^{13}C (heavy) atoms and an acid-cleavable linker added. These new generation cleavable ICAT (cICAT) reagents gave more precise co-migration of the light- and heavy-tagged peptides in RP-LC, whilst the cleavage strategy eliminated undesired residual fragmentation of biotinylated peptides and improved recovery (12). However, an additional clean-up step was required for removal of the cleaved biotin moiety prior to LC-MS analysis. A subsequent large-scale investigation of the cICAT strategy, which measured protein changes in *Escherichia coli* treated with an inhibitor of fatty acid biosynthesis, quantified more than 24,000 peptides in four independent runs using an ion-trap mass spectrometer (13). Good reproducibility in quantification was reported (median CV of ratios was 18.6%), and on average >450 unique proteins were identified per experiment. However, the method was biased towards the detection of acidic proteins and under-represented small and hydrophobic proteins.

In another multiple-run study examining proteomic changes in lymphoma cells following different ligand and drug treatments, Vaughn et al. reported that the majority of ICAT-labelled proteins were identified by single peptides with only 24–41% of these containing cysteine residues (14). Eighty-five percent of cysteine-containing peptides yielded quantification data, but this represented only 28% of all identified proteins. This identification of proteins from single peptides highlights a significant drawback of cysteine-labelling strategies, where more false-positive identifications are likely to occur and quantification accuracy will be lower than when data from multiple peptides derived from the same protein is combined.

Although not covered in this volume, we refer the reader to several published protocols describing sample labelling with ICAT reagents, chromatographic fractionation of labelled tryptic peptides, protein identification and ICAT quantification using MS/MS (15–17). Variations to the original ICAT strategy have also been reported. These include fluorescent-labelled ICAT (FCAT), allowing absolute quantification by fluorescence measurement and peptide enrichment using an anti-FITC antibody or iminodiacetic acid-coated beads (18); visible isotope-coded affinity tags (VICAT) that contain a visible probe for monitoring chromatographic behaviour and a photo-releasable biotin affinity tag for selective capture and release of labelled peptides (19, 20); acid-labile isotope-coded extractants (ALICE), a class of chemically modified resins that contain a thiol-reactive group to capture cysteine-containing peptides and a heavy or a light isotope-coded, acid-labile linker (21); and metal-coded affinity tags (MeCATs), where different element-coded metal chelates are used for cysteine labelling which can be purified on metal chelate-specific affinity resins (22). The compatibility and robustness of the MeCAT technology for the relative quantification was recently shown using standard LC-MS techniques and offered the unique advantage of absolute quantification via inductively coupled plasma mass spectrometry (23).

The principle of ICAT has also been developed to examine differential protein post-translational modifications. In the phosphoprotein isotope-coded affinity tag (PhIAT) approach, phosphoserine and phosphothreonine residues are derivatised by hydroxyl ion-mediated beta-elimination followed by Michael addition of 1,2-ethanedithiol (EDT). Peptides are captured after labelling the EDT moiety with an isotope-coded biotin affinity tag (24). This idea was further developed such that instead of using biotin–avidin enrichment, EDT peptides were captured and labelled on a solid-phase support in a single step using light ($^{12}C_6$, ^{14}N) or heavy ($^{13}C_6$, ^{15}N) phosphoprotein isotope-coded solid-phase tagging (PhIST) reagents and released from the solid-phase support by UV photo-cleavage (25). Similarly, a beta-elimination and Michael addition-based approach for the relative quantification of *O*-phosphate or *O*-GlcNAc-modified peptides used differential isotopic labelling with normal or deuterated dithiothreitol (26). Thiol chromatography was used for enrichment, whilst the specificity of *O*-phosphate versus *O*-GlcNAc mapping was achieved by enzymatic dephosphorylation or *O*-GlcNAc hydrolysis. Finally, given the fact that cysteine thiols are the targets of oxidative modifications relevant to biological function and disease, the ICAT reagents have been used for quantification of oxidative PTMs by virtue of the fact that cysteine thiol oxidation blocks ICAT labelling (27, 28).

2.3. Amino and Carboxyl Group Differential Isotope Labelling

As the majority of protease-generated peptides (and some proteins) do not contain cysteines, the ICAT strategy has limitations in terms of proteomic coverage, particularly for peptidomic analyses. Thus, alternative methods have been described where the amino groups of the N terminus and lysine residues or the carboxylic acids of the C terminus and aspartic and glutamic acid residues are labelled with either deuterium-free (light) or deuterium-containing (heavy) tags. Reagents that have been used to label amino groups include formaldehyde (29, 30), acetic anhydride (31), propionic anhydride (32, 33), succinic anhydride (34), *N*-acetoxy-succinimide (35, 36), phenyl isocyanate (PIC) (37), *N*-nicotinoyloxy-succinimide (31, 38) and [3-(2,5-dioxopyrrolidin-1-yloxycarbonyl)propyl] trimethyl ammonium chloride (34), whilst peptide carboxylic acids have been differentially labelled by methyl esterification with either D_0- or D_3-methanol (39).

Although relatively inexpensive, formaldehyde, the acetic and succinic anhydrides and PIC convert positively charged amines into neutral (formaldehyde/acetyl/PIC) or negatively charged sites (succinyl), with the effect that some peptides give weak MS signals in the positive ion mode (34). Weak signals may also occur due to partial labelling of lysine ε-amino groups (31). Labelling with *N*-nicotinoyloxy-succinimide for nicotinylation or [3-(2,5-dioxopyrrolidin-1-yloxycarbonyl)propyl]trimethylammonium chloride to give the 4-trimethylammoniumbutyryl (TMAB) modification avoids this problem since the positive charge of the amino group is maintained (38, 40, 41). However, it has been reported that TMAB-labelled peptides can partially decompose during CID, making data interpretation and quantification more complex (34, 41). As with the cysteine-labelling isotopic tags, pairs of peptides labelled with the light and heavy deuterated forms of these tags do not always co-elute in RP-LC (34). Amino-labelling reagents have also been used for N-terminal labelling alone, where lysine side chains are first blocked by guanidination (32). Finally, a method for labelling the amino groups of intact proteins using isotope-coded tags has been reported (42). Termed isotope-coded protein labelling (ICPL), the method used D_0- and D_4-*N*-nicotinoyloxy-succinimide or D_0-, D_3- and D_7-butyric acid-2,5-dioxopyrrolidin-1-yl-ester to label proteins prior to digestion with trypsin or Glu-C. The method was shown to be accurate and reproducible for quantification and circumvents technical variations introduced at the digestion step.

2.4. Differential Proteolytic $^{16}O/^{18}O$ Labelling

Differential stable isotopic labelling can also be achieved enzymatically. The major method relies on the oxygen atom exchange that takes place at the C-terminal carboxyl group of peptides during proteolytic digestion. Here, one or two ^{16}O atoms are replaced by one or two ^{18}O atoms through enzyme-catalysed exchange in the presence of $H_2{}^{18}O$. Since any variability in labelling relies solely on the digestion step, this method is expected to give smaller technical variations than do the two-step chemical labelling methods, where both labelling and digestion are a source of introduced variation. The method was first suggested as a protein quantification tool when ^{18}O-labelled internal standards were generated for absolute quantification by MALDI–MS (43). This was followed by the reporting of conditions for protein labelling (44) and then the first proteomic application of the method, where trypsin was used to incorporate two ^{18}O atoms into the C termini of all tryptic peptides and was applied to compare proteins from two serotypes of adenovirus (3). Subsequent work by the same group showed that Glu-C could also be used for labelling, that ^{18}O-labelled and unlabelled (^{16}O) peptide pairs co-eluted in RP-LC and measurements of isotope ratios by LC-MS were accurate and precise (45). Applications include combining with 2D-LC–FT-ICR and an accurate mass and time (AMT) tag strategy to identify and quantify 429 distinct plasma proteins from an individual prior to and after lipopolysaccharide administration (46); combining with $^{16}O/^{18}O$-methanol esterification, immobilised metal-ion affinity chromatography and RP-LC followed by neutral loss-dependent MS^3 for phosphopeptide identification in the study of lysophosphatidic acid-induced chemotaxis (47); the differential analysis of NF-κB transcription factor complexes following TNF-α stimulation (48); the labelling of a 'universal' reference sample of pooled plasma for spiking into individual unlabelled samples for quantitative analysis across clinical samples (49).

Several drawbacks of the proteolytic labelling strategy are apparent. Only two samples can be compared simultaneously, C-terminal peptides of proteins cannot be quantified and variable incorporation of ^{18}O into peptides can occur (50). There is also a lack of computational tools for accurate quantification of peptide differences and this is exacerbated by the overlap of isotopic envelopes for ^{16}O- and $^{18}O_1$- and $^{18}O_2$-labelled peptides. However, some methods have been reported for correcting $^{16}O/^{18}O$ ratios from overlapping isotopic multiplets (51, 52). Rao et al. (52) have also shown the potential of Lys-N labelling, where conditions were established such that only a single ^{18}O atom was incorporated into the C terminus of each peptide, compared to incorporation of one or two ^{18}O atoms when using trypsin, Lys-C or Glu-C. Finally, a significant degree of chemical back exchange of the carboxyl ^{18}O can occur in $H_2{}^{16}O$ solvent, particularly at extreme pH, although handling recommendations to limit this have been reported (44).

2.5. Isobaric Tagging (iTRAQ, TMT, CILAT)

Isobaric tagging is a multiplex peptide labelling method that relies on the introduction of stable isotope tags that are chemically identical but distinguishable by MS/MS due to their fragmentation into reporter ions of different masses. The most commonly used method of this type has been the 4-plex iTRAQ reagents, which are *N*-hydroxysuccinimidyl esters for the labelling of primary amino groups. A specific reporter group in each tag based on *N*-methylpiperazine generates ions of 114, 115, 116 and 117 *m/z* upon CID fragmentation. These appear in an ion-sparse region of MS/MS spectra and their relative intensities provide the relative abundance of labelled peptides between the samples. The reporter groups in each tag are mass-balanced with a linker group making the tags isobaric. The major advantage of this MS/MS-dependent strategy is that the multiplex labelling does not increase the mass complexity of the samples and only peptides subjected to CID fragmentation are quantified. In addition, higher signal-to-noise ratios can be achieved with MS/MS-based detection versus MS-mode measurements.

In the first reported use of iTRAQ, Ross et al. (1) compared global protein expression of wild-type and mutant yeast strains defective in the nonsense-mediated and 5′ to 3′ mRNA decay pathways using 2D-LC linked to MALDI– and ESI-MS/MS. Under optimised labelling conditions, there was an estimated 97% labelling of N termini and lysine -amino groups, with a minimal degree of unlabelled or tyrosine-labelled peptides. Lysine-derivatised peptides were more frequently identified, possibly due to their higher ionisation efficiency versus arginine-terminated peptides. Peptide ratios were averaged for each protein and 685 proteins were quantified in all three yeast strains using two or more significant scoring peptides. A high degree of reproducibility (CV 15–17%) for individual peptides contributing to any given protein was reported. This study also determined the absolute levels of a target protein after spiking with a synthetic peptide standard labelled with one of the isobaric tags. An 8-plex version of iTRAQ has also recently been commercialised, generating a spectrum of eight unique reporter ions at 113, 114, 115, 116, 117, 118, 119 and 121 *m/z* increasing sample throughput for complex differential analyses (53–55). **Chapter 12** describes a protocol for iTRAQ labelling of multiple samples with subsequent 2D-LC-MS/MS-based quantification.

Use of the iTRAQ strategy has increased rapidly in the last few years with numerous applications including the study of transformation-dependent protein changes (56), protein complex formation (57), post-translational modifications (57–64), temporal changes in cellular signalling and protein localisation (65–68), the effects of drug treatments (53, 69, 70) and in disease biomarker discovery (71, 72). In one of these studies, tryptic peptides from four different EGFR stimulation time points were labelled with the 4-plex iTRAQ reagents, the samples mixed and tyrosine-phosphorylated peptides enriched by immunoprecipitation and IMAC prior to LC-MS/MS analysis (65). In other studies, iTRAQ was used to label and hence identify newly formed N termini as endogenous proteolytic cleavage sites after blocking lysine amino groups by guanidination (62), whilst differentially carbonylated proteins were identified by combining biotin hydrazide labelling of carbonyl groups and avidin affinity chromatography (63). Finally, Quaglia et al. (73) have reported the use of iTRAQ for absolute protein quantification using synthetic peptides as standards.

An alternative 6-plex isobaric tagging method known as tandem mass tags (TMTs) was developed and reported prior to the 4-plex iTRAQ strategy (74), but was only recently commercialised. The labels are used in the same way as the iTRAQ reagents to label primary amino groups on digested protein samples and generate reporter ions of 126, 127, 128, 129, 130 and 131 *m/z* after fragmentation. The TMT method was recently used to compare protein fractions from post-mortem and ante-mortem human cerebrospinal fluid samples after immunoaffinity depletion and SCX chromatography (75). After RP-LC separation, peptides were identified and quantified by MALDI–TOF/TOF and ESI-Q-TOF analysis, revealing putative brain damage biomarkers and demonstrating the validity and robustness of the TMT approach. The TMT strategy has also been combined with ETD to identify and quantify PTMs on bovine alpha-crystallin (76).

Isobaric cysteine-tagging reagents have also been reported, known as cleavable isobaric-labelled affinity tags (CILATs) (77). The original CILAT reagents were then further developed to combine the benefits of both ICAT and iTRAQ and allow the quantification of up to 12 samples in a single run. The tags are comprised of a thiol capture group, reporter and balance groups and a cleavable alkyne tag. The reporter groups incorporate ^{2}H, ^{13}C and ^{15}N in various combinations, giving reporter ions spanning from 130 to 142 Da (12 masses excluding 136 Da; the mass of the tyrosine immonium ion). The cleavable alkyne tag allows the solid-phase-based enrichment of labelled peptides onto beads coated with azide via cyclo-addition (click chemistry). Tagged peptides are then released from beads by acid treatment for LC-MS/MS analysis (6). Also of note, a non-isobaric version of iTRAQ known as mTRAQ has now been commercialised for absolute quantification of proteins based on multiple reaction monitoring (MRM) (see below). The mTRAQ methodology relies on MRM to target tryptic peptides from the protein of interest in the test sample, while the tags enable quantification of these peptides in the test sample versus spiked internal standard peptides through unique MRM transitions conferred by the labels (78).

There are several drawbacks to the isobaric tagging methods when compared to other labelling strategies. The iTRAQ/TMT reagents are expensive, difficult to synthesise and show signals only when peptides are subjected to fragmentation. Thus, the strategy misses peptides not selected for MS/MS, thus lowering proteomic coverage. Dedicated software must also be used for data analysis, although as well as commercially available software (Mascot, Proteome Discoverer, Protein Pilot), free computational tools for iTRAQ quantification and protein identification with details of statistical considerations when analysing iTRAQ data have been reported (79–87). It is also evident that the low collision energies used in ion traps and some Q-TOF platforms can result in low iTRAQ reporter ion abundances and hence less accurate quantification data. Thus, higher energy CID methods have been employed, such as pulsed Q dissociation available on the popular LTQ linear ion-trap instruments or 'higher energy CID' available on the LTQ-Orbitrap. However, it is apparent that careful tuning is required for optimal fragmentation (70, 88, 89). Electron transfer dissociation (ETD) is also possible in iTRAQ experiments, although it is best combined with CID to give the full complement of reporter ions (54, 55). Finally, isotope purity correction in measured peak areas needs to be applied for each batch of reagents used and there is reported evidence of the compression of the dynamic range of ratios determined by the iTRAQ technique (78).

2.6. Stable Isotope Labelling of Amino Acids in Culture (SILAC)

As mentioned previously, SILAC is an in vivo stable isotope labelling method which uses heavy and light versions of essential amino acids that are added to the growth media of metabolically active cells. The idea behind the strategy originates from comparative proteomic experiments of simple model organisms (90–93) and plants (93) which can be grown in either medium containing ^{14}N at natural abundance (99.6%) or the same medium enriched in ^{15}N. The method was adapted for the analysis of mammalian cell culture systems where deuterated leucine ([D_3]Leu) was supplemented into the growth media of cells in one state for comparison with unlabelled cells in another state (4). Cells were harvested and equal amounts of cells or protein lysate mixed prior to fractionation and LC-MS. As mentioned above, this has the benefit of reducing technical variability since the samples for comparison are mixed at the earliest possible stage in the workflow and are thus treated as one sample and subjected to the same downstream manipulations. The intensities of all peptides ions containing the label can be compared with their unlabelled counterparts for relative quantification.

The method was improved with the introduction of $[^{13}C_6]$Lys and $[^{13}C_6]$Arg, providing a larger mass difference between light and heavy peptides, giving predominantly C-terminal tagging of tryptic peptides and improving co-elution of heavy and light peptides by RP–LC. Various labels have subsequently been combined for 3-, 4- and 5-plex comparisons ($[^{13}C_6]$Lys (+6 Da); $[^{13}C_6]$Arg (+6 Da); $[^{15}N_4]$Arg (+4 Da); $[D_4]$Lys (+4 Da); $[^{13}C_6,^{15}N_2]$Lys (+8 Da); $[^{13}C_6,^{15}N_4]$Arg (+10 Da); $[^{13}C_6,^{15}N_2,D_9]$Lys (+17 Da); $[^{13}C_6,^{15}N_4,D_7]$Arg (+17 Da)], with the method shown to give reproducible quantitative information (94–96). Whilst this volume describes protocols for SILAC labelling and LC-MS quantification (*see* **Chapters 11**, **15** and **16**), several others have been reported recently (97–99). Software platforms for SILAC-based quantification have also been described, including MaxQuant for LTQ-Orbitrap-acquired data (100).

The SILAC method has mostly been used for the comparison of protein expression and PTMs between two or more samples. As an excellent example, a recent study by Graumann et al. showed that murine embryonic stem (ES) cells could be fully SILAC labelled when grown feeder-free during the last phase of cell culture. In a strategy using parallel 1D gel electrophoresis and isoelectric focusing of peptides from three crude ES cellular fractions, high-resolution analysis on an LTQ-Orbitrap at sub-ppm mass accuracy yielded confident identification and quantification of >5,000 proteins (101). SILAC has also been used to measure rates of protein translation and turnover in pulse-chase style experiments (102–104) and to measure dynamic changes in protein complexes (105). Importantly, the SILAC approach has been used to address specificity issues in affinity purifications, where the association of specific proteins can be distinguished from co-purifying background proteins by comparison of peptide isotope ratios from control and test samples. In this way, SILAC has been used to qualify protein–protein (106–110), DNA–protein (111); small molecule–protein (112, 113) and phosphopeptide–protein (114) interactions.

Internal standard SILAC approaches have also been applied, where equal amounts of individual heavy-labelled samples are mixed and compared with each of the light-labelled samples, thereby improving cross-sample quantitative accuracy (115). Absolute SILAC for protein quantitation is another variation, where recombinant proteins are SILAC labelled during their production and then used as internal standards by mixing directly into cell or tissue lysates (116). An adaptation of the SILAC protein profiling method called heavy methyl SILAC has also been reported, where cells convert $[^{13}C, D_3]$Met to $[^{13}C, D_3]$ *S*-adenosyl methionine, the sole biological methyl donor. The heavy methyl groups are incorporated into proteins during in vivo methylation for the labelling and relative quantification of this PTM (117). Finally, a method which combines features of both the SILAC and isobaric labelling strategies has been reported (118). Here, cultures are grown in media containing isobaric forms of amino acids, labelled with either ^{13}C on the carbonyl (C-1) carbon or ^{15}N on the backbone nitrogen. Labelled peptides from the two samples thus have the same mass but generate distinct immonium ions upon fragmentation differing by 1 mass unit. The relative intensities of these immonium ions are used for relative quantification of the parent peptide ions.

The major drawback of the SILAC method is that it can be used only for samples where in vivo labelling is possible. This has restricted its use to simple organisms or cultured cell models, and so the quantitative comparison of proteins derived from tissues or body fluids is not possible. Despite this, a recent study showed that proteins from mice fed on a diet containing [$^{13}C_6$]Lys could be completely labelled with the heavy amino acid without obvious effect on their development, growth or behaviour. MS analysis was used to examine incorporation rates into proteins across generations and in different tissues to measure in vivo protein half-lives and to more accurately determine protein expression changes in knockout versus control animals (119). An additional drawback of the SILAC method is the cellular conversion of isotope-labelled arginine to proline, resulting in dilution of heavy peptide ion signals and hence inaccuracies in quantification. This is of particular concern as it can effect up to half of all peptides in a proteomic experiment. The problem can however be alleviated by reducing the L-arginine concentration (120), by supplementing L-proline into the SILAC media (121) or by mathematical correction (120). An internal correction for arginine conversion has also been applied by using different heavy arginine forms ([$^{15}N_4$]- and [$^{13}C_6$,$^{15}N_4$]Arg) in both the light and heavy conditions (122). Here, heavy proline will be formed at the same rate under both conditions and can be used to correct for the conversion.

3. Label-Free Quantification

3.1. General Considerations

The protein labelling techniques described above can quantify the relative abundance of proteins and peptides across samples. Although powerful for the investigation of cell biological processes, their main drawback is that labelling methods can be used to compare only a limited number of specimens. This feature means that it is often difficult to obtain sufficient data to perform statistical analysis and to infer biological significance of differences observed with these proteomic experiments. This is particularly problematic for clinical studies where a relatively large number of samples need to be evaluated to account for biological variability. In addition, metabolic labelling techniques cannot be used to analyse human primary tissues, and protein derivatisation methods are expensive, difficult to implement, need to be optimised for each type of sample and can result in the introduction of variability at the chemical derivatisation step.

The proteomics community has therefore shown great interest in the development and use of label-free quantitative LC-MS as an alternative (42, 105, 123–145). Particular advantages of label-free quantitative MS are their wide applicability, low expense and their ability to compare unlimited numbers of samples. In addition to its uses for relative quantification, techniques based on the label-free concept may also be used for absolute quantification of proteins on a global scale.

Label-free quantitative methods are defined here as those techniques that do not require chemical derivatisation steps or incorporation of stable isotopes by proteins in metabolically active cells. As introduced above, they can be classified into two groups: methods based on the evaluation of ion currents and those based on spectral counting. These are described in detail in the following sections.

3.2. Methods Based on Ion Currents

LC-MS and LC-MS/MS techniques are routinely used in pharmacological, forensic and environmental laboratories to quantify small molecules such as drugs, natural metabolites and pesticides. Results of such analyses are accepted by regulatory and official bodies and in court. Thus the fact that LC-MS is a quantitative technique is beyond doubt. Quantification by LC-MS involves measuring the elution profile of molecules and correlating this with their concentration in samples. This can be done by calculating peak areas or heights of extracted ion chromatograms (XICs) of the m/z of the molecule to be quantified. The specificity of the quantification is enhanced when this is done by MS/MS, which involves measuring the elution profile of one or more fragment ions produced as a result of fragmentation of the molecular ion by, for example, collision-induced dissociation (CID). This technique thus measures the transition of parent to fragment ion and is commonly referred to as selected reaction monitoring (SRM) or multiple reaction monitoring (MRM), depending on the number of molecules being analysed in an LC-MS run; these are normally carried out in triple quadrupole instruments.

In principle, the same concepts and techniques used to quantify small molecules may be used to quantify peptides and proteins. However, until relatively recently, the consensus in the proteomics community was that labelling approaches were needed to accurately quantify proteins present in complex mixtures (2, 146). The reason for the initial reluctance of the proteomics community to make use of these more straightforward, non-labelling modes of quantification was probably due to the belief that, because of the extreme complexity of the samples, signal suppression was going to adversely affect protein quantification.

Although ion suppression is commonly observed in the LC-MS analysis of small molecules, there is in fact very little experimental evidence to support the notion that ion suppression occurs under the experimental conditions normally encountered in LC-MS/MS settings for proteomics. Indeed, the evidence actually points to the opposite; when experimental conditions for the analysis of small molecules are modified so that these resemble the conditions used in proteomics, signal suppression effects are minimised or completely avoided (147–149). This is because proteomics use techniques for extensive separation of analytes before MS analysis, which reduces sample complexity, and because ionisation by nanoelectrospray (employed in proteomics) is much more efficient than ionisation by conventional electrospray (used in the analysis of small molecules). Indeed, at sufficiently low flow rates, signal suppression by competition for protonation is extremely unlikely (150).

Chelius and Bondarenko (151) was one of the first groups to show that the ratios of peak areas in XICs could be used to quantify relative abundance of proteins across samples. Later it was shown that protein quantification with this technique was also suitable for the analysis of proteins in complex mixtures (127–129, 143, 145, 151). It was also demonstrated that by spiking the samples with non-labelled internal standards at known concentration, the accuracy of quantification could be improved by normalising for protein losses during handling and for differences in the analytical behaviour of the LC-MS/MS instrumentation (127). It was later proposed that introduction of an internal standard (IS) could also be used to quantify proteins in absolute units (143). This technique considers the peak areas of the three most intense peptides derived from a protein. It was found that the intensities of these three peaks are the same regardless of protein identity; thus by comparing the intensities of a known amount of IS with those of the protein to be quantified, one can provide estimates of protein amounts in absolute units with a certain degree of accuracy. These results may be rationalised by considering that proteins produce several peptides upon trypsinisation; thus although peptides with different sequences ionise with different efficiencies, the chances that at least three peptides of the many that are derived from a given protein may have similar ionisation efficiencies is high. A limitation of this technique is that small proteins may not produce a sufficient number of peptides for this to be applicable to all proteins, and therefore absolute quantification by this label-free method may not always be possible. It has also been proposed that the proportion of sum of all peptide ion currents relative to the total ion current can be used as a measure of protein amounts in absolute units. Nonetheless, regardless of its use as an absolute quantification strategy, the usefulness of label-free LC-MS for relative quantification of proteins is now well accepted by the proteomics community and has also been used to quantify phosphorylation sites on proteins (152, 153).

A major problem for the implementation of label-free quantitative LC-MS based on ion currents is that of choosing an appropriate strategy for data analysis. Computationally, it is more difficult to compare LC-MS runs than to evaluate the signals of groups of ions in the same MS trace, as in label-based experiments. This is because even with the most accurate nanoflow LC systems, changes in retention times across samples are unavoidable, especially when nanoflow columns are loaded with concentrated samples containing large amounts of peptides (in these cases of nearly overloading, interactions of peptides may account for differences in retention times in consecutive LC-MS runs). We have occasionally observed that up to 4 min shift in retention time in consecutive LC-MS runs (100-min gradient) when using a modern ultrahigh-pressure liquid chromatography system. This lack of absolute reproducibility is not a problem for label-based techniques because comparison of ion intensities is performed within the same MS data files, but it can affect the results of label-free LC-MS experiments.

Different strategies are currently available to the researcher in order to overcome difficulties in the analysis of label-free data; these include chromatogram alignment, generation of XIC of previously identified peptides, or using an MRM approach.

Comparison of aligned ion chromatograms. A common strategy for comparing LC-MS runs is to align and compare chromatograms resulting from the analysis of the samples to be evaluated (5). After alignment the intensities of mass features identified are compared using statistical tests. The identities of differentially expressed peptide ions are then obtained by including their masses and retention times in an inclusion list for targeted LC-MS/MS analysis. Procedures for alignment of chromatograms are not straightforward because differences in RT across samples are not always linear; RT offsets may be a few minutes in one direction at the beginning of the gradient, while they could be on the other direction at the end of it. This was the same problem that the 2D gel community faced when this technique was in its infancy, and the same solutions that were found to align 2D gels may also be used to align chromatograms. In fact, chromatogram alignment may be less problematic than aligning 2D gels because only RT may vary across samples (m/z values can be calculated with high degree of accuracy by modern mass spectrometers), whereas, because of variability of gel electrophoresis, spots in 2D gels vary in both the isoelectric and molecular mass axes. Several strategies have been published for the alignment of chromatograms (123, 154,155).

The extracted ion chromatogram (XIC) strategy. A different strategy is to target the quantification of peptides previously identified in LC-MS/MS experiments (5). This is achieved by automating the construction of XICs of identified peptides across the samples to be evaluated (128, 152). Normally the XIC have mass and retention time windows that are chosen to match the performance of the LC-MS system. In complex mixtures, several peptides may have the same *m/z* within the specified RT window; therefore narrowing the *m/z* window may help in reducing the probability of selecting the wrong peak for quantification. This is the basis for the accurate mass and time tag (AMT) approach pioneered by Richard Smith, which makes use of FT-ICR MS with a mass accuracy of <1 ppm for accurate selection of LC-MS features for quantification (129, 131, 156–160). The relatively recent availability of user-friendly FT instruments such as the Orbitrap and Q-TOFs with enhanced resolution and mass accuracies in the low ppm range makes this approach applicable in a wider number of laboratories. **Chapter 13** describes a Q-TOF-based data-independent analysis (MS^E), where during MS acquisition the collision energy is alternated such that two channels are collected. The first channel includes the abundance measurements of the intact peptides and the second channel the fragmented peptides for identification.

It is believed that at sufficiently low mass window, the specificity of quantification increases because it is unlikely that any two peptides may share the same *m/z* within a specified RT window. However, we have observed isobaric ions co-eluting within minutes of each other when analysing complex peptide mixtures. Because of the RT variability discussed above, it makes it difficult for computer programs to choose the right peak for quantification. In order to address this limitation, the charge of the peptides to be quantified can be used as a further restriction for selecting the correct peak for quantification. Thus we have found that in occasions isobaric peptides ions may elute in a narrow retention time window, yet the charges of these peptides may be different. Thus by performing XICs of the three first isotopes, it may be possible to more accurately select the right peptide for quantification in cases of co-eluting isobaric peptide ions.

Multiple reaction monitoring. Further specificity may be achieved by MRM. This mode of MS/MS makes use of fast scan triple quadrupoles (or quadrupole linear ion traps, such as the Q-TRAP, operating in triple quadrupole mode) in which a quadrupole mass filter selects for the *m/z* of the ion to be quantified, which is then fragmented in a collision cell by CID. At least one of the resultant fragment ions is then monitored by a second quadrupole. This analytical strategy can result in very accurate quantification, especially when using internal standards labelled with stable isotopes, but even when unlabeled isotopes are not available and structurally related compounds are used instead, SRM and MRM techniques are the gold standard for the analysis of small molecules in analytical chemistry laboratories. The use of MRM in proteomics is increasingly common (161), with some groups contributing to the systematic identification of optimised parent to fragment transitions to be used in large-scale quantification of proteomes either experimentally or computationally (162–166).

Although not canonical, MRM can also be performed in Q-TOF instruments. As with MRM in triple quadrupoles, this involves selecting parent ions for fragmentation and then analysing fragment ions by TOF MS instead of using a second quadrupole. The disadvantage is that up to 80% of ions are lost in the orthogonal acceleration region of the Q-TOF configuration, thus resulting in less sensitivity than when using triple quadrupoles. Also, because longer scan rates are used, Q-TOF instruments may not be able to monitor as many different ions simultaneously as triple quadrupoles can. However, MRM in Q-TOF instruments may also have advantages, including the fact that since all fragment ions are recorded by TOF, it is possible to add the signal of several of these fragments in XICs to increase sensitivity. This may be particularly important in MRM of peptides, where splitting of the signal from the parent ion may translate into a loss in sensitivity. Another advantage is that Q-TOFs can detect fragment ions with much greater mass accuracy than triple quadrupoles. This is advantageous and can also result in enhanced specificity when several isobaric ions co-elute and for reducing background to enhance signal to noise.

Most modern and competitive proteomics laboratories are now equipped with triple quadrupoles for MRM quantification of peptides. This technique is thus being increasingly used for targeted quantification of several peptides and post-translational modifications in validation experiments (167). However, whether or not MRM can be used as a general tool for the quantification of proteomes in discovery experiments is debatable. This is because of duty cycle constraints imposed in scanning mass analysers when switching to the analysis of ions in succession. The limit of ions that may be quantified in a single experiment by MRM may be around 500. If internal standards are used and several peptides per protein are to be monitored, then this number may be even lower. Thus at present MRM is restricted to validation experiments where the aim is to analyse a limited number of molecules (162–166). **Chapters 9** and **10** provide overviews of SRM applied to quantitative proteomics and basic design of MRM experiments, respectively.

3.3. Spectral Counting Techniques

Quantification by spectral counting is based on a limitation well known to practitioners of LC-MS/MS for proteomics, which is that of under-sampling. In complex mixtures, such as a peptide mixture obtained by the digestion of a total cell lysate or the proteome of a biofluid, there is a staggering number of peptides (perhaps in excess of 100,000 different components), and therefore, even fast scanning tandem mass spectrometers cannot possibly select all peptide ions for MS/MS in data-dependent acquisition (DDA) experiments. In this type of experiment, peptide ions are selected for MS/MS as they elute from the LC column when their intensities reach a predefined threshold value in the MS scan. Peptides producing larger ionic intensities are preferentially selected over peptides with less intense signals. Since proteins produce several peptides when digested with a protease, peptides derived from more abundant proteins will be selected for MS/MS more often than peptides derived from less abundant proteins. It has been found that this spectral count (essentially number of times peptides from a given protein are selected for MS/MS) is roughly proportional to protein concentration and has therefore been used as the basis for several protein quantification strategies, which differ in the way spectral counts are normalised and in their scope (168, 169).

Normalisation is essential since large proteins produce more peptides than smaller proteins. There may also be differences in the quality of peptides produced from the digestion of different proteins. Some of these tryptic peptides (when the protease of choice is trypsin) may be more amenable to MS/MS analysis (because they are the right size and chemical characteristics) than others.

4. Conclusion and Perspectives

Several methods now exist for the quantification of peptides, proteins and activities. Quantifying sites of modification and metabolites can be used as a measure of enzymatic and pathway activity. This is an exciting realisation because this implies that it may soon be possible to quantify pathway fluxes with the same depth of analysis as that achieved for the quantification of proteins. LC-MS is likely to continue playing a central role in this regard. Several of the different techniques that can be used for this purpose have been introduced in this chapter. Choosing the most appropriate of these methods should be based on the question that needs to be addressed and the nature of the samples. For example, SILAC may be the technique of choice when the aim is to compare the proteomes of two cell lines amenable to growth in SILAC media, whereas label-free approach may be more appropriate for biomarker discovery projects. Thus an understanding of the pros and cons of the different techniques introduced in this chapter will assist in the most appropriate use of quantitative LC-MS for biomedical research.

References

1. Ross, P. L., Huang, Y. N., Marchese, J. N., Williamson, B., Parker, K., Hattan, S., Khainovski, N., Pillai, S., Dey, S., Daniels, S., Purkayastha, S., Juhasz, P., Martin, S., Bartlet-Jones, M., He, F., Jacobson, A., and Pappin, D. J. (2004) Multiplexed protein quantitation in *Saccharomyces cerevisiae* using amine-reactive isobaric tagging reagents. *Mol. Cell. Proteomics* **3**, 1154–1169.
2. Gygi, S. P., Rist, B., Gerber, S. A., Turecek, F., Gelb, M. H., and Aebersold, R. (1999) Quantitative analysis of complex protein mixtures using isotope-coded affinity tags. *Nat. Biotechnol.* **17**, 994–999.
3. Yao, X., Freas, A., Ramirez, J., Demirev, P. A., and Fenselau, C. (2001) Proteolytic ^{18}O labeling for comparative proteomics: model studies with two serotypes of adenovirus. *Anal. Chem.* **73**, 2836–2842.
4. Ong, S. E., Blagoev, B., Kratchmarova, I., Kristensen, D. B., Steen, H., Pandey, A., and Mann, M. (2002) Stable isotope labeling by amino acids in cell culture, SILAC, as a simple and accurate approach to expression proteomics. *Mol. Cell. Proteomics* **1**, 376–386.
5. Mueller, L. N., Brusniak, M. Y., Mani, D. R., and Aebersold, R. (2008) An assessment of software solutions for the analysis of mass spectrometry based quantitative proteomics data. *J. Proteome Res.* 7, 51–61.
6. Zeng, D., and Li, S. (2009) Improved CILAT reagents for quantitative proteomics. *Bioorg. Med Chem Lett* **19**, 2059–2061.
7. Han, D. K., Eng, J., Zhou, H., and Aebersold, R. (2001) Quantitative profiling of differentiation-induced microsomal proteins using isotope-coded affinity tags and mass spectrometry. *Nat. Biotechnol.* **19**, 946–951.
8. Griffin, T. J., Gygi, S. P., Rist, B., Aebersold, R., Loboda, A., Jilkine, A., Ens, W., and Standing, K. G. (2001) Quantitative proteomic analysis using a MALDI quadrupole time-of-flight mass spectrometer. *Anal. Chem.* **73**, 978–986.
9. Zhang, R., Sioma, C. S., Thompson, R. A., Xiong, L., and Regnier, F. E. (2002) Controlling deuterium isotope effects in comparative proteomics. *Anal. Chem.* **74**, 3662–3669.
10. Zhang, R., Sioma, C. S., Wang, S., and Regnier, F. E. (2001) Fractionation of isotopically labeled peptides in quantitative proteomics. *Anal. Chem.* **73**, 5142–5149.
11. Borisov, O. V., Goshe, M. B., Conrads, T. P., Rakov, V. S., Veenstra, T. D., and Smith, R. D. (2002) Low-energy collision-induced dissociation fragmentation analysis of cysteinyl-modified peptides. *Anal. Chem.* **74**, 2284–2292.
12. Yu, L. R., Conrads, T. P., Uo, T., Issaq, H. J., Morrison, R. S., and Veenstra, T. D. (2004) Evaluation of the acid-cleavable isotope-coded affinity tag reagents: application to camptothecin-treated cortical neurons. *J. Proteome Res.* **3**, 469–477.
13. Molloy, M. P., Donohoe, S., Brzezinski, E. E., Kilby, G. W., Stevenson, T. I., Baker, J. D., Goodlett, D. R., and Gage, D. A. (2005) Large-scale evaluation of quantitative reproducibility and proteome coverage using acid cleavable isotope coded affinity tag mass spectrometry for proteomic profiling. *Proteomics* **5**, 1204–1208.
14. Vaughn, C. P., Crockett, D. K., Lim, M. S., and Elenitoba-Johnson, K. S. (2006) Analytical characteristics of cleavable isotope-coded affinity tag-LC-tandem mass spectrometry for quantitative proteomic studies. *J. Mol. Diagn.* **8**, 513–520.
15. Shiio, Y., and Aebersold, R. (2006) Quantitative proteome analysis using isotope-coded affinity tags and mass spectrometry. *Nat. Protoc* **1**, 139–145.
16. Haqqani, A. S., Kelly, J. F., and Stanimirovic, D. B. (2008) Quantitative protein profiling by mass spectrometry using isotope-coded affinity tags. *Methods Mol. Biol.* **439**, 225–240.

17. Pan, S., and Aebersold, R. (2007) Quantitative proteomics by stable isotope labeling and mass spectrometry. *Methods Mol. Biol.* **367**, 209–218.
18. Rivera-Monroy, Z., Bonn, G. K., and Guttman, A. (2009) Fluorescent isotopecoded affinity tag 2: peptide labeling and affinity capture. *Electrophoresis* **30**, 1111–1118.
19. Bottari, P., Aebersold, R., Turecek, F., and Gelb, M. H. (2004) Design and synthesis of visible isotope-coded affinity tags for the absolute quantification of specific proteins in complex mixtures. *Bioconjug. Chem.* **15**, 380–388.
20. Lu, Y., Bottari, P., Aebersold, R., Turecek, F., and Gelb, M. H. (2007) Absolute quantification of specific proteins in complex mixtures using visible isotope-coded affinity tags. *Methods Mol. Biol.* **359**, 159–176.
21. Qiu, Y., Sousa, E. A., Hewick, R. M., and Wang, J. H. (2002) Acid-labile isotope-coded extractants: a class of reagents for quantitative mass spectrometric analysis of complex protein mixtures. *Anal. Chem.* **74**, 4969–4979.
22. Whetstone, P. A., Butlin, N. G., Corneillie, T. M., and Meares, C. F. (2004) Element-coded affinity tags for peptides and proteins. *Bioconjug. Chem.* **15**, 3–6.
23. Ahrends, R., Pieper, S., Neumann, B., Scheler, C., and Linscheid, M. W. (2009) Metal-coded affinity tag labeling: a demonstration of analytical robustness and suitability for biological applications. *Anal. Chem.* **81**, 2176–2184.
24. Goshe, M. B., Conrads, T. P., Panisko, E. A., Angell, N. H., Veenstra, T. D., and Smith, R. D. (2001) Phosphoprotein isotope-coded affinity tag approach for isolating and quantitating phosphopeptides in proteome-wide analyses. *Anal. Chem.* **73**, 2578–2586.
25. Qian, W. J., Goshe, M. B., Camp, D. G., 2nd, Yu, L. R., Tang, K., and Smith, R. D. (2003) Phosphoprotein isotope-coded solid-phase tag approach for enrichment and quantitative analysis of phosphopeptides from complex mixtures. *Anal. Chem.* **75**, 5441–5450.
26. Vosseller, K., Hansen, K. C., Chalkley, R. J., Trinidad, J. C., Wells, L., Hart, G. W., and Burlingame, A. L. (2005) Quantitative analysis of both protein expression and serine/threonine post-translational modifications through stable isotope labeling with dithiothreitol. *Proteomics* **5**, 388–398.
27. Sethuraman, M., McComb, M. E., Heibeck, T., Costello, C. E., and Cohen, R. A. (2004) Isotope-coded affinity tag approach to identify and quantify oxidant-sensitive protein thiols. *Mol. Cell. Proteomics* **3**, 273–278. Epub 2004 Jan 2015.
28. Hagglund, P., Bunkenborg, J., Maeda, K., and Svensson, B. (2008) Identification of thioredoxin disulfide targets using a quantitative proteomics approach based on isotope-coded affinity tags. *J. Proteome Res.* 7, 5270–5276.
29. Hsu, J. L., Huang, S. Y., Chow, N. H., and Chen, S. H. (2003) Stable-isotope dimethyl labeling for quantitative proteomics. *Anal. Chem.* **75**, 6843–6852.
30. Boersema, P. J., Raijmakers, R., Lemeer, S., Mohammed, S., and Heck, A. J. (2009) Multiplex peptide stable isotope dimethyl labeling for quantitative proteomics. *Nat. Protoc* **4**, 484–494.
31. Lemmel, C., Weik, S., Eberle, U., Dengjel, J., Kratt, T., Becker, H. D., Rammensee, H. G., and Stevanovic, S. (2004) Differential quantitative analysis of MHC ligands by mass spectrometry using stable isotope labeling. *Nat. Biotechnol.* **22**, 450–454.
32. Zappacosta, F., and Annan, R. S. (2004) N-terminal isotope tagging strategy for quantitative proteomics: results-driven analysis of protein abundance changes. *Anal. Chem.* **76**, 6618–6627.
33. Huang, H., Hittle, J., Zappacosta, F., Annan, R. S., Hershko, A., and Yen, T. J. (2008) Phosphorylation sites in BubR1 that regulate kinetochore attachment, tension, and mitotic exit. *J. Cell Biol.* **183**, 667–680.
34. Che, F. Y., and Fricker, L. D. (2005) Quantitative peptidomics of mouse pituitary: comparison of different stable isotopic tags. *J. Mass Spectrom.* **40**, 238–249.
35. Ji, J., Chakraborty, A., Geng, M., Zhang, X., Amini, A., Bina, M., and Regnier, F. (2000) Strategy for qualitative and quantitative analysis in proteomics based on signature peptides. *J. Chromatogr.* **745**, 197–210.
36. Chakraborty, A., and Regnier, F. E. (2002) Global internal standard technology for comparative proteomics. *J. Chromatogr. A* **949**, 173–184.
37. Mason, D. E., and Liebler, D. C. (2003) Quantitative analysis of modified proteins by LC-MS/MS of peptides labeled with phenyl isocyanate. *J. Proteome Res.* **2**, 265–272.

38. Munchbach, M., Quadroni, M., Miotto, G., and James, P. (2000) Quantitation and facilitated de novo sequencing of proteins by isotopic N-terminal labeling of peptides with a fragmentation-directing moiety. *Anal. Chem.* **72**, 4047–4057.
39. Goodlett, D. R., Keller, A., Watts, J. D., Newitt, R., Yi, E. C., Purvine, S., Eng, J. K., von Haller, P., Aebersold, R., and Kolker, E. (2001) Differential stable isotope labeling of peptides for quantitation and de novo sequence derivation. *Rapid Commun. Mass Spectrom.* **15**, 1214–1221.
40. Zhang, R., and Regnier, F. E. (2002) Minimizing resolution of isotopically coded peptides in comparative proteomics. *J. Proteome Res.* **1**, 139–147.
41. Morano, C., Zhang, X., and Fricker, L. D. (2008) Multiple isotopic labels for quantitative mass spectrometry. *Anal. Chem.* **80**, 9298–9309.
42. Schmidt, A., Bisle, B., and Kislinger, T. (2009) Quantitative peptide and protein profiling by mass spectrometry. *Methods Mol. Biol.* **492**, 21–38.
43. Mirgorodskaya, O. A., Kozmin, Y. P., Titov, M. I., Korner, R., Sonksen, C. P., and Roepstorff, P. (2000) Quantitation of peptides and proteins by matrix-assisted laser desorption/ionization mass spectrometry using (18)O-labeled internal standards. *Rapid Commun. Mass Spectrom.* **14**, 1226–1232.
44. Stewart, II, Thomson, T., and Figeys, D. (2001) ^{18}O labeling: a tool for proteomics. *Rapid Commun. Mass Spectrom.* **15**, 2456–2465.
45. Reynolds, K. J., Yao, X., and Fenselau, C. (2002) Proteolytic ^{18}O labeling for comparative proteomics: evaluation of endoprotease Glu-C as the catalytic agent. *J. Proteome Res.* **1**, 27–33.
46. Qian, W. J., Monroe, M. E., Liu, T., Jacobs, J. M., Anderson, G. A., Shen, Y., Moore, R. J., Anderson, D. J., Zhang, R., Calvano, S. E., Lowry, S. F., Xiao, W., Moldawer, L. L., Davis, R. W., Tompkins, R. G., Camp, D. G., 2nd, and Smith, R. D. (2005) Quantitative proteome analysis of human plasma following in vivo lipopolysaccharide administration using $^{16}O/^{18}O$ labeling and the accurate mass and time tag approach. *Mol. Cell. Proteomics* **4**, 700–709.
47. Ding, S. J., Wang, Y., Jacobs, J. M., Qian, W. J., Yang, F., Tolmachev, A. V., Du, X., Wang, W., Moore, R. J., Monroe, M. E., Purvine, S. O., Waters, K., Heibeck, T. H., Adkins, J. N., Camp, D. G., 2nd, Klemke, R. L., and Smith, R. D. (2008) Quantitative phosphoproteome analysis of lysophosphatidic acid induced chemotaxis applying dual-step (18)O labeling coupled with immobilized metal-ion affinity chromatography. *J. Proteome Res.* **7**, 4215–4224.
48. Bantscheff, M., Dumpelfeld, B., and Kuster, B. (2004) Femtomol sensitivity post-digest (18)O labeling for relative quantification of differential protein complex composition. *Rapid Commun. Mass Spectrom.* **18**, 869–876.
49. Qian, W. J., Liu, T., Petyuk, V. A., Gritsenko, M. A., Petritis, B. O., Polpitiya, A. D., Kaushal, A., Xiao, W., Finnerty, C. C., Jeschke, M. G., Jaitly, N., Monroe, M. E., Moore, R. J., Moldawer, L. L., Davis, R. W., Tompkins, R. G., Herndon, D. N., Camp, D. G., and Smith, R. D. (2009) Large-scale multiplexed quantitative discovery proteomics enabled by the use of an (18)O-labeled "universal" reference sample. *J. Proteome Res.* **8**, 290–299.
50. Julka, S., and Regnier, F. (2004) Quantification in proteomics through stable isotope coding: a review. *J. Proteome Res.* **3**, 350–363.
51. Johnson, K. L., and Muddiman, D. C. (2004) A method for calculating $^{16}O/^{18}O$ peptide ion ratios for the relative quantification of proteomes. *J. Am. Soc. Mass Spectrom.* **15**, 437–445.
52. Rao, K. C., Carruth, R. T., and Miyagi, M. (2005) Proteolytic ^{18}O labeling by peptidyl-Lys metalloendopeptidase for comparative proteomics. *J. Proteome Res.* **4**, 507–514.
53. Choe, L., D'Ascenzo, M., Relkin, N. R., Pappin, D., Ross, P., Williamson, B., Guertin, S., Pribil, P., and Lee, K. H. (2007) 8-plex quantitation of changes in cerebrospinal fluid protein expression in subjects undergoing intravenous immunoglobulin treatment for Alzheimer's disease. *Proteomics* **7**, 3651–3660.
54. Pierce, A., Unwin, R. D., Evans, C. A., Griffiths, S., Carney, L., Zhang, L., Jaworska, E., Lee, C. F., Blinco, D., Okoniewski, M. J., Miller, C. J., Bitton, D. A., Spooncer, E., and Whetton, A. D. (2008) Eight-channel iTRAQ enables comparison of the activity of six leukemogenic tyrosine kinases. *Mol. Cell. Proteomics* **7**, 853–863.
55. Phanstiel, D., Unwin, R., McAlister, G. C., and Coon, J. J. (2009) Peptide quantification using 8-plex isobaric tags and electron transfer dissociation tandem mass spectrometry. *Anal. Chem.* **81**, 1693–1698.
56. Unwin, R. D., Pierce, A., Watson, R. B., Sternberg, D. W., and Whetton, A. D.

(2005) Quantitative proteomic analysis using isobaric protein tags enables rapid comparison of changes in transcript and protein levels in transformed cells. *Mol. Cell. Proteomics* **22**, 22.

57. Pflieger, D., Junger, M. A., Muller, M., Rinner, O., Lee, H., Gehrig, P. M., Gstaiger, M., and Aebersold, R. (2008) Quantitative proteomic analysis of protein complexes: concurrent identification of interactors and their state of phosphorylation. *Mol. Cell. Proteomics* 7, 326–346.

58. Butler, G. S., Dean, R. A., Smith, D., and Overall, C. M. (2009) Membrane protease degradomics: proteomic identification and quantification of cell surface protease substrates. *Methods Mol. Biol.* **528**, 159–176.

59. Chen, Y., Choong, L. Y., Lin, Q., Philp, R., Wong, C. H., Ang, B. K., Tan, Y. L., Loh, M. C., Hew, C. L., Shah, N., Druker, B. J., Chong, P. K., and Lim, Y. P. (2007) Differential expression of novel tyrosine kinase substrates during breast cancer development. *Mol. Cell. Proteomics* **6**, 2072–2087.

60. Chiappetta, G., Corbo, C., Palmese, A., Marino, G., and Amoresano, A. (2009) Quantitative identification of protein nitration sites. *Proteomics* **9**, 1524–1537.

61. Dean, R. A., and Overall, C. M. (2007) Proteomics discovery of metalloproteinase substrates in the cellular context by iTRAQ labeling reveals a diverse MMP-2 substrate degradome. *Mol. Cell. Proteomics* **6**, 611–623.

62. Enoksson, M., Li, J., Ivancic, M. M., Timmer, J. C., Wildfang, E., Eroshkin, A., Salvesen, G. S., and Tao, W. A. (2007) Identification of proteolytic cleavage sites by quantitative proteomics. *J. Proteome Res.* **6**, 2850–2858.

63. Meany, D. L., Xie, H., Thompson, L. V., Arriaga, E. A., and Griffin, T. J. (2007) Identification of carbonylated proteins from enriched rat skeletal muscle mitochondria using affinity chromatography-stable isotope labeling and tandem mass spectrometry. *Proteomics* 7, 1150–1163.

64. Wang, X., and Huang, L. (2008) Identifying dynamic interactors of protein complexes by quantitative mass spectrometry. *Mol. Cell. Proteomics* 7, 46–57.

65. Zhang, Y., Wolf-Yadlin, A., Ross, P. L., Pappin, D. J., Rush, J., Lauffenburger, D. A., and White, F. M. (2005) Time-resolved mass spectrometry of tyrosine phosphorylation sites in the epidermal growth factor receptor signaling network reveals dynamic modules. *Mol. Cell. Proteomics* **4**, 1240–1250.

66. Keshamouni, V. G., Jagtap, P., Michailidis, G., Strahler, J. R., Kuick, R., Reka, A. K., Papoulias, P., Krishnapuram, R., Srirangam, A., Standiford, T. J., Andrews, P. C., and Omenn, G. S. (2009) Temporal quantitative proteomics by iTRAQ 2D-LC-MS/MS and corresponding mRNA expression analysis identify post-transcriptional modulation of actin-cytoskeleton regulators during TGF-beta-induced epithelial–mesenchymal transition. *J. Proteome Res.* **8**, 35–47.

67. Yan, W., Hwang, D., and Aebersold, R. (2008) Quantitative proteomic analysis to profile dynamic changes in the spatial distribution of cellular proteins. *Methods Mol. Biol.* **432**, 389–401.

68. Chen, X., and Andrews, P. C. (2008) Purification and proteomics analysis of pancreatic zymogen granule membranes. *Methods Mol. Biol.* **432**, 275–287.

69. Dwivedi, R. C., Dhindsa, N., Krokhin, O. V., Cortens, J., Wilkins, J. A., and El-Gabalawy, H. S. (2009) The effects of infliximab therapy on the serum proteome of rheumatoid arthritis patients. *Arthritis Res. Ther.* **11**, R32.

70. Bantscheff, M., Boesche, M., Eberhard, D., Matthieson, T., Sweetman, G., and Kuster, B. (2008) Robust and sensitive iTRAQ quantification on an LTQ Orbitrap mass spectrometer. *Mol. Cell. Proteomics* 7, 1702–1713.

71. Ralhan, R., Desouza, L. V., Matta, A., Chandra Tripathi, S., Ghanny, S., Dattagupta, S., Thakar, A., Chauhan, S. S., and Siu, K. W. (2009) iTRAQ-multidimensional liquid chromatography and tandem mass spectrometry-based identification of potential biomarkers of oral epithelial dysplasia and novel networks between inflammation and premalignancy. *J. Proteome Res.* **8**, 300–309.

72. Ho, J., Kong, J. W., Choong, L. Y., Loh, M. C., Toy, W., Chong, P. K., Wong, C. H., Wong, C. Y., Shah, N., and Lim, Y. P. (2009) Novel breast cancer metastasis-associated proteins. *J. Proteome Res.* **8**, 583–594.

73. Quaglia, M., Pritchard, C., Hall, Z., and O'Connor, G. (2008) Amine-reactive isobaric tagging reagents: requirements for absolute quantification of proteins and peptides. *Anal. Biochem.* **379**, 164–169.

74. Thompson, A., Schafer, J., Kuhn, K., Kienle, S., Schwarz, J., Schmidt, G., Neumann, T., Johnstone, R., Mohammed, A. K., and Hamon, C. (2003) Tandem

mass tags: a novel quantification strategy for comparative analysis of complex protein mixtures by MS/MS. *Anal. Chem.* **75**, 1895–1904.

75. Dayon, L., Hainard, A., Licker, V., Turck, N., Kuhn, K., Hochstrasser, D. F., Burkhard, P. R., and Sanchez, J. C. (2008) Relative quantification of proteins in human cerebrospinal fluids by MS/MS using 6-plex isobaric tags. *Anal. Chem.* **80**, 2921–2931.
76. Viner, R. I., Zhang, T., Second, T., and Zabrouskov, V. (2009) Quantification of post-translationally modified peptides of bovine alpha-crystallin using tandem mass tags and electron transfer dissociation. *J. Proteomics* **72**, 874–885.
77. Li, S., and Zeng, D. (2007) CILATa new reagent for quantitative proteomics. *Chem. Commun. (Cambridge, England)*, 2181–2183.
78. DeSouza, L. V., Taylor, A. M., Li, W., Minkoff, M. S., Romaschin, A. D., Colgan, T. J., and Siu, K. W. (2008) Multiple reaction monitoring of mTRAQ-labeled peptides enables absolute quantification of endogenous levels of a potential cancer marker in cancerous and normal endometrial tissues. *J. Proteome Res.* 7, 3525–3534.
79. Boehm, A. M., Putz, S., Altenhofer, D., Sickmann, A., and Falk, M. (2007) Precise protein quantification based on peptide quantification using iTRAQ. *BMC Bioinformatics* **8**, 214.
80. D'Ascenzo, M., Choe, L., and Lee, K. H. (2008) iTRAQPak: an R based analysis and visualization package for 8-plex isobaric protein expression data. *Brief Funct. Genomic Proteomic* 7, 127–135.
81. Hundertmark, C., Fischer, R., Reinl, T., May, S., Klawonn, F., and Jansch, L. (2009) MS-specific noise model reveals the potential of iTRAQ in quantitative proteomics. *Bioinformatics (Oxford, England)* **25**, 1004–1011.
82. Lacerda, C. M., Xin, L., Rogers, I., and Reardon, K. F. (2008) Analysis of iTRAQ data using Mascot and peaks quantification algorithms. *Brief Funct. Genomic Proteomic* 7, 119–126.
83. Laderas, T., Bystrom, C., McMillen, D., Fan, G., and McWeeney, S. (2007) TandTRAQ: an open-source tool for integrated protein identification and quantitation. *Bioinformatics (Oxford, England)* **23**, 3394–3396.
84. Lin, W. T., Hung, W. N., Yian, Y. H., Wu, K. P., Han, C. L., Chen, Y. R., Chen, Y. J., Sung, T. Y., and Hsu, W. L. (2006) Multi-Q: a fully automated tool for multiplexed protein quantitation. *J. Proteome Res.* **5**, 2328–2338.
85. Shadforth, I. P., Dunkley, T. P., Lilley, K. S., and Bessant, C. (2005) i-Tracker: for quantitative proteomics using iTRAQ. *BMC Genomics* **6**, 145.
86. Yu, C. Y., Tsui, Y. H., Yian, Y. H., Sung, T. Y., and Hsu, W. L. (2007) The Multi-Q web server for multiplexed protein quantitation. *Nucleic Acids Res.* **35**, W707–W712.
87. Hill, E. G., Schwacke, J. H., Comte-Walters, S., Slate, E. H., Oberg, A. L., Eckel-Passow, J. E., Therneau, T. M., and Schey, K. L. (2008) A statistical model for iTRAQ data analysis. *J. Proteome Res.* 7, 3091–3101.
88. Griffin, T. J., Xie, H., Bandhakavi, S., Popko, J., Mohan, A., Carlis, J. V., and Higgins, L. (2007) iTRAQ reagent-based quantitative proteomic analysis on a linear ion trap mass spectrometer. *J. Proteome Res.* **6**, 4200–4209.
89. Guo, T., Gan, C. S., Zhang, H., Zhu, Y., Kon, O. L., and Sze, S. K. (2008) Hybridization of pulsed-Q dissociation and collision-activated dissociation in linear ion trap mass spectrometer for iTRAQ quantitation. *J. Proteome Res.* 7, 4831–4840.
90. Oda, Y., Huang, K., Cross, F. R., Cowburn, D., and Chait, B. T. (1999) Accurate quantitation of protein expression and site-specific phosphorylation. *Proc. Natl. Acad. Sci. USA* **96**, 6591–6596.
91. Conrads, T. P., Alving, K., Veenstra, T. D., Belov, M. E., Anderson, G. A., Anderson, D. J., Lipton, M. S., Pasa-Tolic, L., Udseth, H. R., Chrisler, W. B., Thrall, B. D., and Smith, R. D. (2001) Quantitative analysis of bacterial and mammalian proteomes using a combination of cysteine affinity tags and ^{15}N-metabolic labeling. *Anal. Chem.* **73**, 2132–2139.
92. Washburn, M. P., Koller, A., Oshiro, G., Ulaszek, R. R., Plouffe, D., Deciu, C., Winzeler, E., and Yates, J. R., 3rd. (2003) Protein pathway and complex clustering of correlated mRNA and protein expression analyses in *Saccharomyces cerevisiae*. *Proc. Natl. Acad. Sci. USA* **100**, 3107–3112.
93. Bindschedler, L. V., Palmblad, M., and Cramer, R. (2008) Hydroponic isotope labelling of entire plants (HILEP) for quantitative plant proteomics; an oxidative stress case study. *Phytochemistry* **69**, 1962–1972.

94. Pan, C., Gnad, F., Olsen, J. V., and Mann, M. (2008) Quantitative phosphoproteome analysis of a mouse liver cell line reveals specificity of phosphatase inhibitors. *Proteomics* **8**, 4534–4546.
95. Molina, H., Parmigiani, G., and Pandey, A. (2005) Assessing reproducibility of a protein dynamics study using in vivo labeling and liquid chromatography tandem mass spectrometry. *Anal. Chem.* 77, 2739–2744.
96. Molina, H., Yang, Y., Ruch, T., Kim, J. W., Mortensen, P., Otto, T., Nalli, A., Tang, Q. Q., Lane, M. D., Chaerkady, R., and Pandey, A. (2009) Temporal profiling of the adipocyte proteome during differentiation using a five-plex SILAC based strategy. *J. Proteome Res.* **8**, 48–58.
97. Harsha, H. C., Molina, H., and Pandey, A. (2008) Quantitative proteomics using stable isotope labeling with amino acids in cell culture. *Nat. Protoc* **3**, 505–516.
98. Gruhler, S., and Kratchmarova, I. (2008) Stable isotope labeling by amino acids in cell culture (SILAC). *Methods Mol. Biol.* **424**, 101–111.
99. Ong, S. E., and Mann, M. (2007) Stable isotope labeling by amino acids in cell culture for quantitative proteomics. *Methods Mol. Biol.* **359**, 37–52.
100. Cox, J., Matic, I., Hilger, M., Nagaraj, N., Selbach, M., Olsen, J. V., and Mann, M. (2009) A practical guide to the MaxQuant computational platform for SILAC-based quantitative proteomics. *Nat. Protoc* **4**, 698–705.
101. Graumann, J., Hubner, N. C., Kim, J. B., Ko, K., Moser, M., Kumar, C., Cox, J., Scholer, H., and Mann, M. (2008) Stable isotope labeling by amino acids in cell culture (SILAC) and proteome quantitation of mouse embryonic stem cells to a depth of 5,111 proteins. *Mol. Cell. Proteomics* 7, 672–683.
102. Schwanhausser, B., Gossen, M., Dittmar, G., and Selbach, M. (2009) Global analysis of cellular protein translation by pulsed SILAC. *Proteomics* **9**, 205–209.
103. Doherty, M. K., Hammond, D. E., Clague, M. J., Gaskell, S. J., and Beynon, R. J. (2009) Turnover of the human proteome: determination of protein intracellular stability by dynamic SILAC. *J. Proteome Res.* **8**, 104–112.
104. Mintz, M., Vanderver, A., Brown, K. J., Lin, J., Wang, Z., Kaneski, C., Schiffmann, R., Nagaraju, K., Hoffman, E. P., and Hathout, Y. (2008) Time series proteome profiling to study endoplasmic reticulum stress response. *J. Proteome Res.* 7, 2435–2444.
105. Wang, M., You, J., Bemis, K. G., Tegeler, T. J., and Brown, D. P. (2008) Label-free mass spectrometry-based protein quantification technologies in proteomic analysis. *Brief Funct. Genomic Proteomic* 7, 329–339.
106. Trinkle-Mulcahy, L., Boulon, S., Lam, Y. W., Urcia, R., Boisvert, F. M., Vandermoere, F., Morrice, N. A., Swift, S., Rothbauer, U., Leonhardt, H., and Lamond, A. (2008) Identifying specific protein interaction partners using quantitative mass spectrometry and bead proteomes. *J. Cell Biol.* **183**, 223–239.
107. Mousson, F., Kolkman, A., Pijnappel, W. W., Timmers, H. T., and Heck, A. J. (2008) Quantitative proteomics reveals regulation of dynamic components within TATA-binding protein (TBP) transcription complexes. *Mol. Cell. Proteomics* 7, 845–852.
108. Dobreva, I., Fielding, A., Foster, L. J., and Dedhar, S. (2008) Mapping the integrin-linked kinase interactome using SILAC. *J. Proteome Res.* 7, 1740–1749.
109. Guerrero, C., Tagwerker, C., Kaiser, P., and Huang, L. (2006) An integrated mass spectrometry-based proteomic approach: quantitative analysis of tandem affinity-purified in vivo cross-linked protein complexes (QTAX) to decipher the 26 S proteasome-interacting network. *Mol. Cell. Proteomics* **5**, 366–378.
110. Blagoev, B., Kratchmarova, I., Ong, S. E., Nielsen, M., Foster, L. J., and Mann, M. (2003) A proteomics strategy to elucidate functional protein–protein interactions applied to EGF signaling. *Nat. Biotechnol.* **21**, 315–318.
111. Mittler, G., Butter, F., and Mann, M. (2009) A SILAC-based DNA protein interaction screen that identifies candidate binding proteins to functional DNA elements. *Genome Res.* **19**, 284–293.
112. Ong, S. E., Schenone, M., Margolin, A. A., Li, X., Do, K., Doud, M. K., Mani, D. R., Kuai, L., Wang, X., Wood, J. L., Tolliday, N. J., Koehler, A. N., Marcaurelle, L. A., Golub, T. R., Gould, R. J., Schreiber, S. L., and Carr, S. A. (2009) Identifying the proteins to which small-molecule probes and drugs bind in cells. *Proc. Natl. Acad. Sci. USA* **106**, 4617–4622.
113. Oppermann, F. S., Gnad, F., Olsen, J. V., Hornberger, R., Greff, Z., Keri, G., Mann, M., and Daub, H. (2009) Large-scale proteomics analysis of the human kinome. *Mol. Cell. Proteomics* **8**, 1751–1764.

114. Hanke, S., and Mann, M. (2009) The phosphotyrosine interactome of the insulin receptor family and its substrates IRS-1 and IRS-2. *Mol. Cell. Proteomics* **8**, 519–534.
115. Yan, Y., Weaver, V. M., and Blair, I. A. (2005) Analysis of protein expression during oxidative stress in breast epithelial cells using a stable isotope labeled proteome internal standard. *J. Proteome Res.* **4**, 2007–2014.
116. Hanke, S., Besir, H., Oesterhelt, D., and Mann, M. (2008) Absolute SILAC for accurate quantitation of proteins in complex mixtures down to the attomole level. *J. Proteome Res.* 7, 1118–1130.
117. Ong, S. E., Mittler, G., and Mann, M. (2004) Identifying and quantifying in vivo methylation sites by heavy methyl SILAC. *Nat. Methods* **1**, 119–126.
118. Colzani, M., Schutz, F., Potts, A., Waridel, P., and Quadroni, M. (2008) Relative protein quantification by isobaric SILAC with immonium ion splitting (ISIS). *Mol. Cell. Proteomics* 7, 927–937.
119. Kruger, M., Moser, M., Ussar, S., Thievessen, I., Luber, C. A., Forner, F., Schmidt, S., Zanivan, S., Fassler, R., and Mann, M. (2008) SILAC mouse for quantitative proteomics uncovers kindlin-3 as an essential factor for red blood cell function. *Cell* **134**, 353–364.
120. Ong, S. E., Kratchmarova, I., and Mann, M. (2003) Properties of ^{13}C-substituted arginine in stable isotope labeling by amino acids in cell culture (SILAC). *J. Proteome Res.* **2**, 173–181.
121. Bendall, S. C., Hughes, C., Stewart, M. H., Doble, B., Bhatia, M., and Lajoie, G. A. (2008) Prevention of amino acid conversion in SILAC experiments with embryonic stem cells. *Mol. Cell. Proteomics* 7, 1587–1597.
122. Van Hoof, D., Pinkse, M. W., Oostwaard, D. W., Mummery, C. L., Heck, A. J., and Krijgsveld, J. (2007) An experimental correction for arginine-to-proline conversion artifacts in SILAC-based quantitative proteomics. *Nat. Methods* **4**, 677–678.
123. America, A. H., and Cordewener, J. H. (2008) Comparative LC-MS: a landscape of peaks and valleys. *Proteomics* **8**, 731–749.
124. Braisted, J. C., Kuntumalla, S., Vogel, C., Marcotte, E. M., Rodrigues, A. R., Wang, R., Huang, S. T., Ferlanti, E. S., Saeed, A. I., Fleischmann, R. D., Peterson, S. N., and Pieper, R. (2008) The APEX quantitative proteomics tool: generating protein quantitation estimates from LC-MS/MS proteomics results. *BMC Bioinformatics* **9**, 529.
125. Cheng, F. Y., Blackburn, K., Lin, Y. M., Goshe, M. B., and Williamson, J. D. (2009) Absolute protein quantification by LC/MS(E) for global analysis of salicylic acid-induced plant protein secretion responses. *J. Proteome Res.* **8**, 82–93.
126. Choi, H., Fermin, D., and Nesvizhskii, A. I. (2008) Significance analysis of spectral count data in label-free shotgun proteomics. *Mol. Cell. Proteomics* 7, 2373–2385.
127. Cutillas, P. R., Biber, J., Marks, J., Jacob, R., Stieger, B., Cramer, R., Waterfield, M., Burlingame, A. L., and Unwin, R. J. (2005) Proteomic analysis of plasma membrane vesicles isolated from the rat renal cortex. *Proteomics* **5**, 101–112.
128. Cutillas, P. R., and Vanhaesebroeck, B. (2007) Quantitative profile of five murine core proteomes using label-free functional proteomics. *Mol. Cell. Proteomics* **6**, 1560–1573.
129. Fang, R., Elias, D. A., Monroe, M. E., Shen, Y., McIntosh, M., Wang, P., Goddard, C. D., Callister, S. J., Moore, R. J., Gorby, Y. A., Adkins, J. N., Fredrickson, J. K., Lipton, M. S., and Smith, R. D. (2006) Differential label-free quantitative proteomic analysis of *Shewanella oneidensis* cultured under aerobic and suboxic conditions by accurate mass and time tag approach. *Mol. Cell. Proteomics* **5**, 714–725.
130. Fraterman, S., Zeiger, U., Khurana, T. S., Wilm, M., and Rubinstein, N. A. (2007) Quantitative proteomics profiling of sarcomere associated proteins in limb and extraocular muscle allotypes. *Mol. Cell. Proteomics* **6**, 728–737.
131. Fu, X., Gharib, S. A., Green, P. S., Aitken, M. L., Frazer, D. A., Park, D. R., Vaisar, T., and Heinecke, J. W. (2008) Spectral index for assessment of differential protein expression in shotgun proteomics. *J. Proteome Res.* 7, 845–854.
132. Govorukhina, N., Horvatovich, P., and Bischoff, R. (2008) Label-free proteomics of serum. *Methods Mol. Biol.* **484**, 67–77.
133. Haqqani, A. S., Kelly, J. F., and Stanimirovic, D. B. (2008) Quantitative protein profiling by mass spectrometry using label-free proteomics. *Methods Mol. Biol.* **439**, 241–256.
134. Higgs, R. E., Knierman, M. D., Gelfanova, V., Butler, J. P., and Hale, J. E. (2005) Comprehensive label-free method for the relative quantification of proteins from biological samples. *J. Proteome Res.* **4**, 1442–1450.

135. Higgs, R. E., Knierman, M. D., Gelfanova, V., Butler, J. P., and Hale, J. E. (2008) Label-free LC-MS method for the identification of biomarkers. *Methods Mol. Biol.* **428**, 209–230.
136. Meng, F., Wiener, M. C., Sachs, J. R., Burns, C., Verma, P., Paweletz, C. P., Mazur, M. T., Deyanova, E. G., Yates, N. A., and Hendrickson, R. C. (2007) Quantitative analysis of complex peptide mixtures using FTMS and differential mass spectrometry. *J. Am. Soc. Mass Spectrom.* **18**, 226–233.
137. Negishi, A., Ono, M., Handa, Y., Kato, H., Yamashita, K., Honda, K., Shitashige, M., Satow, R., Sakuma, T., Kuwabara, H., Omura, K., Hirohashi, S., and Yamada, T. (2009) Large-scale quantitative clinical proteomics by label-free liquid chromatography and mass spectrometry. *Cancer Sci.* **100**, 514–519.
138. Ono, M., Shitashige, M., Honda, K., Isobe, T., Kuwabara, H., Matsuzuki, H., Hirohashi, S., and Yamada, T. (2006) Label-free quantitative proteomics using large peptide data sets generated by nanoflow liquid chromatography and mass spectrometry. *Mol. Cell. Proteomics* **5**, 1338–1347.
139. Rinner, O., Mueller, L. N., Hubalek, M., Muller, M., Gstaiger, M., and Aebersold, R. (2007) An integrated mass spectrometric and computational framework for the analysis of protein interaction networks. *Nat. Biotechnol.* **25**, 345–352.
140. Ru, Q. C., Zhu, L. A., Silberman, J., and Shriver, C. D. (2006) Label-free semiquantitative peptide feature profiling of human breast cancer and breast disease sera via two-dimensional liquid chromatography–mass spectrometry. *Mol. Cell. Proteomics* **5**, 1095–1104.
141. Sardiu, M. E., Cai, Y., Jin, J., Swanson, S. K., Conaway, R. C., Conaway, J. W., Florens, L., and Washburn, M. P. (2008) Probabilistic assembly of human protein interaction networks from label-free quantitative proteomics. *Proc. Natl. Acad. Sci. USA* **105**, 1454–1459.
142. Schmidt, M. W., Houseman, A., Ivanov, A. R., and Wolf, D. A. (2007) Comparative proteomic and transcriptomic profiling of the fission yeast *Schizosaccharomyces pombe*. *Mol. Syst. Biol.* **3**, 79.
143. Silva, J. C., Denny, R., Dorschel, C., Gorenstein, M. V., Li, G. Z., Richardson, K., Wall, D., and Geromanos, S. J. (2006) Simultaneous qualitative and quantitative analysis of the *Escherichia coli* proteome: a sweet tale. *Mol. Cell. Proteomics* **5**, 589–607.
144. Tabata, T., Sato, T., Kuromitsu, J., and Oda, Y. (2007) Pseudo internal standard approach for label-free quantitative proteomics. *Anal. Chem.* **79**, 8440–8445.
145. Wiener, M. C., Sachs, J. R., Deyanova, E. G., and Yates, N. A. (2004) Differential mass spectrometry: a label-free LC-MS method for finding significant differences in complex peptide and protein mixtures. *Anal. Chem.* **76**, 6085–6096.
146. Mann, M. (1999) Quantitative proteomics? *Nat. Biotechnol.* **17**, 954–955.
147. Gangl, E. T., Annan, M. M., Spooner, N., and Vouros, P. (2001) Reduction of signal suppression effects in ESI–MS using a nanosplitting device. *Anal. Chem.* **73**, 5635–5644.
148. Shen, J. X., Motyka, R. J., Roach, J. P., and Hayes, R. N. (2005) Minimization of ion suppression in LC-MS/MS analysis through the application of strong cation exchange solid-phase extraction (SCX-SPE). *J. Pharm. Biomed. Anal.* **37**, 359–367.
149. Weaver, R., and Riley, R. J. (2006) Identification and reduction of ion suppression effects on pharmacokinetic parameters by polyethylene glycol 400. *Rapid Commun. Mass Spectrom.* 20, 2559–2564.
150. Wilm, M., and Mann, M. (1994) Electrospray and Taylor-Cone theory, Dole's beam of macromolecules at last? *Int. J. Mass Spectrom. Ion Process.* **136**, 167–180.
151. Chelius, D., and Bondarenko, P. V. (2002) Quantitative profiling of proteins in complex mixtures using liquid chromatography and mass spectrometry. *J. Proteome Res.* **1**, 317–323.
152. Cutillas, P. R., Geering, B., Waterfield, M. D., and Vanhaesebroeck, B. (2005) Quantification of gel-separated proteins and their phosphorylation sites by LC-MS using unlabeled internal standards: analysis of phosphoprotein dynamics in a B cell lymphoma cell line. *Mol. Cell. Proteomics* **4**, 1038–1051.
153. Steen, H., Jebanathirajah, J. A., Springer, M., and Kirschner, M. W. (2005) Stable isotope-free relative and absolute quantitation of protein phosphorylation stoichiometry by MS. *Proc. Natl. Acad. Sci. USA* **102**, 3948–3953.
154. America, A. H., Cordewener, J. H., van Geffen, M. H., Lommen, A., Vissers, J. P., Bino, R. J., and Hall, R. D. (2006) Alignment and statistical difference analysis of complex peptide data sets generated by multidimensional LC-MS. *Proteomics* **6**, 641–653.

155. Zhang, B., VerBerkmoes, N. C., Langston, M. A., Uberbacher, E., Hettich, R. L., and Samatova, N. F. (2006) Detecting differential and correlated protein expression in label-free shotgun proteomics. *J. Proteome Res.* **5**, 2909–2918.
156. Hixson, K. K. (2009) Label-free relative quantitation of prokaryotic proteomes using the accurate mass and time tag approach. *Methods Mol. Biol.* **492**, 39–63.
157. Jaitly, N., Monroe, M. E., Petyuk, V. A., Clauss, T. R., Adkins, J. N., and Smith, R. D. (2006) Robust algorithm for alignment of liquid chromatography–mass spectrometry analyses in an accurate mass and time tag data analysis pipeline. *Anal. Chem.* **78**, 7397–7409.
158. Luo, Q., Hixson, K. K., Callister, S. J., Lipton, M. S., Morris, B. E., and Krumholz, L. R. (2007) Proteome analysis of *Desulfovibrio desulfuricans* G20 mutants using the accurate mass and time (AMT) tag approach. *J. Proteome Res.* **6**, 3042–3053.
159. Strittmatter, E. F., Ferguson, P. L., Tang, K., and Smith, R. D. (2003) Proteome analyses using accurate mass and elution time peptide tags with capillary LC time-of-flight mass spectrometry. *J. Am. Soc. Mass Spectrom.* **14**, 980–991.
160. Zimmer, J. S., Monroe, M. E., Qian, W. J., and Smith, R. D. (2006) Advances in proteomics data analysis and display using an accurate mass and time tag approach. *Mass Spectrom. Rev.* **25**, 450–482.
161. Kuster, B., Schirle, M., Mallick, P., and Aebersold, R. (2005) Scoring proteomes with proteotypic peptide probes. *Nat. Rev. Mol. Cell. Biol.* **6**, 577–583.
162. Kitteringham, N. R., Jenkins, R. E., Lane, C. S., Elliott, V. L., and Park, B. K. (2008) Multiple reaction monitoring for quantitative biomarker analysis in proteomics and metabolomics. *J. Chromatogr. B Anal. Technol. Biomed. Life Sci.* **877**, 1229–1239.
163. Mallick, P., Schirle, M., Chen, S. S., Flory, M. R., Lee, H., Martin, D., Ranish, J., Raught, B., Schmitt, R., Werner, T., Kuster, B., and Aebersold, R. (2007) Computational prediction of proteotypic peptides for quantitative proteomics. *Nat. Biotechnol.* **25**, 125–131.
164. Mead, J. A., Bianco, L., Ottone, V., Barton, C., Kay, R. G., Lilley, K. S., Bond, N. J., and Bessant, C. (2008) MRMaid: the webbased tool for designing multiple reaction monitoring (MRM) transitions. *Mol. Cell. Proteomics* **8**, 696–705.
165. Walsh, G. M., Lin, S., Evans, D. M., Khosrovi-Eghbal, A., Beavis, R. C., and Kast, J. (2008) Implementation of a data repository-driven approach for targeted proteomics experiments by multiple reaction monitoring. *J Proteomics* **72**, 838–852.
166. Fusaro, V. A., Mani, D. R., Mesirov, J. P., and Carr, S. A. (2009) Prediction of high-responding peptides for targeted protein assays by mass spectrometry. *Nat. Biotechnol.* **27**, 190–198.
167. Rifai, N., Gillette, M. A., and Carr, S. A. (2006) Protein biomarker discovery and validation: the long and uncertain path to clinical utility. *Nat. Biotechnol.* **24**, 971–983.
168. Ishihama, Y., Oda, Y., Tabata, T., Sato, T., Nagasu, T., Rappsilber, J., and Mann, M. (2005) Exponentially modified protein abundance index (emPAI) for estimation of absolute protein amount in proteomics by the number of sequenced peptides per protein. *Mol. Cell. Proteomics* **4**, 1265–1272.
169. Lu, P., Vogel, C., Wang, R., Yao, X., and Marcotte, E. M. (2007) Absolute protein expression profiling estimates the relative contributions of transcriptional and translational regulation. *Nat. Biotechnol.* **25**, 117–124.

Chapter 3

Instrumentation for LC-MS/MS in Proteomics

Robert Chalkley

Abstract

Mass spectrometers now have sufficient sensitivity and acquisition rates to allow analysis of complex proteomic samples on a chromatographic timescale. In this chapter the different instrument options for protein and peptide analysis will be presented, along with their relative strengths and weaknesses for producing different types of information, such as protein identification, modification characterization, or reporting quantitative measurements.

Key words: Quadrupole, Orbitrap, Fourier transform, mass spectrometry, proteomics, metabolomics.

1. Introduction

The two major types of ionization for analysis of biomolecules are electrospray ionization (ESI) (1) and matrix-assisted laser desorption/ionization (MALDI) (2). One of the fundamental differences between these methods is that MALDI is employed on samples in a solid state, whereas ESI is employed on samples in a liquid state. Hence, interfacing liquid chromatography with ESI is relatively straightforward, whereas LC–MALDI analysis is an offline process, where fractions are collected and then analyzed by MALDI at a later stage. Partly as a result of this, ESI is the dominant ionization process for analysis of samples separated by liquid chromatography and will be the emphasis of instrumentation discussed from hereon.

P.R. Cutillas, J.F. Timms (eds.), *LC-MS/MS in Proteomics*, Methods in Molecular Biology 658,
DOI 10.1007/978-1-60761-780-8_3,

2. Liquid Chromatography Instrumentation

Although electrospray starts from the solution phase, the mass spectrometer is ultimately detecting gaseous ions. Hence, during the ionization process, all the solvent has to be removed. The process of ESI converts a solution into a mist of charged droplets. These droplets shrink as the solvent is evaporated and when the charge density in the droplet reaches a critical level, coulombic repulsion causes desorption of charged gaseous ions from the droplet (3, 4). This process continues; as solvent continues to evaporate, droplets get smaller and more ions are ejected into the gaseous phase. Unsurprisingly, this conversion of liquid sample into gaseous ions is more efficient when less solvent is present. Hence, electrospray is referred to as a concentration-sensitive process. This means that the smaller the volume of sample introduced, the better the efficiency and sensitivity of the process. Increasing sample concentration can be achieved in two ways using liquid chromatography. First, use of a narrower column and lower flow rate will cause elution in smaller volumes. Second, by improving the resolution of separation, the same amount of sample will elute in a narrower profile, giving a higher concentration at the maxima of peak elution.

Nanospray is more sensitive than ESI approaches at higher flow rates (5); so sub-microliter flow rates are typically used for proteomic analyses. Many chromatographic systems cannot natively produce reliable gradients at these low flow rates due to the presence of solvent-mixing chambers in the plumbing that have too large volumes in comparison to the solvent flow rate, leading to inconsistent solvent mixing and irreproducible gradients. Hence, pressure-based flow-splitting systems are commonly employed and built into the chromatography system. Generally, a split in the range of 1:100 to 1:1,000 is used post-pump but prior to the separation column, allowing efficient mixing of solvents prior to chromatography to produce consistent separations, albeit with large amounts of solvent waste. Recently, some systems have been developed that use air pressure as the pumping mechanism and allow splitless delivery of reproducible nanoliter per minute flow rates.

Different stationary phases in chromatography columns provide variable levels of resolution. Reverse-phase chromatography is highly compatible with subsequent mass spectrometric analysis due to the lack of salts in the buffers and provides relatively high-resolution separation, so is the dominant separation method in use for proteomic analysis. Most reverse-phase stationary phases for LC-MS analysis consist of silica beads of 3–5 μm in diameter with alkyl chains of either eight or eighteen carbons in length (C8 or C18) attached. The resolution of separation can be increased through the use of smaller particle-size resins, e.g., 2-μm-diameter beads, which leads to higher efficiency but also higher pressure separation; as a result this is sometimes referred to as ultrahigh-pressure liquid chromatography (UPLC) and generally requires specialized chromatography instrumentation to cope with the associated higher pressures (6, 7). Another form of UPLC involves the use of long columns (up to 80-cm-long columns have been reported for ultrahigh-pressure LC by the group of Richard Smith), which also results in high-resolution and high peak capacity separations (8) .

3. Mass Spectrometry Instrumentation

For identification of peptides in complex mixtures, measurement of peptide mass alone is not sufficiently informative. While the combination of accurate mass and retention time can be employed for identification in well-defined samples (9), the most flexible and generally applicable approach involves fragmentation analysis of components. In an initial scan, the masses of intact components are measured, and then in a subsequent scan/s, individual components are isolated in the mass spectrometer and then fragmented. A selection of different instrument configurations can be employed to perform these two steps of analysis.

3.1. Ion Traps

An ion trap is the only analyzer that can be used for measurement of peptides, isolation of selected components, and subsequent fragmentation analysis in the same chamber. There are three different types of ion traps employed in mass spectrometers: the quadrupole ion trap, the Penning ion trap, and the Kingdon trap (orbitrap).

3.1.1. Quadrupole Ion Traps

A quadrupole ion trap, sometimes called a Paul trap after Wolfgang Paul, who received the Nobel Prize in 1989 for its development, consists of a ring electrode and electrostatic end caps that create a box-like chamber to trap ions. The classical version of these (sometimes referred to as a 3D trap) uses two linear rods as end caps and a single ring electrode, but more recently a linear or 2D equivalent has been developed and made commercially available, which uses a set of four linear rods and two end-cap electrodes (10, 11). The 2D and 3D refer to the number of dimensions in which the ions are being oscillated during the trapping mechanism: in a 3D quadrupole trap, ions are sequentially "squeezed" and "stretched" in all three dimensions, whereas in a linear trap, the end caps are purely performing trapping and so their voltage is not changing and ions are oscillated only radially. The main advantages of linear ion traps over 3D traps are their more efficient ion trapping and larger ion capacity, allowing them to produce spectra with higher signal-to-noise and better dynamic range (11).

The equations describing the stability of ions inside a quadrupole ion trap are not simple. However, commercial quadrupole ion trap instruments are operated in a way that maintains stability in the axial direction; so manipulation of stability in the radial direction, by altering a parameter known as the q_z, is used for determining whether an ion is trapped or not. The parameter q_z is defined by a combination of the m/z of the ion, the ion trap radius, and the frequency and amplitude of the RF voltage applied. Scans are performed by changing the amplitude of the RF voltage to sequentially eject ions radially, normally from low to high mass (12). As ions are ejected, they are detected, and by correlating the time of detection to the RF amplitude, it is possible to determine their mass-to-charge ratio (m/z).

Ions of a given m/z in the quadrupole ion trap will be moving in a motion at a certain frequency. By applying an AC voltage on the end-cap electrodes at the same frequency as this motion, it is possible to resonantly excite ions of that m/z. These ions collide with gas molecules in the ion trap and can cause fragmentation through collision-induced dissociation (CID). Unfortunately, low-mass ions (below about one-third of the precursor m/z) formed from CID analysis will have a q_z value too high to be trapped, meaning the low-mass region of the fragmentation spectrum is missing. By resonantly exciting for a very short period of time, it is possible to still trap a few low-mass ions (a technique called pulsed Q dissociation or PQD). However, this compromises the fragmentation efficiency. By applying a higher amount of resonant excitation, it is possible to eject ions from the trap without fragmentation, and this provides an alternative method for isolating a particular ion to scanning ions out at the extremes of the q_z stability range.

Quadrupole ion traps are the most common type of ion trap in use, and almost invariably if someone states that they are using an ion trap mass spectrometer, they mean a quadrupole ion trap instrument. The instruments are known for their high sensitivity. Their resolving power (ability to distinguish between components of similar masses) is dependent on the rate at which they scan the mass range. However, as one cannot detect new ions until the previous ions have been scanned/ejected out of the trap, fast scan rates are typically employed to maximize sensitivity and speed. As a result, ion trap data are typically of low resolution and relatively poor mass accuracy.

If too many ions are isolated in any type of ion trap, the charge density causes ions to interact with each other (referred to as space-charging effects), causing distortion of the electric field and subsequent mass spectrum. Hence, the number of ions trapped is usually regulated by a process known as automatic gain control (AGC), where the number of ions is monitored over a very short period and then the same ion flux is assumed to produce a target number of ions isolated in the trap (13).

Some quadrupole ion trap instruments now have a chemical ionization source either attached to the back end of the instrument or interfaced to a quadrupole region before the quadrupole ion trap. These allow introduction of anions that can be used for electron transfer dissociation (ETD) fragmentation (14) and can also be used to introduce chemicals for proton transfer reactions that allow manipulation of the charge state of ions (15). ETD will be discussed later in this chapter.

3.1.2. Penning Ion Traps

By placing an ion chamber of similar geometry to a Paul trap in the center of a magnetic cylinder, it is possible to trap ions using a combination of magnetic and electrostatic fields, and this type of ion trapping was first used as part of a mass spectrometer by Alan Marshall (16). Ions are trapped in the center of the magnetic field and through application of an oscillating electric field in the same direction as the magnetic field, the ions are moved into a cyclotron motion. The rate at which they cycle through this motion is dependent on the *m/z* ratio of the ion. Ions can be detected as an image current as they pass a pair of plates on either side of the chamber in a motion referred to as ion cyclotron resonance. The sum of all the image currents for all ions trapped can be converted into an *m/z* scale using Fourier transform mathematics; so this technique is generally referred to as Fourier transform ion cyclotron resonance (FT-ICR) mass spectrometry.

Fragmentation can be performed in the Penning trap using a wide variety of methods, including variants of collision-induced dissociation (17, 18), infrared multiphoton dissociation (IRMPD) (19), and radical-based fragmentation using electron capture dissociation (20). For a thorough review of FT-ICR mass spectrometry, *see* (21).

FT-ICR mass spectrometry provides the highest mass precision and resolution, and can be used for determining elemental composition of components. These facets are useful for peptide analysis but are particularly beneficial for analysis of large peptides and small proteins (22).

3.1.3. Kingdon Ion Traps/Orbitraps

An orbitrap consists of an outer electrode of a barrel shape surrounding a co-axial inner electrode (**Fig. 3.1**) (23). Through electrostatic forces the ions are attracted toward the central electrode, but these forces are balanced by centrifugal forces that cause the ions to cycle around the inner electrode. The frequency of this oscillation is related to the *m/z* of the ion, and the ions

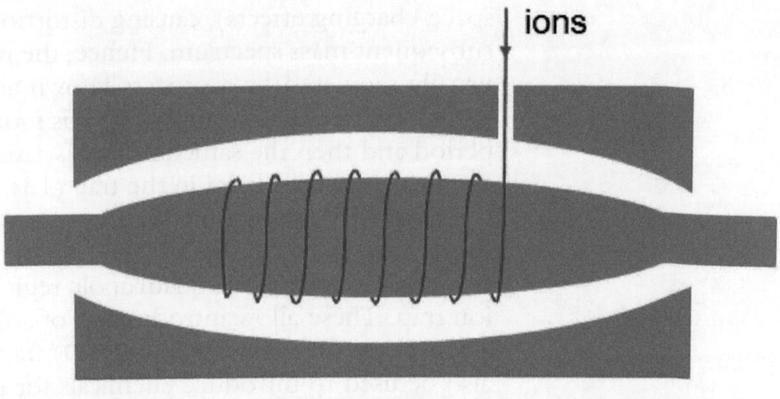

Fig. 3.1. Schematic of an orbitrap mass analyzer. Ions are injected into the orbitrap off-center and with a velocity perpendicular to the long axis of the orbitrap. The off-center injection also gives the ions potential energy parallel to the length of the orbitrap, causing them to oscillate along the plane of the orbitrap in an analogous fashion to a swinging pendulum as they cycle around the central electrode.

can be detected as an image current and converted to m/z using Fourier transformation in the same way as in FT-ICR.

An orbitrap provides high-resolution and mass accuracy detection, and has a higher dynamic range than a Penning trap has (24), meaning low-level components can still be detected in the background of more abundant co-eluting compounds.

3.2. Hyphenated Instruments

Many of the most popular mass spectrometric instruments used for proteomic analysis employ a combination of more than one analyzer type. As different analyzers are better for certain types of experiments than others, by combining analyzers together, the user can sometimes get "the best of both worlds."

3.2.1. Triple Quadrupoles

A quadrupole is a set of four linear rods, similar to a linear ion trap, but without the end caps. Variations exist with six (hexapole) or eight (octupole) rods, but they are all functionally the same. DC and RF AC currents are applied in a similar way to in a quadrupole ion trap and at given voltages, only ions of a certain m/z successfully traverse the chamber. A quadrupole can be operated in a mass selection mode, where all ions of a certain m/z can be separated from the others, or in a scanning mode, where voltages are ramped across a range of values and at a given time point, only ions of a given m/z traverse the chamber.

The triple quadrupole is a comparatively cheap mass spectrometer for fragmentation analysis. As the name suggests, it consists of three sequential quadrupoles. For measurement of intact components, the first two quadrupoles function essentially as ion transmission devices allowing all ions through and the third quadrupole is operated in a scanning mode to measure the m/z of all components (**Fig. 3.2a**). For fragmentation analysis, the first

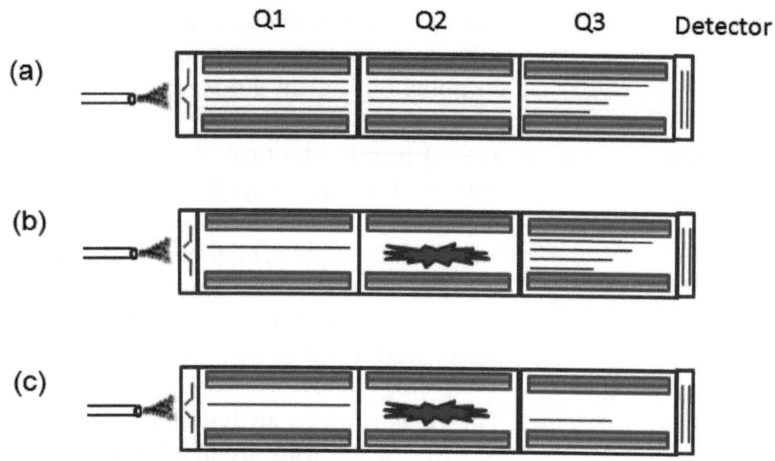

Fig. 3.2. Different operating modes of a triple quadrupole mass spectrometer. (**a**) For measurement of intact components, both quadrupoles Q1 and Q2 transmit all ions and Q3 is operated in a scanning mode where only ions of a given *m*/*z* pass through the chamber at a given time point during the scan. (**b**) For measurement of fragmentation spectra of a component, Q1 is operated in a mass selective mode to transmit only ions of the *m*/*z* of the desired precursor, Q2 is used as a collision cell, and Q3 is operated in a scanning mode to measure the fragment ion *m*/*z*. (**c**) For selective reaction monitoring, Q1 transmits only the *m*/*z* of the precursor ion, Q2 is used as a collision cell, and Q3 measures only one fragment *m*/*z*.

quadrupole is operated in a mass selection mode to isolate a single *m*/*z* component for subsequent analysis. The second quadrupole is used as a collision chamber where applied energy and introduction of low levels of a collision gas (often nitrogen, but sometimes helium or air) cause collisions and form fragments that can then be measured using the third quadrupole (**Fig. 3.2b**).

Quadrupoles are low-resolution and low mass accuracy analyzers. In a mass selection mode, they exhibit high sensitivity, but in a scanning mode, they are insensitive compared to other analyzers. Hence, triple quadrupoles are generally not the instruments of choice for discovery of peptide identifications through fragmentation analysis. However, they are powerful in studies where you already know how a component fragments and you want to either detect this component at very low levels and/or quantify its level. In experiments known as selected reaction monitoring (SRM), a single mass can be monitored in the first quadrupole, components are fragmented, and then the mass of a single fragment ion is monitored in the final quadrupole (**Fig. 3.2c**). By monitoring a single reaction (given mass precursor forms a given mass fragment), components can be detected at high sensitivity and specificity, even in the background of a very complex mixture. It is possible to cycle between a series of these SRM experiments within an LC-MS analysis (a technique often known as multiple reaction monitoring or MRM), allowing high-sensitivity and high-specificity monitoring of a set of components (25), and this approach has great applicability for monitoring biomarkers in complex biological fluids (26).

3.2.2. Q-Traps

The Q-trap is identical to a triple quadrupole instrument except that the final quadrupole has been replaced by a quadrupole ion trap. This dramatically improves the sensitivity of the instrument for scanning mass ranges and makes the instrument more suitable for analyzing peptide fragmentation spectra. The final ion trap can also be used as a quadrupole; so SRM studies can also be performed using this instrument (26, 27).

3.2.3. QqTOF

The basis of mass analysis by time of flight (TOF) resides in the fact that when ions are accelerated with the same kinetic energy, ions of higher m/z will have a lower velocity. Hence, by measuring the time ions take to travel along a flight path to a detector, it is possible to determine their m/z. Most TOF mass spectrometers, rather than having a simple linear flight path to the detector, employ an ion mirror or a reflectron that "bounces" the ions back to a detector toward the other end of the flight tube (28). This is a mechanism that allows partial compensation for any variation in initial kinetic energy that ions may have had; if two ions of the same m/z had slightly different initial kinetic energies, then the one with higher kinetic energy will have slightly higher velocity and so will penetrate further into the ion reflectron and take more time to be reflected. By positioning the detector at the correct focus point, the two ions of slightly different initial kinetic energies but same m/z will be detected at the same time. The use of a reflectron improves the resolution and mass accuracy of TOF mass spectrometers, giving performance in between a quadrupole ion trap and an orbitrap or FT-ICR mass spectrometer. In a TOF analyzer, all accelerated ions are measured. Hence, because of their short duty cycle, TOF instruments are relatively sensitive when compared to quadrupoles.

QqTOF instruments employ a first quadrupole for precursor isolation and a second as a fragmentation chamber in the same way as triple quadrupole and Q-trap instruments. However, the final analyzer is a TOF with a flight path orthogonal to the axis of the quadrupoles (29), as shown in **Fig. 3.3**. For TOF analysis, ions need to start with no initial kinetic energy, then a packet of ions are accelerated together, and flight times are measured. Ions passing through the quadrupoles in a QqTOF have kinetic energy in the plane of the quadrupole axes but none in the orthogonal direction and so are suitable for TOF measurement perpendicular to their flight path.

QqTOF instruments are good compromise instruments between sensitivity, mass accuracy, and resolution; they are more

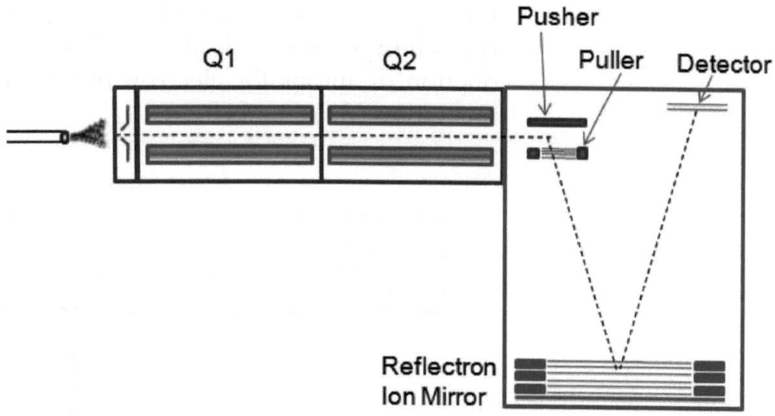

Fig. 3.3. Schematic of a QqTOF geometry instrument. Q1 can either allow all ions through or select a single *m/z* for subsequent fragmentation analysis. Q2 is used as a collision cell when CID analysis is desired. Ions entering the pusher/puller region have no velocity perpendicular to the path through the quadrupoles and so can be accelerated in pulses and their time of flight measured as they are reflected off the ion mirror and back to an ion detector.

sensitive than FT-ICR instruments but have better mass accuracy and resolution than do quadrupole ion traps.

3.2.4. Quadrupole Ion Trap–Orbitrap

As previously mentioned, quadrupole ion traps are able to measure ions, perform ion chemistry reactions to fragment components, and measure their masses, all in the same chamber. However, their routine low resolution and mass accuracy is a problem for peptide identification using database-searching strategies. Hence, there are clear advantages to an instrument where there is the alternative to measure the ions at high mass accuracy and resolution. The LTQ-Orbitrap is an instrument that combines a linear quadrupole ion trap and an orbitrap (24). All ion isolations and manipulations are performed in the quadrupole ion trap, but then the user has a choice of either measuring ions at high sensitivity but low resolution and mass accuracy in the quadrupole ion trap or transferring ions to the orbitrap (through a chamber known as the C-trap) for high-resolution and mass accuracy detection. Unfortunately there is some ion loss during transmission of ions from the linear quadrupole trap to the orbitrap; so there is a sensitivity hit in making this measurement. For identification of peptides by database-searching strategies, mass accuracy of the intact peptide is significantly more important than for fragment ions. Hence, the typical mode of operation for a LTQ-Orbitrap is to measure intact masses in the orbitrap but measure the fragments at higher sensitivity in the linear quadrupole trap.

The LTQ-Orbitrap XL (schematic shown in **Fig. 3.4**) is the same as the LTQ-Orbitrap described above, except that it has a hexapole located after the C-trap. This allows quadrupole-type

CID spectra to be acquired (although they must be measured in the orbitrap) (30). It is also used as a transmission cell for introduction of anions for electron transfer dissociation (ETD) fragmentation (31) (see below).

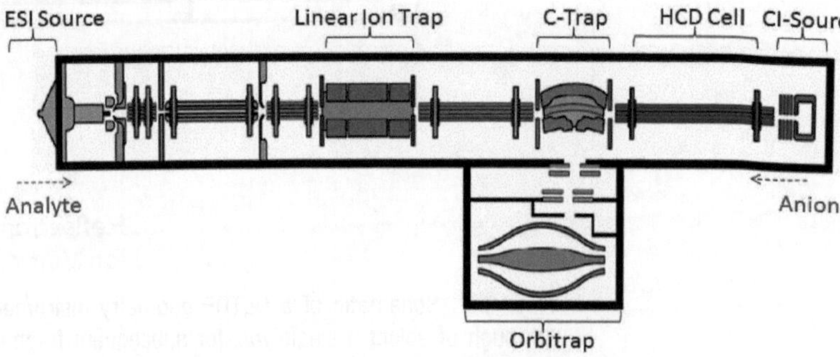

Fig. 3.4. Schematic of an LTQ-Orbitrap XL enabled with ETD. Analyte ions are introduced on the left of the instrument and pass through into the linear quadrupole ion trap. Ions can be measured in the linear trap at high sensitivity or passed out of the trap and diverted into the orbitrap through the C-trap for high mass accuracy and resolution measurement. For ETD analysis, anions are produced in the CI source at the right of the instrument and are passed through the HCD cell and C-trap to the linear ion trap, where ETD fragmentation occurs. Yet again, ions can then be measured in either the linear ion trap or orbitrap. Quadrupole (rather than quadrupole ion trap) fragmentation analysis can be performed in the HCD cell but can be measured only in the orbitrap.

3.2.5. Quadrupole Ion Trap–FT

Interfacing a quadrupole ion trap with an FT-ICR gives very similar functionality to an LTQ-Orbitrap, with the ability to measure ions either at high sensitivity or with high mass accuracy (32). FT-ICR can produce higher resolution measurement than can an orbitrap, but for proteomic applications, both have more resolution than strictly required. The advantages of the orbitrap are better dynamic range and sensitivity; the transfer distance between quadrupole trap and orbitrap is significantly shorter than into the cell inside an FT-ICR magnet; so there is less sample loss during ion transfer. However, the FT instrument does provide the opportunity to employ different fragmentation mechanisms.

3.3. Fragmentation Alternatives

CID is the major fragmentation approach employed in mass spectrometers but is not the only approach employed for proteomic studies, and particularly the use of ETD is likely to increase in the near future. Also, CID in a quadrupole ion trap produces slightly different fragmentation to CID performed in a quadrupole and the two methods are found to be complementary (see below).

3.3.1. CID in a Quadrupole Ion Trap vs in a Quadrupole

CID fragmentation in a quadrupole ion trap is produced by resonant excitation of a specific *m*/*z*. Hence, once the component has been fragmented, its products are not further excited. This means that the majority of the fragments observed are the products of a single-bond cleavage and so mainly b and y ions are observed (33). It also means that if there is a labile bond in the peptide, for example, the O-phosphate linkage in a phosphopeptide, then the fragmentation spectrum is dominated by a single fragment ion (34), sometimes precluding the ability to identify the peptide.

In quadrupole CID, all ions, precursors and fragments included, are excited. Hence, products formed by secondary fragmentation are also formed. Two backbone cleavages can form internal ions. In addition, b and y ions can be further fragmented. The b ion structure is relatively unstable and so readily fragments further to form smaller b ions (35). However, the b2 ion is stable. Hence, in quadrupole CID spectra, a large percentage of b ions are fragmented down to a b2 ion giving relatively intense b2 and a2 ions.

It should also be pointed out that resonant fragmentation spectra in a quadrupole ion trap lose the low-mass region of the spectrum, whereas quadrupole CID spectra contain a full mass range spectrum. This has important implications for certain quantitation strategies based on MS/MS peaks, e.g., iTRAQ (36), but this will be discussed in a later chapter.

3.3.2. Electron Capture Dissociation and Electron Transfer Dissociation

Electron capture dissociation (ECD) (20) and electron transfer dissociation (ETD) (14) provide orthogonal alternatives to CID for fragmentation analysis (37). Both of these approaches form unstable radical ions that then fragment at sites that are not the weakest bonds in the molecular structure. These techniques produce fragmentation spectra that are less dependent on the peptide sequence (with the exception of the inability to cleave N-terminal to proline residues).

ECD involves firing a beam of electrons at the trapped cloud of sample ions (20). Electron capture by the analyte produces a radical ion, which is unstable and fragments to produce predominantly c and z. ions from peptides. ECD is almost exclusively performed in FT-ICR instruments.

ETD uses anions, most commonly fluoranthene ions, to transfer electrons to the analyte, forming radical ions that then fragment similarly to ECD (14). ETD can be performed in quadrupole ion traps, making the technique much more sensitive and affordable than ECD in an FT-ICR instrument.

Both of these fragmentation approaches are more effective on highly charged and charge-dense precursors. As electron capture or transfer by definition reduces the charge of the precursor, one cannot produce charged fragments from singly charged components; so ETD/ECD MS/MS analysis of singly charged species cannot be performed. Also, the efficiency of fragmentation of doubly charged species is not as high as triply charged or greater precursors (38). Hence, these approaches are not as generally applicable as CID, but as they are effective on higher charged species that generally are less efficiently fragmented by CID, they can provide useful complementary data. Indeed, a decision tree that decides whether to perform CID or ETD based on the precursor *m/z* and charge has been shown to provide more comprehensive results than does either fragmentation approach alone (39). An area where these radical-based fragmentations are very powerful is in the analysis of labile post-translational modifications, where sites of modification can be determined on peptides bearing modifications that are highly labile under CID conditions (40–43). They also have the potential to be used for LC-MS analysis of large peptides and small intact proteins.

In conclusion, the correct choice of mass spectrometer to use for proteomic studies is somewhat dependent on the type of analysis to be performed. With any selection a trade-off has to be made between sensitivity versus mass accuracy and resolution, but hybrid instruments provide more flexibility in workflows to suit different types of experiments. The availability of different fragmentation mechanisms on the same instrument also allows a wider range of sample analysis strategies, and this can be particularly important for the analysis of protein and peptide modifications.

References

1. Fenn, J. B, Mann, M., Meng, C. K., Wong, S. F., and Whitehouse, C. M. (1989) Electrospray ionization for mass spectrometry of large biomolecules. *Science* **246**, 64–71.
2. Hillenkamp. F., and Karas, M. (1990) Mass spectrometry of peptides and proteins by matrix-assisted ultraviolet laser desorption/ionization. *Methods Enzymol.* **193**, 280–295.
3. Iribarne, J. V., and Thomson, B. A. (1976) On the evaporation of small ions from charged droplets. *J. Chem. Phys.* **64**, 2287–2294.
4. Nguyen, S., and Fenn, J. B. (2007) Gas-phase ions of solute species from charged droplets of solutions. *Proc. Natl. Acad. Sci. USA* **104**. 1111–1117.
5. Wilm, M., Shevchenko, A., Houthaeve, T. et al. (1996) Femtomole sequencing of proteins from polyacrylamide gels by nano-electrospray mass spectrometry. *Nature* **379**, 466–469.
6. MacNair, J. E., Patel, K. D., and Jorgenson, J. W. (1999) Ultrahigh-pressure reversed-phase capillary liquid chromatography: isocratic and gradient elution using columns packed with 1.0-micron particles. *Anal. Chem.* **71**, 700–708.
7. Plumb, R., Castro-Perez, J., Granger, J., Beattie, I., Joncour, K., and Wright, A. (2004) Ultra-performance liquid chromatography coupled to quadrupole-orthogonal time-of-flight mass spectrometry. *Rapid Commun. Mass Spectrom.* **18**, 2331–2337.
8. Shen, Y., Zhao, R., Berger, S. J., Anderson, G. A., Rodriguez, N., and Smith, R. D. (2002) High-efficiency nanoscale liquid chromatography coupled on-line with mass spectrometry using nanoelectrospray ionization for proteomics. *Anal. Chem.* **74**, 4235–4249.
9. Strittmatter, E. F., Ferguson, P. L., Tang, K., and Smith, R. D. (2003) Proteome analyses using accurate mass and elution time peptide tags with capillary LC time-of-flight mass spectrometry. *J. Am. Soc. Mass Spectrom.* **14**, 980–991.
10. Douglas, D. J., Frank, A. J., and Mao, D. (2005) Linear ion traps in mass spectrometry. *Mass Spectrom. Rev.* **24**, 1–29.
11. Schwartz, J. C., Senko, M. W., and Syka, J. E. (2002) A two-dimensional quadrupole ion trap mass spectrometer. *J. Am. Soc. Mass Spectrom.* **13**, 659–669.
12. Stafford, G. C., Kelley, P. E., Syka, J. E. P., Reynolds, W. E., and Todd, J. F. J. (1984)

Recent improvements in and analytical applications of advanced ion trap technology. *Int. J. Mass Spectrom. Ion Process.* **60**, 85–98.

13. Schwartz, J. C., Zhou, X., Bier, M. E., inventors (1996) Method and apparatus of increasing dynamic range and sensitivity of a mass spectrometer, U.S. Patent 5572022.
14. Coon, J. J., Shabanowitz, J., Hunt, D. F., and Syka, J. E. (2005) Electron transfer dissociation of peptide anions. *J. Am. Soc. Mass Spectrom.* **16**, 880–882.
15. Pitteri, S. J., and McLuckey, S. A. (2005) Recent developments in the ion/ion chemistry of high-mass multiply charged ions. *Mass Spectrom. Rev.* **24**, 931–958.
16. Comisarow, M. B., and Marshall, A. G. (1974) Fourier transform ion cyclotron resonance spectroscopy. *Chem. Phys. Lett.* **25**, 282–283.
17. Cody, R. B., Burnier, R. C., and Freiser, B. S. (1982) Collision-induced dissociation with Fourier transform mass spectrometry. *Anal. Chem.* **54**, 96–101.
18. Gauthier, J. W., Trautman, T. R., and Jacobsen, D. B. (1991) Sustained off-resonance irradiation for CAD involving FTMS. CAD technique that emulates infrared multiphoton dissociation. *Anal. Chim. Acta* **246**, 211–225.
19. Little, D. P., Speir, J. P., Senko, M. W, O'Connor, P. B., and McLafferty, F. W. (1994) Infrared multiphoton dissociation of large multiply charged ions for biomolecule sequencing. *Anal. Chem.* **66**, 2809–2815.
20. Zubarev, R. A. (2004) Electron-capture dissociation tandem mass spectrometry. *Curr. Opin. Biotechnol.* **15**, 12–16.
21. Marshall, A. G., Hendrickson, C. L., and Jackson, G. S. (1998) Fourier transform ion cyclotron resonance mass spectrometry: a primer. *Mass Spectrom. Rev.* **17**, 1–35.
22. Meng, F., Forbes, A. J., Miller, L. M., and Kelleher, N. L. (2005) Detection and localization of protein modifications by high resolution tandem mass spectrometry. *Mass Spectrom. Rev.* **24**, 126–134.
23. Hu, Q., Noll, R. J., Li, H., Makarov, A., Hardman, M., and Graham Cooks, R. (2005) The Orbitrap: a new mass spectrometer. *J. Mass Spectrom.* **40**, 430–443.
24. Makarov, A., Denisov, E., Kholomeev, A. et al. (2006) Performance evaluation of a hybrid linear ion trap/orbitrap mass spectrometer. *Anal. Chem.* **78**, 2113–2120.
25. Lange, V., Picotti, P., Domon, B., and Aebersold, R. (2008) Selected reaction monitoring for quantitative proteomics: a tutorial. *Mol. Syst. Biol.* **4**, 222.
26. Stahl-Zeng, J., Lange, V., Ossola, R. et al. (2007) High sensitivity detection of plasma proteins by multiple reaction monitoring of N-glycosites. *Mol. Cell. Proteomics* **6**, 1809–1817.
27. Le Blanc, J. C., Hager, J. W., Ilisiu, A. M., Hunter, C., Zhong, F., and Chu, I. (2003) Unique scanning capabilities of a new hybrid linear ion trap mass spectrometer (Q TRAP) used for high sensitivity proteomics applications. *Proteomics* **3**, 859–869.
28. Mamyrin, B. A., Karataev, V. I., Shmikk, D. V., and Zagulin, V. A. (1973) *Sov. Phys. JETP* **37**, 45.
29. Chernushevich, I. V., Loboda, A. V., and Thomson, B. A. (2001) An introduction to quadrupole-time-of-flight mass spectrometry. *J. Mass Spectrom.* **36**, 849–865.
30. Olsen, J. V., Macek, B., Lange, O., Makarov, A., Horning, S., and Mann, M. (2007) Higher-energy C-trap dissociation for peptide modification analysis. *Nat. Methods* **4**, 709–712.
31. McAlister, G. C., Phanstiel, D., Good, D. M., Berggren, W. T., and Coon, J. J. Implementation of electron-transfer dissociation on a hybrid linear ion trap–orbitrap mass spectrometer. *Anal. Chem.* **79**, 3525–3534.
32. Syka, J. E., Marto, J. A., Bai, D. L. et al. (2004) Novel linear quadrupole ion trap/FT mass spectrometer: performance characterization and use in the comparative analysis of histone H3 post-translational modifications. *J. Proteome Res.* **3**, 621–626.
33. Biemann, K. (1990) Appendix 5. Nomenclature for peptide fragment ions (positive ions). *Methods Enzymol.* **193**, 886–888.
34. McLachlin, D. T., and Chait, B. T. (2001) Analysis of phosphorylated proteins and peptides by mass spectrometry. *Curr. Opin. Chem. Biol.* **5**, 591–602.
35. Paizs, B., and Suhai, S. (2005) Fragmentation pathways of protonated peptides. *Mass Spectrom. Rev.* **24**, 508–548.
36. Ross, P. L., Huang, Y. N., Marchese, J. N. et al. (2004) Multiplexed protein quantitation in *Saccharomyces cerevisiae* using amine-reactive isobaric tagging reagents. *Mol. Cell. Proteomics* **3**, 1154–1169.
37. Bateman, R. H., Carruthers, R., Hoyes, J. B. et al. (2002) A novel precursor ion discovery method on a hybrid quadrupole orthogonal acceleration time-of-flight (Q-TOF) mass spectrometer for studying protein phosphorylation. *J. Am. Soc. Mass Spectrom.* **13**, 792–803.
38. Good, D. M., Wirtala, M., McAlister, G. C., and Coon, J. J. (2007) Performance characteristics of electron transfer dissociation

mass spectrometry. *Mol. Cell. Proteomics* **6**, 1942–1951.

39. Swaney, D. L., McAlister, G. C., Coon, J. J. (2008) Decision tree-driven tandem mass spectrometry for shotgun proteomics. *Nat. Methods* **5**, 959–964.
40. Kelleher, N. L., Zubarev, R. A., Bush, K. et al. (1999) Localization of labile posttranslational modifications by electron capture dissociation: the case of gamma-carboxyglutamic acid. *Anal. Chem.* **71**, 4250–4253.
41. Mirgorodskaya, E., Roepstorff, P., and Zubarev, R. A. (1999) Localization of O-glycosylation sites in peptides by electron capture dissociation in a Fourier transform mass spectrometer. *Anal. Chem.* **71**, 4431–4436.
42. Vosseller, K., Trinidad, J. C., Chalkley, R. J. et al. (2006) O-Linked *N*-acetylglucosamine proteomics of postsynaptic density preparations using lectin weak affinity chromatography and mass spectrometry. *Mol. Cell. Proteomics* **5**, 923–934.
43. Chi, A., Huttenhower, C., Geer, L. Y. et al. (2007) Analysis of phosphorylation sites on proteins from *Saccharomyces cerevisiae* by electron transfer dissociation (ETD) mass spectrometry. Proc. Natl. Acad. Sci. USA **104**, 2193–2198.

Chapter 4

Bioinformatics for LC-MS/MS-Based Proteomics

Richard J. Jacob

Abstract

Mass spectrometry instrumentation has continued to develop rapidly in the last two decades, enabled in part by advances in microelectronic hardware controllers and computerized control and data acquisition systems. The wealth and complexity of data produced by a modern instrument is such that the data can no longer be analyzed manually. Computerized data analysis has become *de rigueur* and the bioinformatics field has expanded to provide software applications for all aspects of the data analysis needed by LC-MS/MS. The bioinformatics field is evolving rapidly and software applications are continually being improved or replaced for existing applications as well as developed to support new types of experiments and analysis enabled by modern instrumentation. Entire books have been written on MS data analysis in proteomics but this review will be necessarily brief. In this chapter we will review the bioinformatics software applications available for different LC-MS/MS analysis tasks.

Key words: Proteomics, bioinformatics, database searching.

1. Introduction

Before the advent of genomic databases and modern MS instruments, sequences could only be obtained from MS/MS of highly abundant endogenous peptides and proteins and the sequencing techniques were manual and slow. All the data analyses were done by de novo sequencing of MS/MS spectra. It was rare that a complete sequence of a peptide could be determined from an MS/MS spectrum. The first computer programs for protein and peptide analyses were written to aid in the interpretation of spectra (1, 2), perform in silico digestions of proteins (3), and map identified peptides on to a protein sequence (4), amongst other tasks. With the advent of BLAST (5), a peptide sequence

P.R. Cutillas, J.F. Timms (eds.), *LC-MS/MS in Proteomics*, Methods in Molecular Biology 658,
DOI 10.1007/978-1-60761-780-8_4, © Springer Science+Business Media, LLC 2010

that was obtained by MS/MS could now be searched against a protein database and hopefully identify a protein. In 1990 the public genomic and protein databases were relatively small with only 40,000 entries. Developments in DNA sequencing led to an explosion of DNA sequence data, basically doubling the number of entries in a database every 12–15 months (6). The increasing number of entries in the database and the improvements in MS and MS/MS instrumentation, data quality, and automation in the 1990s led to the initial development of three different database-searching techniques for identifying proteins from MS and MS/MS data: (1) peptide mass fingerprinting (PMF) (7), (2) uninterpreted MS/MS (8), and (3) sequence tags (9).

Both rapid advances in modern microelectronics and the dramatic increase in computer processing power have enabled the development of mass spectrometers and allowed computers to keep pace with the volume of MS/MS data generated with the growth of the databases. The sheer amount of data generated makes it impossible to manually analyze, de novo sequence, and search all the MS/MS spectra from an LC-MS/MS run. This makes computerized data analysis routine and indispensable. There are now hundreds of software applications covering all aspects of LC-MS/MS data analysis, developed by an active community of software engineers, developers, and scientists. Many software applications are written for and used by individual labs. However, a considerable number of them are made available to the general public under a number of different licensing agreements but typically distributed free. Some of the software applications are free and open source software (FOSS) applications which allow users to study, change, and improve their design through the availability of its source code. There is also a smaller community of commercial companies that sell or distribute software applications for LC-MS/MS analysis. The suppliers of commercial software applications are split between the MS instrument vendors themselves and a mixture of smaller private and publicly held companies.

The bioinformatics field is developing rapidly and some of the software applications mentioned here will be superseded or no longer supported over time. New applications are being continually released and these, for obvious reasons, will not be mentioned here. Some of the applications developed by the MS instrument vendors are limited to working only with data produced by their own instruments, and for this reason the vendors' software applications have not been extensively reviewed. Finally, due to the large number of publicly available applications, only a small subset have been described here and are used as examples of the genre. Fortunately, there are a number of Web sites that list and collate applications that are currently available. Along with the current literature, these Web sites are good sources of links to the actual applications or documentation about them. Two recent book publications cover the theory and application of the software applications mentioned in the different sections of this review in far more detail and are highly recommended (10, 11).

Bioinformatic software can be used to theoretically analyze a proteome before even stepping into the laboratory, which can aid in the design of an experiment. Once the LC-MS/MS data has been acquired, it needs to be processed and converted into a peak list before database searching and peptide and protein identification. The LC-MS/MS data search results can be visualized and compared before generating reports for use in the laboratory or for publication. Recently, there has been a lot of focus on quantitation of proteins by MS/MS which requires specialized software. There are a number of reasons why a large proportion of the MS/MS spectra remain unidentified by a database search. A second-pass search can be used to identify spectra of peptides with unexpected post-translational modifications or that are a product of non-specific enzyme cleavages. De novo sequencing can be used to identify the peptides from some of the remaining spectra. Once a list of all the proteins in a sample have been determined, bioinformatics software can be used to query databases for biological information about the proteins, their function, localization, and known interaction partners. Information about LC-MS/MS experiments, the data, and search results can be tracked, organized, and stored in Laboratory Information Management Systems (LIMS) and made available for data mining. Consideration has to be made for the long-term storage of LC-MS/MS data on computer servers. Raw data and search results may need to be converted into standardized formats for data exchange and submittal to public repositories.

2. Theoretical Analysis of Proteomes

As the technologies used in experimental proteomics have multiplied, the need to model the effect of these technologies on theoretical analysis has grown. The choice of technologies used in an experiment has an impact on the number of proteins that can be identified. Proteomic experiments targeting specific proteins will need careful choice of technologies to optimize the possibility that the proteins of interest are present and identified. Theoretical proteomic analysis provides a description of some aspects of the proteome (for example, the proportion of the amino acids and the molecular weight and p*I* of the proteins), which can be used to design an experiment that targets proteins normally missed in typical analyses (12). A more complicated analysis involves theoretical modeling of the proteolytic processing of a proteome to determine the effect of proteolytic agents on the nature and properties of the resulting peptides.

There have been a number of theoretical analyses of whole proteome digests as part of the development of the accurate mass (AM) and the accurate mass and time (AMT) tag concepts (see later) and by other groups exploring the effects of chemical derivatization (13), interpreting peptide mass spectra (14) and peptide isoelectric point (15) or carrying out basic descriptive and statistical analysis (16). These groups have developed software programs to perform the analysis, some of which have been released to the public (**Table 4.1**). Virtual Expert Mass Spectrometrist (VEMS) is probably the most developed program with a lot of additional functionalities and has been well described (10, 17).

Table 4.1
Software for the theoretical analysis of proteomes

VEMS (10, 17)	http://personal.cicbiogune.es/rmatthiesen/
Proteogest (16)	http://www.utoronto.ca/emililab/proteogest.htm
ProteomeXplorer	http://pc4-133.ludwig.ucl.ac.uk/

3. Raw Data Processing

Each mass spectrometry manufacturer uses a number of different proprietary file formats to store the results of MS experiments. The data are acquired and stored as profile or centroid data. The structure of the data varies by instrument and experiment type; the files typically consist of MS spectra interleaved with multiple MS/MS spectra. The spectra may be in high or low resolution or a mixture depending on the mass spectrometer and the method. Low-resolution instruments can use zoom scans to improve the charge state determination of the precursor ions. Experimental methods can be programmed to acquire MS^n or MS^E spectra or use other targeted acquisitions like multiple reaction monitoring (MRM). The vendors typically include data analysis software that can process the raw data by smoothing and noise filtering followed by calibration, deisotoping, and peak picking. The MS vendor's software package can include their own database-searching software or may integrate with other commercial packages. Raw files can be very large for a long LC-MS/MS acquisition on a high-resolution instrument. The protein identification search engines cannot search the raw files directly; so they need to be converted into peak lists first.

Producing a high-quality peak list from the raw data is crucial to producing a high-quality protein database search result. For best results, the peak list should be calibrated and deisotoped with just the ^{12}C ions reported at the correct charge state. The ideal peak list would contain the peptide fragment ions for the MS/MS spectra and no noise peaks or extraneous masses but this can be hard to achieve in practice and not always necessary as the database search engines can cope with a fair amount of noise by design. Some of the raw data-processing tools included with the Trans-Proteomic Pipeline (TPP) can produce peak lists for raw data as does the free toolkit ProteoWizard. A number of search engines and LIMS include peak-picking tools and there is at least one commercial program, Mascot Distiller (Matrix Science Ltd, London, UK), that can open and process raw data files from most instruments. The resulting peak list from all these tools is normally a text file formatted to match the input specifications of a search engine. It is easy to convert these peak lists from one format to another.

A modern LC-MS/MS system can analyze tens of samples a day, far too much to interpret manually. To efficiently analyze the output of one or more instruments, a certain level of automation is required. The raw data files are often processed through a software pipeline (**Table 4.2**). The vendor's software may have some automation features or a third-party automation client might be used. The automation client is often included with the search engine software; for example, Matrix Science's Mascot Daemon is included with the Mascot Server software or the automation scripts in the TPP. The data analysis pipelines may be built into a Laboratory Information Management System that can automate the data processing from acquisition to final reporting.

Table 4.2
Raw data-processing and format conversion software

Trans-Proteomic Pipeline (77) mzXML	http://tools.proteomecenter.org/wiki/index.php?title=Formats:mzXML
ProteoWizard (78)	http://proteowizard.sourceforge.net/
Mascot Distiller	http://www.matrixscience.com/distiller.html
Spectrum Mill	http://www.home.agilent.com
Rosetta Elucidator	http://www.rosettabio.com/products/elucidator

Depending on the quality and number of the spectra in the peak list, it can be beneficial to clean up the data before submitting it to a search engine. Spectra can be checked for quality and noise, those failing the tests being rejected, thereby reducing the overall size and search time for the analysis. Finally the peak lists are submitted to the search engine or multiple search engines of choice.

4. Peptide and Protein Identification

A protein or a complex mixture of proteins is digested with a proteolytic agent and subjected to MS and MS/MS analysis. The mixture of peptides is separated and individual peptides are selected for MS/MS analysis. The resulting raw data file can contain many thousands of MS/MS experiments. The result of this experimental design is that the relationship between peptide and protein is lost and there may be some ambiguity as to which protein a peptide should be assigned to. There are a number of different methods for identifying the proteins in the sample, and the most frequently used is the searching of the uninterpreted data.

4.1. Uninterpreted MS/MS Database Searching

Searching uninterpreted MS/MS data from a single peptide or from a complete LC-MS/MS run was pioneered by Yates and Eng (18) and uses software to match lists of fragment ion mass and intensity values, without any manual sequencing to peptides, in a database (**Table 4.3**). The precursor mass is used as a filter to find a list of candidate peptide sequences from the theoretical digest of the database. A variety of different systems are used to score the experimental MS/MS spectrum against spectra predicted from the candidate peptide sequences. The scoring algorithms range from simple filters to cross-correlation algorithms to true probabilistic methods, and an identity threshold is used to determine which peptides are significant. As the results of an MS/MS database search normally represent a mixture of proteins, they can be quite complicated to report. Each MS/MS spectrum may match to one or more peptides and each peptide match may be assignable to multiple proteins. Using the principle of parsimony, or Occam's razor, the minimum number of proteins that can account for the observed peptides is reported (19). Although searches can be slow, they do allow identification of peptides from poor-quality spectra.

4.2. Sequence Tag Searching

An alternative to uninterpreted MS/MS searching is the use of sequence tags. In a sequence tag, the mass information for a peptide is combined with amino acid sequence or composition

Table 4.3
Peptide and protein identification software

Uninterpreted MS/MS database searching	
ProbID (79)	http://sashimi.sourceforge.net/software_mi.html
PepProbe	http://bart.scripps.edu/public/search/pep_probe/search.jsp
OMSSA (80)	http://Pubchem.ncbi.nlm.nih.gov/omssa
Phenyx (81)	http://www.genebio.com/products/phenyx/
Spectrum Mill	http://www.home.agilent.com
X!Tandem (82)	http://human.thegpm.org/tandem/thegpm_tandem.html
MS-Tag (22)	http://Prospector.ucsf.edu
Mascot Server (19)	http://www.matrixscience.com/
Protein Pilot/Paragon (83)	https://products.appliedbiosystems.com/ab/en/US/adirect/ab?cmd=catNavigate2&catID=601680&tab=DetailInfo
SEQUEST (8)	http://www.thermo.com/com/cda/product/detail/0,,16483,00.html?CA=bioworks
XProteo	http://xproteo.com:2698/
MyriMatch (84)	http://www.mc.vanderbilt.edu/msrc/bioinformatics/myrimatch/index.php
Sequence tag searching	
GutenTag (85)	http://fields.scripps.edu/GutenTag/
InsPecT (86)	http://bix.ucsd.edu/
MultiIdent (87)	http://www.expasy.org/tools/multiident/
Spider (88)	http://proteome.sharcnet.ca:8080/spider.htm
MS-Seq (22)	http://Prospector.ucsf.edu
Mascot Server (19)	http://www.matrixscience.com/
PepFrag (89)	http://prowl.rockefeller.edu/prowl/pepfrag.html
Accurate mass and time tags	
VIPER (24)	http://omics.pnl.gov/software/VIPER.php
msInspect/AMT (90)	http://proteomics.fhcrc.org/CPL/amt/
Spectral matching	
P3	http://www.thegpm.org/PPP/index.html
SpectraST (91)	http://www.peptideatlas.org/spectrast/
BiblioSpec (92)	http://proteome.gs.washington.edu/software/bibliospec/documentation/index.html
NoDup	fields.scripps.edu/nodupe/index.html
MS-Clustering (93)	http://bix.ucsd.edu/

(Continued)

Table 4.3 (Continued)

Machine-learning algorithms	
Gist (26)	svm.sdsc.edu
Percolator (27, 28)	http://noble.gs.washington.edu/proj/percolator/
Mascot Percolator (28)	http://www.sanger.ac.uk/Software/analysis/MascotPercolator/

data. The original method was developed by Mann and Wilm (9). Because it is very rare to obtain complete sequence coverage in an MS/MS spectrum, only a partial interpretation is performed identifying three or more amino acids in a sequence. This short stretch of sequence is combined with the fragment ion mass values which enclose it, the peptide mass, and the enzyme specificity in a database search (**Table 4.3**). This can be enough information to unambiguously identify a protein. A sequence tag acts as a filter on the database. A more powerful approach to using sequence tags is an error-tolerant tag. Relaxing the specificity by removing the peptide molecular mass constraint allows the tag to float within the candidate sequence so that a match is possible even if there is a difference in the calculated mass to one side or the other side of the tag. This enables a sequence tag to match a peptide when there is an unsuspected modification or a variation in the primary amino acid sequence. The sequence tag approach boasts rapid search times as it is essentially a filter and can be error tolerant allowing for the matching of unknown modifications or single-nucleotide polymorphisms (SNPs) but requires correct interpretation of the MS/MS spectrum, although ambiguity is acceptable.

4.3. Accurate Mass and Time Tags

Depending upon the complexity of a digested proteome and the accuracy of the mass measurements of the resulting peptides, many proteins can be identified by a single peptide which has a unique mass termed an accurate mass tag (AM) (20). This idea was then extended to utilize the liquid chromatography retention time in a later paper (21) and termed an accurate mass and time tag (AMT). The accurate mass and time tag idea is fairly new, although the basis can be traced back to the original protein identification search engines (22). Because of the complexity of protease-treated proteomes, only mass spectrometers with sufficiently high mass accuracy, such as FT-ICR, Orbitrap, and very accurate QTOF instruments are able to produce data suitable for AM and AMT tag analysis. There are now a number of search engines that use the AMT tag approach to identify peptides (23, 24) (**Table 4.3**).

4.4. Spectral Matching

The searching of unknown MS spectra against databases of spectra from known compounds has been used for a long time in all forms of analytical spectroscopy including mass spectrometry. Until recently there were no large databases of identified and annotated peptides available. As the size of the publicly available annotated databases has increased, so has the feasibility of spectral searching. Spectral matching software evaluates the overlap between observed and predicted fragments in the spectra and scores them using probability functions or spectral similarity metrics (**Table 4.3**). Spectral matching is considerably faster than uninterpreted MS/MS searching but is normally reliant on the peptides of interest being present in the annotated database. Peptides from low abundance proteins, rarely seen alternative splice forms or weakly fragmented under MS/MS conditions, will be poorly represented. An alternative approach is to populate a database with simulated spectra of all the theoretical peptides for a proteome (25).

4.5. Machine-Learning Algorithms

Machine-learning algorithms, which improve their performance over time, can also be used to analyze data or improve results (**Table 4.3**). The algorithms are classified into different types depending on their design. The most commonly used machine-learning algorithms in proteomics are supervised learning algorithms, which learn from a training set of manually curated examples, and semi-supervised learning algorithms, which learn from both manually curated examples and uninterpreted examples. A support vector machine, Gist (a supervised learning algorithm), was used to discriminate between positive and negative identifications by SEQUEST using the scoring systems reported by SEQUEST and other calculated factors (26), while a semi-supervised learning algorithm, Percolator, was used to discriminate between correct and decoy spectrum identifications (27, 28).

5. Visualization of LC-MS/MS Data, Validating, Comparing, and Reporting Results

The software applications described in this section are all multifunctional but can be grouped around the general tasks of visualization, validation, comparing, and reporting results (**Table 4.4**). The applications interact with the raw LC-MS/MS data and the peptide search results generated by specialized algorithms described in **Section 4**.

Table 4.4
Visualization of LC-MS/MS data, validating, comparing, and reporting results software

Visualization of LC-MS/MS data	
MSight (94)	http://www.expasy.ch/MSight/
Pep3D (95)	http://tools.proteomecenter.org/wiki/index.php?title=Software:Pep3D
DeCyder MS	http://www5.gelifesciences.com/aptrix/upp01077.nsf/Content/Products?OpenDocument&parentid=976024&moduleid=166070&zone=Elpho
Prequips (96)	http://prequips.sourceforge.net/
Validating and comparing search results	
Scaffold (30)	http://www.proteomesoftware.com/
PROTEOIQ	https://www.bioinquire.com/
Proteome Discoverer	http://www.thermo.com/com/cda/product/detail/1,1055,10142329,00.html
Phenyx (81)	http://www.genebio.com/products/phenyx/
Trans-Proteomic Pipeline (*77*)	http://tools.proteomecenter.org/wiki/index.php?title=Software:TPP
IDPicker (97)	http://fenchurch.mc.vanderbilt.edu/bumbershoot/idpicker/index.html
DTASelect (98)	http://fields.scripps.edu/DTASelect/
Reporting search results	
Excel (Microsoft)	http://office.microsoft.com/en-us/excel/default.aspx
Mascot Parser	http://www.matrixscience.com/msparser.html
MASCOT HTML and XML parser (32)	http://www.ccbm.jhu.edu
MascotDatfile (33)	http://genesis.ugent.be/MascotDatfile/
mres2x (34)	http://grosse-coosmann.de/~florian/Overview.html
Scaffold (30)	http://www.proteomesoftware.com/
PROTEOIQ	https://www.bioinquire.com/
Multiplierz (99)	http://blais.dfci.harvard.edu/multiplierz/
MassSieve (100)	http://www.ncbi.nlm.nih.gov/staff/slottad/MassSieve/
MASPECTRAS (101)	https://maspectras.genome.tugraz.at/maspectras/
VEMS (17)	http://personal.cicbiogune.es/rmatthiesen/
MascotDatfile (33)	http://genesis.ugent.be/MascotDatfile/

(Continued)

Table 4.4 (Continued)

Complete pipelines	
Proteios Software Environment (102)	http://www.proteios.org/
Trans-Proteomic Pipeline (77)	http://tools.proteomecenter.org/wiki/index.php?title=Software:TPP
Elucidator System	http://www.rosettabio.com/products/elucidator
Spotfire	http://spotfire.tibco.com/

5.1. Visualization of LC-MS/MS Data

LC-MS/MS data are a fairly complex three-dimensional array of structured data. The exact structure depends on the MS instrument used to acquire the data and the analysis method. Visualization of LC-MS/MS data with the precursors that lead to peptide identifications can be very useful in assessing the quality of the data. Data are typical plotted with m/z vs time and intensity in the third dimension with interactive controls that allow visual zooming and panning and data processing. Data-processing options range from feature detection, background subtraction, and charge state deconvolution to quantitation and multiple sample alignment and comparison or differential analysis. Precursors that were analyzed by MS/MS may be marked in a different color as are those that lead to a significant peptide identification.

5.2. Validating and Comparing Search Results

Not all the peptide and protein identification applications use a true probabilistic scoring system; so an additional level of validation is often desired. LC-MS/MS data might be searched with multiple search engines to gain further confidence in the peptide assignments or to identify additional matches to those identified by only one of the search engines. The applications often provide some level of false discovery rate (FDR) determination. The results from multiple search engines can then be compared and combined for the final interpretation and reporting. Statistical analysis is also used to validate the database search engine's spectra to peptide sequence and peptide to protein assignments (29). Finally, further statistical analysis can be used to compare and combine search results from different fractions or between different search engine algorithms (30). This allows for the analysis of large data sets containing multiple samples that have been analyzed using the MudPIT (31) protocol or one-dimensional gel electrophoresis in combination with LC-MS/MS (GeLC-MS/MS). The complexity of the statistical analysis and combined search results is such that the results are often visualized and viewed in an interactive browser (**Table 4.4**).

5.3. Reporting Search Results

Many of the LC-MS/MS data search algorithms store their search results in a format that is not normally human readable. Therefore, to view the results, collaborators may need access to the search engine. The level of information presented by the search engine may be more or less than the collaborators' need and the format of the results may not be amenable to further analysis. Most search engines can export the results to either industry standard formats like pepXML or to simple comma separated value (CSV) files. pepXML can be used as an input format for the TPP, other applications, and relational databases. pepXML is written in structured plain text but is not designed for manual reading or editing. Excel is frequently used to view, edit, and analyze data in laboratories and can open CSV files. Excel contains a number of analysis tools but it does not offer a very rich view of the search results. In recent years, a number of tools have been developed that provide a more interactive interface to the search results (**Table 4.4**). These applications allow analysts to quickly drill down into the data, reviewing peptide matches to MS/MS data, searching and filtering results, etc. For more complex data analysis, the data often need to be imported into a relational database. Although a pepXML file is a good starting point, it does not contain all the results produced by a search engine. In this case, a specialized result parser can be used to extract all the data so that it can be formatted and inserted into a database or used with a custom application. Some search engine manufacturers provide parsers for their result files and there are also many FOSS parsers developed by the community (32–34). With the publication of guidelines for proteomics-based papers by the journals *Molecular and Cellular Proteomics* (35) and *Proteomics* (36), some of these applications have started to produce guideline-compliant reports.

6. Quantitation

LC-MS/MS has long been used for small molecule quantitation but has only more recently been widely used for peptide and protein quantitation (37). Despite the profusion of different methods, they can be broken down into six fundamental protocols:

1. Reporter
2. Precursor
3. Multiplex
4. Replicate
5. Spectral counting
6. Average

Reporter ion-based quantitation methods use the relative intensities of specific fragment peaks at fixed *m/z* values generated from chemically derivatized peptides that break down under MS/MS conditions (38, 39). Precursor quantitation uses the relative intensities of extracted ion chromatograms (XICs) for precursors within a single data set. Any chemical derivatization that creates a large enough precursor mass shift can be used. The more popular techniques include ICAT (40), ^{18}O labeling (41, 42), spiking in an isotopically enriched peptide (43), and growing under isotopically enriched conditions (44) amongst other methods. For multiplex quantitation, the two samples are differentially labeled at a terminal creating a small mass shift and both peptides are passed through the MS/MS transmission window for fragmentation. The relative intensities of sequence ion fragment peaks are used for the quantitation (45). A protocol that is receiving a lot of attention is the replicate protocol because it is label free (46–51). The quantitation is based on the relative intensities of XICs for the precursors in multiple data sets. Spectral counting is another label-free quantitation protocol and is based on the number of spectra observed per protein. A variation is the emPAI method that compares the number of observed spectra against the number of expected peptides (52). The average quantitation protocol is also a label-free method for quantifying proteins in a mixture based on the identified peptide's XIC intensities (53).

There are quite a few software packages available for quantitation (**Table 4.5**). Many of them are restricted to data from a particular instrument so that they can generate XICs, and this is

Table 4.5
Quantitation software

MSQuant (103)	http://msquant.sourceforge.net/
MaxQuant (104, 105)	http://www.biochem.mpg.de/en/rd/maxquant/index.html
Mascot Distiller	http://www.matrixscience.com/distiller.html
Trans-Proteomic Pipeline (77)	http://tools.proteomecenter.org/wiki/index.php?title=Software:TPP
iTracker (106)	http://www.cranfield.ac.uk/health/researchareas/bioinformatics/page6801.jsp
MFPaQ (107)	http://mfpaq.sourceforge.net/
DAnTE (108)	http://omics.pnl.gov/software/DAnTE.php
RAAMS (109)	http://informatics.mayo.edu/svn/trunk/mprc/raams/index.html
StatQuant (110)	https://gforge.nbic.nl/projects/statquant/
ZoomQuant (41)	http://proteomics.mcw.edu/zoomquant

particularly the case for software from mass spectrometry instrument companies. Additionally, most of the quantitation applications support only one custom method based on one of the six protocols listed above. The application provides a solution to the method of choice for the developing laboratory and keeps the complexity of the application to a minimum.

7. Post-translational Modification Identification and De Novo Sequencing

The percentage of MS/MS queries that can be identified from an LC-MS/MS run typically runs from 5 to 30% under standard search conditions. This is not particularly high and represents a huge waste in instrument analysis time and data processing. There are a number of reasons why so many queries remain unidentified. The main reasons can be grouped as follows:

1. Underestimated mass measurement error.
2. Incorrect determination of precursor charge or incorrect assignment of the ^{12}C isotope.
3. The peptide has an unexpected modification.
4. Enzyme non-specificity.
5. The peptide sequence is not in the databases.
6. Multiple peptides with overlapping retention times and masses that fall into the same precursor selection window have been co-fragmented.
7. The data are of poor quality and the spectra do not contain enough information to identify it with any kind of statistical significance.

7.1. Mass Error and Deisotoping

If mass measurement error has been underestimated, it should be apparent from graphs plotting the differences between the calculated and measured mass values for the precursor and fragment matches. Researching the data with corrected tolerances will solve the problem. The software algorithms used during the acquisition of the MS data have very little time to correctly determine either the charge state or the ^{12}C isotope. Incorrect determination of precursor charge can be dealt with post-acquisition during peak detection. If it is not possible to determine the precursor charge reliably due to poor resolution, one option is to generate peak lists for all probable charge states. The ^{12}C isotope peak can be optionally re-determined and corrected during the peak-picking procedure or, if the search engine supports it, multiple precursor mass tolerance windows can be used that are spaced one carbon isotope apart.

7.2. Post-translational Modifications

Choosing too many variable modifications for a standard database search will lead to a combinatorial explosion in the size of the search space resulting in long searches and potential lower sensitivity for correct peptide identifications. To work around this problem, many search engines have a second-pass search function built in where the proteins identified in the first-round search are exhaustively searched with all the variable modifications known to the program (54). An alternative method to the second-pass search is spectral alignment with a delta mass shift corresponding to the expected modification (55–57). For some modifications, particularly phosphorylation, the peptides as well as unmodified peptides under standard CID conditions do not fragment. This, combined with having multiple fragmentation channels, means that the peptides are hard to identify and the site of modification may not be localized. If the peptide is identified, it can be rescored using more of the available peaks to localize the modification. An example of this method is the AScore (58) algorithm (**Table 4.6**). Finally, POSTMan can be used for identifying post-translational modified peptides for targeted MS/MS analysis at a later date (59).

7.3. Non-specific Cleavages

Peptides with non-specific cleavages or from the truncated terminals of the protein can be identified with no-enzyme searches. However, this is not very efficient as the searches take a long time and the sensitivity is often reduced. Semi-specific (where one terminus conforms to the proteolytic cleavage rules and one is free) or no-enzyme specificity can be incorporated into the second-pass search strategy instead. In some cases, a no-enzyme search is the only option for endogenous or major histocompatibility complex (MHC) peptides, for example.

7.4. De Novo Sequencing

A number of spectra will have excellent peptide fragmentation information but are not identified in a database search because the sequence is not in the database. The proteins could come from a species that has not yet had its genome sequenced. They could also come from an area of the genome that was not predicted to be a coding region or from proteins with alternative splice sites or multiple SNPs or other genomic modifications. If the fragmentation information is extensive enough, a complete peptide sequence solution can be generated by de novo sequencing. Otherwise, partial sequences or sequence tags that contain ambiguous amino acid sequences, pairs of amino acids in an unknown order, extra masses, and gaps can be produced. Due to the scale of an LC-MS/MS experiment, it is not always possible to de novo analyze every prospective spectra by hand. Additionally, manual de novo analysis is a skilled task that requires familiarity with the peptide fragmentation pathways under the MS/MS conditions

Table 4.6
PTM and de novo sequencing software

PTM software	
deltaMasses (111)	http://www.detectorvision.com/deltaMasses.html
ModifiComb (57)	http://www.medicwave.com/pas_modificomb.php
Spectral networks (55)	http://bix.ucsd.edu/
Modiro	http://www.modiro.com:8080/licenseserver/home.seam
POSTMan (59)	http://www.probe.uib.no/software.php
Ascore (58)	http://ascore.med.harvard.edu/
MS-Alignment (112)	http://proteomics.ucsd.edu/
PhosCalc (113)	http://www.ayeaye.tsl.ac.uk/PhosCalc/
SPIDER (88)	http://www.bioinformaticssolutions.com/products/peaks/spider.php
Popitam (114)	www.expasy.org/tools/popitam
De novo software	
Mascot Distiller	http://www.matrixscience.com/distiller.html
Lutefisk (115)	http://www.hairyfatguy.com/Lutefisk/
PEAKS (116)	http://www.bioinformaticssolutions.com/products/peaks/db_bsipaper.php
NovoHMM (117)	http://people.inf.ethz.ch/befische/proteomics/
Audens (118)	http://www.ti.inf.ethz.ch/pw/software/audens/
RAId (119)	yyu@ncbi.nlm.nih.gov
SHERENGA (120)	http://code.google.com/p/sherenga/
De novo search engines	
PEAKS (116)	http://www.bioinformaticssolutions.com/products/peaks/db_bsipaper.php
PepNovo (121)	http://proteomics.ucsd.edu/Software/PepNovo.html
Sequence-based search programs	
MS-BLAST (122)	dove.embl-heidelberg.de/Blast2/msblast.html
FASTS (123)	http://fasta.bioch.virginia.edu/fasta_www2/fasta_www.cgi?rm=select&pgm=fs
CIDentify (115)	http://faculty.virginia.edu/wrpearson/fasta/OLD/CIDentify/
Multiple peptides from one spectra	
ProbIDtree (124)	http://sourceforge.net/project/showfiles.php?group_id=69281&package_id=173843
Hardklör (125)	http://proteome.gs.washington.edu/software/hardklor/

used to acquire the data and a fair amount of experience. It is also a difficult computational task and there have been many different attempts at solving the de novo sequencing problem. The majority of the programs use a spectrum graph and scoring system approach but alternative methods using constraint satisfaction, genetic algorithms, or divide-and-conquer approaches have also been used (**Table 4.6**). The de novo sequencing programs can process a set of spectra and automatically output the solutions and partial solutions with ambiguities. There are two main approaches to searching the de novo solutions: general sequence search programs such as BLAST (5) and FASTS (60) that have been modified to cope with MS-specific sequence ambiguities and peptide sequence tag search algorithms (9).

7.5. Multiple Precursors

MS/MS spectra that are derived from multiple peptides with a similar retention time and *m/z* will often fail to be identified by a search engine that is only expecting fragment ions from a single peptide. Such parallel fragmentation is used intentionally with the MS^E technique, where the mass spectrometer alternates between MS and CID of all the eluting peptides (53, 61, 62).

Of the unidentified MS/MS spectra remaining, they typically contain data too poor to identify a peptide in an uninterpreted MS/MS database search. A number of these spectra can be identified by using different identification techniques such as spectral matching or machine-learning algorithms (**Table 4.6**). There have been a number of studies to try and assign all the unidentified spectra remaining after uninterpreted MS/MS database searching. A combination of manual analysis of the spectra was combined with the techniques described above (63, 64). However, the quality of the vast majority of spectra in the remaining group is too poor to be identified by any existing methods.

8. Post-identification Informatics

Information about proteins is stored in a variety of sources around the world. The information varies from the DNA or amino acid sequence to a protein's three-dimensional structure, known post-translational modifications, function, localization, and interaction partners of a protein. In the 1990s, researchers started to collate all the different sources into more accessible cross-referenced databases. For example, the SwissProt-curated protein sequence database has a high level of annotation, such as the description of the function of a protein, its domain structure, post-translational modifications, and variants; it has a minimal level of redundancy and a high level of integration with other databases. There is also a very easy-to-use Web site that displays this information and direct links are made to it from the protein database search reports. Other databases were initiated with the goals to organize the information about a protein's function, localization, and interactions (BIND) (65). The main databases and organizations are the Gene Ontology Annotation (GOA) project (66) at the European Bioinformatics Institute (EBI), the Kyoto Encyclopedia of Genes and Genomes (KEGG) (67), and individual model organism groups (for example, the *Schizosaccharomyces pombe* genome project at the Sanger Center (68) and the FlyBase (69) team are currently assigning annotations for individual genomes). There are a lot of applications for linking the results of microarray analysis to additional information in part because a microarray experiment uses a defined set of proteins. However, the number of applications that can handle proteomics data has been growing rapidly (**Table 4.7**). The applications either interact directly with the database search engine results or take a list of accession numbers as input. The applications offer a variety of information varying from integrated views of complete genomes and proteomes to links to the GO and KEGG annotations and physiochemical properties.

Table 4.7
Post-ID informatics software

ProteinCenter (126)	http://www.proxeon.com/productrange/data_interpretation/introduction/index.html
InforSense KDE	http://www.inforsense.com/
Integr8 (127)	http://www.ebi.ac.uk/integr8/EBI-Integr8-HomePage.do
Multi-Protein Survey System (128)	http://www.scbit.org/mpss/
PIGOK (126)	http://pc4-133.ludwig.ucl.ac.uk/pigok.html
GOEx (129)	http://pcarvalho.com/patternlab/goex.shtml
QuickGo (130)	http://www.ebi.ac.uk/QuickGO/

9. Data Mining with Laboratory Information Management Systems

Laboratory Information Management Systems (LIMS) were developed in the 1970s specifically for tracking sample and data in analytical laboratories. LIMS were developed in-house by many organizations to manage the various laboratory data from sample logging to reporting of the results. The first commercial LIMS was released in the early 1980s and provided automated reporting features and ran on mini-computers. These LIMS were often developed by analytical instrument manufacturers to organize the data their instruments produced. Commercial LIMS were built on proprietary architectures and targeted at specific industries. They required considerable customization at each laboratory making them very expensive. Today LIMS use open system architectures and platforms using client/server communication via the World Wide Web (**Table 4.8**). The systems are very flexible and offer a lot of advanced functionality. The three major functions of a LIMS are sample tracking, data storage, and reporting. The key actions that form the core of a LIMS can be broken down into the following:

1. Data capture or entry into the system.
2. Data analysis and organization.
3. Data reporting.
4. Sample tracking and workload assignment.
5. Audit trails and archival functions.

Table 4.8
Laboratory Information Management System

LIMS providers	
Biotracker	http://www.ocimumbio.com/lims2/
Computational Proteomics Analysis System (CPAS)	https://www.labkey.org/project/home/begin.view
LabWare LIMS	http://www.labware.com/lwweb.nsf
Mascot Integra	http://www.matrixscience.com/integra.html
Nautilus LIMS	http://www.thermoscientific.com/wps/portal/ts/products/detail?productId=11962427&groupType=PRODUCT&searchType=0
Proteomics Analyzer Software System (PASS)	http://www.i-a-inc.com/Products/pass.htm
Proteus	http://www.genologics.com/proteomics
Sapphire	http://www.labvantage.com/index.html
WinLIMS.NET	http://www.lims-software.com/index.php?title=Main_Page
ms_lims	http://genesis.ugent.be/ms_lims/
PRIME	http://www.proteomeconsortium.org/prime.html
Proteios software environment	http://www.proteios.org/
SBEAMS – Proteomics	http://tools.proteomecenter.org/wiki/index.php?title=Software:SBEAMS

Additional actions might include invoicing and accounting functions for lab management, instrument interfacing, sample scheduling, and quality control.

A recent review compared the major proteomics commercial and FOSS LIMS providers (70) (**Table 4.8**). All the systems were based on relational databases and could use one- or two-way communication with a variety of laboratory equipments and software applications. Once the experiment and LC-MS/MS search results have been captured by a LIMS and represented in tables of data, it is possible to build sophisticated queries in Standard Query Language (SQL). These queries can be used to search subsets of the database, all the results from one experiment for example, or the complete database. The results from the SQL queries can then be reported in standardized or custom reports or visualized in an interactive viewer. A LIMS can be used to keep quality control statistics on the different instruments in the lab, perform in-depth analysis of individual runs, search for all the occurrences of a peptide or a protein, perform large comparisons across multiple samples and much more.

10. Data Exchange and Long-Term Storage

A single modern mass spectrometer might generate roughly two to three TB of LC-MS/MS data per year. Worldwide the proteomics community is generating in the order of petabytes of data per year. Laboratories need to have infrastructure in place to be able to store the data, and potentially the derived data too, for the period of a grant, preparation of a publication, or longer. The LC-MS/MS data are acquired on half a dozen different makes of instrument and processed with a similar number of different peak-picking applications and search engines. Standardized data formats have been developed for each of the three types of data: raw, peak list, and search results. The HUPO/PSI standardized data formats use a controlled vocabulary or ontology, specified by the Proteomics Standards Initiative (PSI)(71), to consistently describe the data. The use of standardized data formats allows for simplified application development that does not require access to proprietary libraries, enhances interoperability, and increases accessibility by collaborators and the general scientific community. Peak lists are derived from the raw LC-MS/MS data and by their very nature do not contain all the information of a raw data file. For this reason when there is a need to make the data available, the raw data are still preferred (72). There are now a number of publicly accessible data repositories available that enable collaboration and publication of raw data, peak lists, and search results.

10.1. Long-Term Storage

A laboratory engaging in LC-MS/MS analysis of complex samples can generate a lot of data over the lifetime of an instrument. Many organizations have data retention policies that require the storage of the MS data from three to five years or occasionally indefinitely. Although this is not strictly a bioinformatics problem, it is something that a laboratory has to be aware of and needs to budget for. Fortunately, every year the cost of storage decreases and the capacity increases. Computers equipped with large RAID or dedicated network-attached storage systems are the preferred choice for long-term data storage and can easily store between 1 and 10 TB of data. However, it should be stressed that these systems are not infallible as multiple disk failures can coincide or the RAID controller hardware can fail potentially corrupting the data. Therefore backing up to offline tape storage or removable hard disks as well are highly recommended.

10.2. Raw Data Standards

Because the raw data formats used by the manufacturers are not interchangeable, a number of open standards have been devised by the proteomics community. mzData was the first format that was initiated by the HUPO/PSI and designed by committee over a 2-year period. While the mzData format was in development, the mzXML format was quickly developed by the Institute of Systems Biology for use in the TPP suite of applications (**Table 4.9**). Finally, a revised format that merged the mzData and mzXML formats, called mzML, was released in June 2008. The mzML format was developed by a working group consisting of the HUPO/PSI committee, SPC/ISB, instrument vendors, and other proteomics software groups. There are a number of FOSS and commercial applications available that can convert raw data from all the popular instruments into mzData, mzXML, and mzML formats. However, many of these tools still rely on software libraries that are a part of the MS vendor software. There is a free viewer available from Insilicos that can read and display data in all three open formats. If the mzData, mzXML, and mzML formatted data files contain profile data, they still need to undergo peak picking before database searching.

10.3. Experiment Result Standards

In a similar fashion to the raw data standards, there are multiple standards for database search results. The HUPO/PSI standard mzidentML (née AnalysisXML) is currently in review and will start to be supported by applications from the summer of 2009 onward. mzidentML conforms to both the minimum information about a proteomics experiment (MIAPE) (73) and the MCP guidelines (35). While the HUPO/PSI standard was being designed, the pepXML and protXML standards were developed as part of TPP (**Table 4.9**). PepXML can be used only for MS/MS data and includes only the “raw” peptide match data;

Table 4.9
Data exchange applications

Raw data converts and standards	
mzXML	http://tools.proteomecenter.org/wiki/index.php?title=Formats:mzXML
Search result to pepXML and protXML converters	
pepXML	http://tools.proteomecenter.org/wiki/index.php?title=Formats:pepXML
Scaffold (30)	http://www.proteomesoftware.com/
Public data repositories	
PeptideAtlas (75)	http://www.peptideatlas.org/
Open Proteomics Database (131)	http://apropos.icmb.utexas.edu/OPD/
Human Proteinpedia (132)	http://www.humanproteinpedia.org/index_html
Proteome commons	https://proteomecommons.org/
PRIDE (76)	http://www.ebi.ac.uk/pride/
PRIDE Converter	http://code.google.com/p/pride-converter/

so an additional tool (such as TPP's ProteinProphet) is needed to infer the protein identities in the sample. ProtXML, again only for MS/MS data, only includes the protein identifications created from MS/MS-derived peptide sequence data. Most of the major database search engines or reporting applications (e.g., Scaffold) can export to either pepXML or protXML directly or there is a conversion program available that can perform the transformation.

10.4. Repositories

With the development of data exchange formats and their adoption by the proteomics community, it became possible to build public repositories to hold LC-MS/MS data (**Table 4.9**). The data repositories act as central locations to store and distribute LC-MS/MS data and search results for inter-lab collaboration and as supporting evidence for publication. The data repositories are split between those that just accept raw MS data, a mixture of data and identification information, and just identification information. An example of a repository that is mainly used for raw data is the ProteomeCommons Tranche data repository (74).

Tranche uses a distributed file system across multiple servers replicating the data to minimize the chance of file loss. Data are distributed from the multiple servers at once, resulting in a faster and more reliable download connection suitable for large data sets. The PeptideAtlas repository is a public compendium of peptides identified in a large set of tandem mass spectrometry proteomics experiments that stores both raw data and identification results (75). The proteomics identifications database (PRIDE) is a centralized, standards compliant, public data repository for proteomics data that provides a searchable repository of protein and peptide identifications (76). Data sets that are uploaded to the PRIDE repository need to report additional experimental information about the sample, including species, tissue, sub-cellular location, and disease state, as well as literature references.

11. General Resources

There are far more FOSS or commercial applications available than can be covered here in this short review. The Web sites at the end of this section either collate applications from multiple sources and act as repositories of applications or link to them or are from organizations that have released substantial software applications covering many of the topics listed above (**Table 4.10**). As mentioned at the beginning of this review, the field is continually and rapidly changing. Applications covered here may be withdrawn or support may be discontinued by their authors; they may also be superseded by better applications that are released and of course there will be many more applications released in the future.

Table 4.10
General resource Web sites for proteomics

Proteome commons	https://proteomecommons.org/
TPP Web site	http://tools.proteomecenter.org/wiki/index.php?title=Software:TPP
PNNL	http://ncrr.pnl.gov/software/
NBIC	https://wiki.nbic.nl/index.php/Proteomics_data_management_and_analysis

References

1. Thirkettle, C., and Morris, H. R. (1980) Computer-assisted sequencing of peptide mass spectra [proceedings]. *Biochem. Soc. Trans.* **8**, 176–177.
2. Johnson, R. S., and Biemann, K. (1989) Computer program (SEQPEP) to aid in the interpretation of high-energy collision tandem mass spectra of peptides. *Biomed. Environ. Mass Spectrom.* **18**, 945–957.
3. Pillay, T. S. (1988) TURBOLYTIK: a peptide cleavage program for personal computers. *Int. I. Bio-med. Comput.* **22**, 259–264.
4. Lee, T. D., and Vemuri, S. (1990) MacProMass: a computer program to correlate mass spectral data to peptide and protein structures. *Biomed. Environ. Mass Spectrom.* **19**, 639–645.
5. Altschul, S. F., Gish, W., Miller, W., Myers, E. W., and Lipman, D. J. (1990) Basic local alignment search tool. *J. Mol. Biol.* **215**, 403–410.
6. (2009) GenBank release notes. National Center for Biotechnology Information. ftp://ftp.ncbi.nih.gov/genbank/release.notes/
7. Stults, J. T., Lai, J., McCune, S., and Wetzel, R. (1993) Simplification of high-energy collision spectra of peptides by amino-terminal derivatization. *Anal. Chem.* **65**, 1703–1708.
8. Yates, J. R., 3rd, Eng, J. K., McCormack, A. L., and Schieltz, D. (1995) Method to correlate tandem mass spectra of modified peptides to amino acid sequences in the protein database. *Anal. Chem.* **67**, 1426–1436.
9. Mann, M., and Wilm, M. (1994) Error-tolerant identification of peptides in sequence databases by peptide sequence tags. *Anal. Chem.* **66**, 4390–4399.
10. Matthiesen, R., (ed.) (2007) Mass spectrometry data analysis in proteomics. Vol. 367. Humana Press, Totowa, NJ, USA.
11. Eidhammer, I., Flikka, K., Martens, L., and Mikalsen, S. (2007) Computational methods for mass spectrometry proteomics. Wiley, Chichester, UK.
12. Pedersen, S. K., Harry, J. L., Sebastian, L., Baker, J., Traini, M. D., McCarthy, J. T., Manoharan, A., Wilkins, M. R., Gooley, A. A., Righetti, P. G., Packer, N. H., Williams, K. L., and Herbert, B. R. (2003) Unseen proteome: mining below the tip of the iceberg to find low abundance and membrane proteins. *J. Proteome Res.* **2**, 303–311.
13. Sidhu, K. S., Sangvanich, P., Brancia, F. L., Sullivan, A. G., Gaskell, S. J., Wolkenhaue, O., Oliver, S. G., and Hubbard, S. J. (2001) *Proteomics* **1**, 1368–1377.
14. Matthiesen, R., Lundsgaard, M., Welinder, K. G., and Bauw, G. (2003) Interpreting peptide mass spectra by VEMS. *Bioinformatics (Oxford, England)* **19**, 792–793.
15. Cargile, B. J., and Stephenson, J. L., Jr. (2004) An alternative to tandem mass spectrometry: isoelectric point and accurate mass for the identification of peptides. *Anal. Chem.* **76**, 267–275.
16. Cagney, G., Amiri, S., Premawaradena, T., Lindo, M., and Emili, A. (2003) In silico proteome analysis to facilitate proteomics experiments using mass spectrometry. *Proteome Sci.* **1**, 5.
17. Matthiesen, R., Trelle, M. B., Hojrup, P., Bunkenborg, J., and Jensen, O. N. (2005) VEMS 3.0: algorithms and computational tools for tandem mass spectrometry based identification of post-translational modifications in proteins. *J. Proteome Res.* **4**, 2338–2347.
18. Eng, J. K., McCormack, A. L., and Yates, J. R. (1994) An approach to correlate tandem mass spectral data of peptides with amino acid sequences in a protein database. *Am. Soc. Mass Spectrom.* **5**, 976–989.
19. Perkins, D. N., Pappin, D. J., Creasy, D. M., and Cottrell, J. S. (1999) Probability-based protein identification by searching sequence databases using mass spectrometry data. *Electrophoresis* **20**, 3551–3567.
20. Conrads, T. P., Anderson, G. A., Veenstra, T. D., Pasa-Tolic, L., and Smith, R. D. (2000) Utility of accurate mass tags for proteome-wide protein identification. *Anal. Chem.* **72**, 3349–3354.
21. Strittmatter, E. F., Ferguson, P. L., Tang, K., and Smith, R. D. (2003) Proteome analyses using accurate mass and elution time peptide tags with capillary LC time-of-flight mass spectrometry. *J. Am. Soc. Mass Spectrom.* **14**, 980–991.
22. Clauser, K. R., Baker, P., and Burlingame, A. L. (1999) Role of accurate mass measurement (+/−10 ppm) in protein identification strategies employing MS or MS/MS and database searching. *Anal. Chem.* **71**, 2871–2882.
23. Mueller, L. N., Brusniak, M. Y., Mani, D. R., and Aebersold, R. (2008) An assessment of software solutions for the analysis of mass spectrometry based quantitative proteomics data. *J. Proteome Res.* **7**, 51–61.
24. Monroe, M. E., Tolic, N., Jaitly, N., Shaw, J. L., Adkins, J. N., and Smith, R. D. (2007)

VIPER: an advanced software package to support high-throughput LC-MS peptide identification. *Bioinformatics (Oxford, England)* **23**, 2021–2023.
25. Yen, C. Y., Meyer-Arendt, K., Eichelberger, B., Sun, S., Houel, S., Old, W. M., Knight, R., Ahn, N. G., Hunter, L. E., and Resing, K. A. (2009) A simulated MS/MS library for spectrum-to-spectrum searching in large scale identification of proteins. *Mol. Cell. Proteomics* **8**, 857–869.
26. Anderson, D. C., Li, W., Payan, D. G., and Noble, W. S. (2003) A new algorithm for the evaluation of shotgun peptide sequencing in proteomics: support vector machine classification of peptide MS/MS spectra and SEQUEST scores. *J. Proteome Res.* **2**, 137–146.
27. Kall, L., Canterbury, J. D., Weston, J., Noble, W. S., and MacCoss, M. J. (2007) Semi-supervised learning for peptide identification from shotgun proteomics datasets. *Nat. Methods* **4**, 923–925.
28. Brosch, M., Yu, L., Hubbard, T., and Choudhary, J. (2009) Accurate and sensitive peptide identification with Mascot Percolator. *J. Proteome Res.*
29. Keller, A., Nesvizhskii, A. I., Kolker, E., and Aebersold, R. (2002) Empirical statistical model to estimate the accuracy of peptide identifications made by MS/MS and database search. *Anal. Chem.* **74**, 5383–5392.
30. Searle, B. C., Turner, M., and Nesvizhskii, A. I. (2008) Improving sensitivity by probabilistically combining results from multiple MS/MS search methodologies. *J. Proteome Res.* 7, 245–253.
31. Washburn, M. P., Wolters, D., and Yates, J. R., 3rd (2001) Large-scale analysis of the yeast proteome by multidimensional protein identification technology. *Nat. Biotechnol.* **19**, 242–247.
32. Yang, C. G., Granite, S. J., Van Eyk, J. E., and Winslow, R. L. (2006) MASCOT HTML and XML parser: an implementation of a novel object model for protein identification data. *Proteomics* **6**, 5688–5693.
33. Helsens, K., Martens, L., Vandekerckhove, J., and Gevaert, K. (2007) MascotDatfile: an open-source library to fully parse and analyse MASCOT MS/MS search results. *Proteomics* 7, 364–366.
34. Grosse-Coosmann, F., Boehm, A. M., and Sickmann, A. (2005) Efficient analysis and extraction of MS/MS result data from Mascot result files. *BMC Bioinformatics* **6**, 290.
35. Bradshaw, R. A., Burlingame, A. L., Carr, S., and Aebersold, R. (2006) Reporting protein identification data: the next generation of guidelines. *Mol. Cell. Proteomics* **5**, 787–788.
36. Wilkins, M. R., Appel, R. D., Van Eyk, J. E., Chung, M. C., Gorg, A., Hecker, M., Huber, L. A., Langen, H., Link, A. J., Paik, Y. K., Patterson, S. D., Pennington, S. R., Rabilloud, T., Simpson, R. J., Weiss, W., and Dunn, M. J. (2006) Guidelines for the next 10 years of proteomics, *Proteomics 6*, 4–8.
37. Ong, S. E., and Mann, M. (2005) Mass spectrometry-based proteomics turns quantitative. *Nat. Chem. Biol.* **1**, 252–262.
38. Ross, P. L., Huang, Y. N., Marchese, J. N., Williamson, B., Parker, K., Hattan, S., Khainovski, N., Pillai, S., Dey, S., Daniels, S., Purkayastha, S., Juhasz, P., Martin, S., Bartlet-Jones, M., He, F., Jacobson, A., and Pappin, D. J. (2004) Multiplexed protein quantitation in *Saccharomyces cerevisiae* using amine-reactive isobaric tagging reagents. *Mol. Cell. Proteomics* **3**, 1154–1169.
39. Thompson, A., Schafer, J., Kuhn, K., Kienle, S., Schwarz, J., Schmidt, G., Neumann, T., Johnstone, R., Mohammed, A. K., and Hamon, C. (2003) Tandem mass tags: a novel quantification strategy for comparative analysis of complex protein mixtures by MS/MS. *Anal. Chem.* **75**, 1895–1904.
40. Han, D. K., Eng, J., Zhou, H., and Aebersold, R. (2001) Quantitative profiling of differentiation-induced microsomal proteins using isotope-coded affinity tags and mass spectrometry. *Nat. Biotechnol.* **19**, 946–951.
41. Hicks, W. A., Halligan, B. D., Slyper, R. Y., Twigger, S. N., Greene, A. S., and Olivier, M. (2005) Simultaneous quantification and identification using 18O labeling with an ion trap mass spectrometer and the analysis software application "ZoomQuant". *J. Am. Soc. Mass Spectrom.* **16**, 916–925.
42. Miyagi, M., and Rao, K. C. (2007) Proteolytic 18O-labeling strategies for quantitative proteomics. *Mass Spectrom. Rev.* **26**, 121–136.
43. Gerber, S. A., Rush, J., Stemman, O., Kirschner, M. W., and Gygi, S. P. (2003) Absolute quantification of proteins and phosphoproteins from cell lysates by tandem MS. *Proc. Natl. Acad. Sci. USA 100*, 6940–6945.
44. Ong, S. E., Blagoev, B., Kratchmarova, I., Kristensen, D. B., Steen, H., Pandey, A.,

and Mann, M. (2002) Stable isotope labeling by amino acids in cell culture, SILAC, as a simple and accurate approach to expression proteomics. *Mol. Cell. Proteomics* **1**, 376–386.

45. Zhang, G., and Neubert, T. A. (2006) Automated comparative proteomics based on multiplex tandem mass spectrometry and stable isotope labeling. *Mol. Cell. Proteomics* **5**, 401–411.

46. Wang, W., Zhou, H., Lin, H., Roy, S., Shaler, T. A., Hill, L. R., Norton, S., Kumar, P., Anderle, M., and Becker, C. H. (2003) Quantification of proteins and metabolites by mass spectrometry without isotopic labeling or spiked standards. *Anal. Chem.* **75**, 4818–4826.

47. Chelius, D., Zhang, T., Wang, G., and Shen, R. F. (2003) Global protein identification and quantification technology using two-dimensional liquid chromatography nanospray mass spectrometry. *Anal. Chem.* **75**, 6658–6665.

48. Radulovic, D., Jelveh, S., Ryu, S., Hamilton, T. G., Foss, E., Mao, Y., and Emili, A. (2004) Informatics platform for global proteomic profiling and biomarker discovery using liquid chromatography–tandem mass spectrometry. *Mol. Cell. Proteomics* **3**, 984–997.

49. Silva, J. C., Denny, R., Dorschel, C. A., Gorenstein, M., Kass, I. J., Li, G. Z., McKenna, T., Nold, M. J., Richardson, K., Young, P., and Geromanos, S. (2005) Quantitative proteomic analysis by accurate mass retention time pairs. *Anal. Chem.* 77, 2187–2200.

50. Hughes, M. A., Silva, J. C., Geromanos, S. J., and Townsend, C. A. (2006) Quantitative proteomic analysis of drug-induced changes in mycobacteria. *J. Proteome Res.* **5**, 54–63.

51. Cutillas, P. R., Geering, B., Waterfield, M. D., and Vanhaesebroeck, B. (2005) Quantification of gel-separated proteins and their phosphorylation sites by LC-MS using unlabeled internal standards: analysis of phosphoprotein dynamics in a B cell lymphoma cell line. *Mol. Cell. Proteomics* **4**, 1038–1051.

52. Ishihama, Y., Oda, Y., Tabata, T., Sato, T., Nagasu, T., Rappsilber, J., and Mann, M. (2005) Exponentially modified protein abundance index (emPAI) for estimation of absolute protein amount in proteomics by the number of sequenced peptides per protein. *Mol. Cell. Proteomics* **4**, 1265–1272.

53. Silva, J. C., Gorenstein, M. V., Li, G. Z., Vissers, J. P., and Geromanos, S. J. (2006) Absolute quantification of proteins by LCMSE: a virtue of parallel MS acquisition. *Mol. Cell. Proteomics* **5**, 144–156.

54. Creasy, D. M., and Cottrell, J. S. (2002) Error tolerant searching of uninterpreted tandem mass spectrometry data. *Proteomics* **2**, 1426–1434.

55. Bandeira, N., Tsur, D., Frank, A., and Pevzner, P. A. (2007) Protein identification by spectral networks analysis. *Proc. Natl. Acad. Sci. USA* **104**, 6140–6145.

56. Potthast, F., Gerrits, B., Hakkinen, J., Rutishauser, D., Ahrens, C. H., Roschitzki, B., Baerenfaller, K., Munton, R. P., Walther, P., Gehrig, P., Seif, P., Seeberger, P. H., and Schlapbach, R. (2007) The Mass Distance Fingerprint: a statistical framework for de novo detection of predominant modifications using high-accuracy mass spectrometry. *J. Chromatogr.* **854**, 173–182.

57. Savitski, M. M., Nielsen, M. L., and Zubarev, R. A. (2006) ModifiComb, a new proteomic tool for mapping substoichiometric post-translational modifications, finding novel types of modifications, and fingerprinting complex protein mixtures. *Mol. Cell. Proteomics* **5**, 935–948.

58. Beausoleil, S. A., Villen, J., Gerber, S. A., Rush, J., and Gygi, S. P. (2006) A probability-based approach for high-throughput protein phosphorylation analysis and site localization. *Nat. Biotechnol.* **24**, 1285–1292.

59. Arntzen, M. O., Osland, C. L., Raa, C. R., Kopperud, R., Doskeland, S. O., Lewis, A. E., and D'Santos, C. S. (2009) POSTMan (POST-translational modification analysis), a software application for PTM discovery. *Proteomics* **9**, 1400–1406.

60. Pearson, W. R. (1990) Rapid and sensitive sequence comparison with FASTP and FASTA. *Methods Enzymol.* **183**, 63–98.

61. Hoaglund-Hyzer, C. S., Li, J., and Clemmer, D. E. (2000) Mobility labeling for parallel CID of ion mixtures. *Anal. Chem.* **72**, 2737–2740.

62. Niggeweg, R., Kocher, T., Gentzel, M., Buscaino, A., Taipale, M., Akhtar, A., and Wilm, M. (2006) A general precursor ion-like scanning mode on quadrupole-TOF instruments compatible with chromatographic separation. *Proteomics* **6**, 41–53.

63. Chalkley, R. J., Baker, P. R., Hansen, K. C., Medzihradszky, K. F., Allen, N. P., Rexach, M., and Burlingame, A. L. (2005)

Comprehensive analysis of a multidimensional liquid chromatography mass spectrometry dataset acquired on a quadrupole selecting, quadrupole collision cell, time-of-flight mass spectrometer: I. How much of the data is theoretically interpretable by search engines? *Mol. Cell. Proteomics* **4**, 1189–1193.

64. Vestal, M. L., Campbell, J. M., Hayden, K. M., Chen, X., Strahler, J. R., and Andrews, P. C. (2005) Dynamic range in MALDI TOF–TOF analysis of protein digests. In "*53th ASMS conference on mass spectrometry*, San Antonio, TX, USA".
65. Bader, G. D., Donaldson, I., Wolting, C., Ouellette, B. F., Pawson, T., and Hogue, C. W. (2001) BINDThe Biomolecular Interaction Network Database. *Nucleic Acids Res.* **29**, 242–245.
66. Ashburner, M., Ball, C. A., Blake, J. A., Botstein, D., Butler, H., Cherry, J. M., Davis, A. P., Dolinski, K., Dwight, S. S., Eppig, J. T., Harris, M. A., Hill, D. P., Issel-Tarver, L., Kasarskis, A., Lewis, S., Matese, J. C., Richardson, J. E., Ringwald, M., Rubin, G. M., and Sherlock, G. (2000) Gene ontology: tool for the unification of biology. The Gene Ontology Consortium. *Nat. Genet.* **25**, 25–29.
67. Ogata, H., Goto, S., Sato, K., Fujibuchi, W., Bono, H., and Kanehisa, M. (1999) KEGG: Kyoto Encyclopedia of Genes and Genomes. *Nucleic Acids Res.* **27**, 29–34.
68. Wood, V., Gwilliam, R., Rajandream, M. A., Lyne, M., Lyne, R., Stewart, A., Sgouros, J., Peat, N., Hayles, J., Baker, S., Basham, D., Bowman, S., Brooks, K., Brown, D., Brown, S., Chillingworth, T., Churcher, C., Collins, M., Connor, R., Cronin, A., Davis, P., Feltwell, T., Fraser, A., Gentles, S., Goble, A., Hamlin, N., Harris, D., Hidalgo, J., Hodgson, G., Holroyd, S., Hornsby, T., Howarth, S., Huckle, E. J., Hunt, S., Jagels, K., James, K., Jones, L., Jones, M., Leather, S., McDonald, S., McLean, J., Mooney, P., Moule, S., Mungall, K., Murphy, L., Niblett, D., Odell, C., Oliver, K., O'Neil, S., Pearson, D., Quail, M. A., Rabbinowitsch, E., Rutherford, K., Rutter, S., Saunders, D., Seeger, K., Sharp, S., Skelton, J., Simmonds, M., Squares, R., Squares, S., Stevens, K., Taylor, K., Taylor, R. G., Tivey, A., Walsh, S., Warren, T., Whitehead, S., Woodward, J., Volckaert, G., Aert, R., Robben, J., Grymonprez, B., Weltjens, I., Vanstreels, E., Rieger, M., Schafer, M., Muller-Auer, S., Gabel, C., Fuchs, M., Dusterhoft, A., Fritzc, C., Holzer, E., Moestl, D., Hilbert, H., Borzym, K., Langer, I., Beck, A., Lehrach, H., Reinhardt, R., Pohl, T. M., Eger, P., Zimmermann, W., Wedler, H., Wambutt, R., Purnelle, B., Goffeau, A., Cadieu, E., Dreano, S., Gloux, S., Lelaure, V., Mottier, S., Galibert, F., Aves, S. J., Xiang, Z., Hunt, C., Moore, K., Hurst, S. M., Lucas, M., Rochet, M., Gaillardin, C., Tallada, V. A., Garzon, A., Thode, G., Daga, R. R., Cruzado, L., Jimenez, J., Sanchez, M., del Rey, F., Benito, J., Dominguez, A., Revuelta, J. L., Moreno, S., Armstrong, J., Forsburg, S. L., Cerutti, L., Lowe, T., McCombie, W. R., Paulsen, I., Potashkin, J., Shpakovski, G. V., Ussery, D., Barrell, B. G., and Nurse, P. (2002) The genome sequence of *Schizosaccharomyces pombe. Nature* **415**, 871–880.
69. Drysdale, R. A., and Crosby, M. A. (2005) FlyBase: genes and gene models. *Nucleic Acids Res.* **33**, D390–D395.
70. Piggee, C. (2008) LIMS and the art of MS proteomics. *Anal. Chem.* **80**, 4801–4806.
71. Orchard, S., Hermjakob, H., and Apweiler, R. (2003) The proteomics standards initiative. *Proteomics* **3**, 1374–1376.
72. Martens, L., Nesvizhskii, A. I., Hermjakob, H., Adamski, M., Omenn, G. S., Vandekerckhove, J., and Gevaert, K. (2005) Do we want our data raw? Including binary mass spectrometry data in public proteomics data repositories. *Proteomics* **5**, 3501–3505.
73. Taylor, C. F., Paton, N. W., Lilley, K. S., Binz, P. A., Julian, R. K., Jr., Jones, A. R., Zhu, W., Apweiler, R., Aebersold, R., Deutsch, E. W., Dunn, M. J., Heck, A. J., Leitner, A., Macht, M., Mann, M., Martens, L., Neubert, T. A., Patterson, S. D., Ping, P., Seymour, S. L., Souda, P., Tsugita, A., Vandekerckhove, J., Vondriska, T. M., Whitelegge, J. P., Wilkins, M. R., Xenarios, I., Yates, J. R., 3rd, and Hermjakob, H. (2007) The minimum information about a proteomics experiment (MIAPE). *Nat. Biotechnol.* **25**, 887–893.
74. Falkner, J. A., and Andrews, P. C. (2006) Open access, peer reviewed, peer-to-peer based proteomics data dissemination and archival system. In *ABRF*, Long Beach, CA, USA.
75. Desiere, F., Deutsch, E. W., King, N. L., Nesvizhskii, A. I., Mallick, P., Eng, J., Chen, S., Eddes, J., Loevenich, S. N., and Aebersold, R. (2006) The PeptideAtlas project. *Nucleic Acids Res.* **34**, D655–D658.
76. Martens, L., Hermjakob, H., Jones, P., Adamski, M., Taylor, C., States, D.,

Gevaert, K., Vandekerckhove, J., and Apweiler, R. (2005) PRIDE: the proteomics identifications database. *Proteomics* **5**, 3537–3545.
77. Keller, A., Eng, J., Zhang, N., Li, X. J., and Aebersold, R. (2005) A uniform proteomics MS/MS analysis platform utilizing open XML file formats. *Mol. Syst. Biol.* **1**, 2005.0017.
78. Kessner, D., Chambers, M., Burke, R., Agus, D., and Mallick, P. (2008) ProteoWizard: open source software for rapid proteomics tools development. *Bioinformatics (Oxford, England)* **24**, 2534–2536.
79. Zhang, N., Aebersold, R., and Schwikowski, B. (2002) ProbID: a probabilistic algorithm to identify peptides through sequence database searching using tandem mass spectral data. *Proteomics* **2**, 1406–1412.
80. Geer, L. Y., Markey, S. P., Kowalak, J. A., Wagner, L., Xu, M., Maynard, D. M., Yang, X., Shi, W., and Bryant, S. H. (2004) Open mass spectrometry search algorithm. *J. Proteome Res.* **3**, 958–964.
81. Heller, M., Ye, M., Michel, P. E., Morier, P., Stalder, D., Junger, M. A., Aebersold, R., Reymond, F., and Rossier, J. S. (2005) Added value for tandem mass spectrometry shotgun proteomics data validation through isoelectric focusing of peptides. *J. Proteome Res.* **4**, 2273–2282.
82. Craig, R., and Beavis, R. C. (2004) TANDEM: matching proteins with tandem mass spectra. *Bioinformatics (Oxford, England)* **20**, 1466–1467.
83. Shilov, I. V., Seymour, S. L., Patel, A. A., Loboda, A., Tang, W. H., Keating, S. P., Hunter, C. L., Nuwaysir, L. M., and Schaeffer, D. A. (2007) The Paragon Algorithm, a next generation search engine that uses sequence temperature values and feature probabilities to identify peptides from tandem mass spectra. *Mol. Cell. Proteomics* **6**, 1638–1655.
84. Tabb, D. L., Fernando, C. G., and Chambers, M. C. (2007) MyriMatch: highly accurate tandem mass spectral peptide identification by multivariate hypergeometric analysis. *J. Proteome Res.* **6**, 654–661.
85. Tabb, D. L., Saraf, A., and Yates, J. R., 3rd (2003) GutenTag: high-throughput sequence tagging via an empirically derived fragmentation model. *Anal. Chem.* **75**, 6415–6421.
86. Tanner, S., Shu, H., Frank, A., Wang, L. C., Zandi, E., Mumby, M., Pevzner, P. A., and Bafna, V. (2005) InsPecT: identification of posttranslationally modified peptides from tandem mass spectra. *Anal. Chem.* **77**, 4626–4639.
87. Wilkins, M. R., Gasteiger, E., Wheeler, C. H., Lindskog, I., Sanchez, J. C., Bairoch, A., Appel, R. D., Dunn, M. J., and Hochstrasser, D. F. (1998) Multiple parameter cross-species protein identification using MultiIdenta world-wide web accessible tool. *Electrophoresis* **19**, 3199–3206.
88. Han, Y., Ma, B., and Zhang, K. (2004) SPIDER: software for protein identification from sequence tags with de novo sequencing error. *Proc./IEEE Comput. Syst. Bioinformatics Conf. CSB*, 206–215.
89. Qin, J., Fenyo, D., Zhao, Y., Hall, W. W., Chao, D. M., Wilson, C. J., Young, R. A., and Chait, B. T. (1997) A strategy for rapid, high-confidence protein identification. *Anal. Chem.* **69**, 3995–4001.
90. May, D., Fitzgibbon, M., Liu, Y., Holzman, T., Eng, J., Kemp, C. J., Whiteaker, J., Paulovich, A., and McIntosh, M. (2007) A platform for accurate mass and time analyses of mass spectrometry data. *J. Proteome Res.* **6**, 2685–2694.
91. Lam, H., Deutsch, E. W., Eddes, J. S., Eng, J. K., King, N., Stein, S. E., and Aebersold, R. (2007) Development and validation of a spectral library searching method for peptide identification from MS/MS. *Proteomics* **7**, 655–667.
92. Frewen, B., and MacCoss, M. J. (2007) Using BiblioSpec for creating and searching tandem MS peptide libraries. *Current protocols in bioinformatics/editorial board, Andreas D. Baxevanis ... [et al Chapter 13*, Unit 13.7].
93. Frank, A. M., Bandeira, N., Shen, Z., Tanner, S., Briggs, S. P., Smith, R. D., and Pevzner, P. A. (2008) Clustering millions of tandem mass spectra. *J. Proteome Res.* **7**, 113–122.
94. Palagi, P. M., Walther, D., Quadroni, M., Catherinet, S., Burgess, J., Zimmermann-Ivol, C. G., Sanchez, J. C., Binz, P. A., Hochstrasser, D. F., and Appel, R. D. (2005) MSight: an image analysis software for liquid chromatography–mass spectrometry. *Proteomics* **5**, 2381–2384.
95. Li, X. J., Pedrioli, P. G., Eng, J., Martin, D., Yi, E. C., Lee, H., and Aebersold, R. (2004) A tool to visualize and evaluate data obtained by liquid chromatography–electrospray ionization–mass spectrometry. *Anal. Chem.* **76**, 3856–3860.
96. Gehlenborg, N., Yan, W., Lee, I. Y., Yoo, H., Nieselt, K., Hwang, D., Aebersold, R., and Hood, L. (2009) Prequipsan extensible

software platform for integration, visualization and analysis of LC-MS/MS proteomics data. *Bioinformatics (Oxford, England)* **25**, 682–683.

97. Zhang, B., Chambers, M. C., and Tabb, D. L. (2007) Proteomic parsimony through bipartite graph analysis improves accuracy and transparency. *J. Proteome Res.* **6**, 3549–3557.
98. Tabb, D. L., McDonald, W. H., and Yates, J. R., 3rd (2002) DTASelect and Contrast: tools for assembling and comparing protein identifications from shotgun proteomics. *J. Proteome Res.* **1**, 21–26.
99. Askenazi, M., Parikh, J. R., and Marto, J. A. (2009) mzAPI: a new strategy for efficiently sharing mass spectrometry data. *Nat. Methods* **6**, 240–241.
100. Slotta, D. J., McFarland, M., Makusky, A., and Markey, S. (2007) P18-T MassSieve: a new tool for mass spectrometry-based proteomics. *J. Biomol. Tech.* **18**, 7.
101. Hartler, J., Thallinger, G. G., Stocker, G., Sturn, A., Burkard, T. R., Korner, E., Rader, R., Schmidt, A., Mechtler, K., and Trajanoski, Z. (2007) MASPECTRAS: a platform for management and analysis of proteomics LC-MS/MS data. *BMC Bioinformatics* **8**, 197.
102. Hakkinen, J., Vincic, G., Mansson, O., Warell, K., and Levander, F. (2009) The Proteios Software Environment: an extensible multiuser platform for management and analysis of proteomics data. *J. Proteome Res.*
103. Andersen, J. S., Wilkinson, C. J., Mayor, T., Mortensen, P., Nigg, E. A., and Mann, M. (2003) Proteomic characterization of the human centrosome by protein correlation profiling. *Nature* **426**, 570–574.
104. Cox, J., Matic, I., Hilger, M., Nagaraj, N., Selbach, M., Olsen, J. V., and Mann, M. (2009) A practical guide to the MaxQuant computational platform for SILAC-based quantitative proteomics. *Nat. Protoc* **4**, 698–705.
105. Cox, J., and Mann, M. (2008) MaxQuant enables high peptide identification rates, individualized p.p.b.-range mass accuracies and proteome-wide protein quantification. *Nat. Biotechnol.* **26**, 1367–1372.
106. Shadforth, I. P., Dunkley, T. P., Lilley, K. S., and Bessant, C. (2005) i-Tracker: for quantitative proteomics using iTRAQ. *BMC Genomics* **6**, 145.
107. Bouyssie, D., Gonzalez de Peredo, A., Mouton, E., Albigot, R., Roussel, L., Ortega, N., Cayrol, C., Burlet-Schiltz, O., Girard, J. P., and Monsarrat, B. (2007) Mascot file parsing and quantification (MFPaQ), a new software to parse, validate, and quantify proteomics data generated by ICAT and SILAC mass spectrometric analyses: application to the proteomics study of membrane proteins from primary human endothelial cells. *Mol. Cell. Proteomics* **6**, 1621–1637.
108. Polpitiya, A. D., Qian, W. J., Jaitly, N., Petyuk, V. A., Adkins, J. N., Camp, D. G., 2nd, Anderson, G. A., and Smith, R. D. (2008) DAnTE: a statistical tool for quantitative analysis of -omics data. *Bioinformatics (Oxford, England)* **24**, 1556–1558.
109. Mason, C. J., Therneau, T. M., Eckel-Passow, J. E., Johnson, K. L., Oberg, A. L., Olson, J. E., Nair, K. S., Muddiman, D. C., and Bergen, H. R., 3rd (2007) A method for automatically interpreting mass spectra of 18O-labeled isotopic clusters. *Mol. Cell. Proteomics* **6**, 305–318.
110. van Breukelen, B., van den Toorn, H. W., Drugan, M. M., and Heck, A. J. (2009) StatQuant: A post quantification analysis toolbox for improving quantitative mass spectrometry. *Bioinformatics (Oxford, England)*.
111. Potthast, F., Ocenasek, J., Rutishauser, D., Pelikan, M., and Schlapbach, R. (2005) Database independent detection of isotopically labeled MS/MS spectrum peptide pairs. *J. Chromatogr.* **817**, 225–230.
112. Pevzner, P. A., Mulyukov, Z., Dancik, V., and Tang, C. L. (2001) Efficiency of database search for identification of mutated and modified proteins via mass spectrometry. *Genome Res.* **11**, 290–299.
113. Maclean, D., Burrell, M. A., Studholme, D. J., and Jones, A. M. (2008) PhosCalc: a tool for evaluating the sites of peptide phosphorylation from mass spectrometer data. *BMC Res. Notes* **1**, 30.
114. Hernandez, P., Gras, R., Frey, J., and Appel, R. D. (2003) Popitam: towards new heuristic strategies to improve protein identification from tandem mass spectrometry data. *Proteomics* **3**, 870–878.
115. Taylor, J. A., and Johnson, R. S. (2001) Implementation and uses of automated de novo peptide sequencing by tandem mass spectrometry. *Anal. Chem.* **73**, 2594–2604.
116. Ma, B., Zhang, K., Hendrie, C., Liang, C., Li, M., Doherty-Kirby, A., and Lajoie, G. (2003) PEAKS: powerful software for peptide de novo sequencing by tandem mass spectrometry. *Rapid Commun. Mass Spectrom.* **17**, 2337–2342.
117. Fischer, B., Roth, V., Roos, F., Grossmann, J., Baginsky, S., Widmayer, P., Gruis-

sem, W., and Buhmann, J. M. (2005) NovoHMM: a hidden Markov model for de novo peptide sequencing. *Anal. Chem.* 77, 7265–7273.

118. Grossmann, J., Roos, F. F., Cieliebak, M., Liptak, Z., Mathis, L. K., Muller, M., Gruissem, W., and Baginsky, S. (2005) AUDENS: a tool for automated peptide de novo sequencing. *J. Proteome Res.* **4**, 1768–1774.

119. Alves, G., and Yu, Y. K. (2005) Robust accurate identification of peptides (RAId): deciphering MS2 data using a structured library search with de novo based statistics. *Bioinformatics (Oxford, England)* **21**, 3726–3732.

120. Dancik, V., Addona, T. A., Clauser, K. R., Vath, J. E., and Pevzner, P. A. (1999) De novo peptide sequencing via tandem mass spectrometry. *J. Comput. Biol.* **6**, 327–342.

121. Frank, A., and Pevzner, P. (2005) PepNovo: de novo peptide sequencing via probabilistic network modeling. *Anal. Chem.* 77, 964–973.

122. Shevchenko, A. (2001) Evaluation of the efficiency of in-gel digestion of proteins by peptide isotopic labeling and MALDI mass spectrometry. *Anal. Biochem.* **296**, 279–283.

123. Mackey, A. J., Haystead, T. A., and Pearson, W. R. (2002) Getting more from less: algorithms for rapid protein identification with multiple short peptide sequences. *Mol. Cell. Proteomics* **1**, 139–147.

124. Zhang, N., Li, X. J., Ye, M., Pan, S., Schwikowski, B., and Aebersold, R. (2005) ProbIDtree: an automated software program capable of identifying multiple peptides from a single collision-induced dissociation spectrum collected by a tandem mass spectrometer. *Proteomics* **5**, 4096–4106.

125. Hoopmann, M. R., Finney, G. L., and MacCoss, M. J. (2007) High-speed data reduction, feature detection, and MS/MS spectrum quality assessment of shotgun proteomics data sets using high-resolution mass spectrometry. *Anal. Chem.* **79**, 5620–5632.

126. Olsen, J. V., Blagoev, B., Gnad, F., Macek, B., Kumar, C., Mortensen, P., and Mann, M. (2006) Global, in vivo, and site-specific phosphorylation dynamics in signaling networks. *Cell* **127**, 635–648.

127. Kersey, P., Bower, L., Morris, L., Horne, A., Petryszak, R., Kanz, C., Kanapin, A., Das, U., Michoud, K., Phan, I., Gattiker, A., Kulikova, T., Faruque, N., Duggan, K., McLaren, P., Reimholz, B., Duret, L., Penel, S., Reuter, I., and Apweiler, R. (2005) Integr8 and Genome Reviews: integrated views of complete genomes and proteomes. *Nucleic Acids Res.* **33**, D297–D302.

128. Hao, P., He, W. Z., Huang, Y., Ma, L. X., Xu, Y., Xi, H., Wang, C., Liu, B. S., Wang, J. M., Li, Y. X., and Zhong, Y. (2005) MPSS: an integrated database system for surveying a set of proteins. *Bioinformatics (Oxford, England)* **21**, 2142–2143.

129. Carvalho, P. C., Fischer, J. S., Chen, E. I., Domont, G. B., Carvalho, M. G., Degrave, W. M., Yates, J. R., 3rd, and Barbosa, V. C. (2009) GO Explorer: a gene-ontology tool to aid in the interpretation of shotgun proteomics data. *Proteome Sci.* 7, 6.

130. Barrell, D., Dimmer, E., Huntley, R. P., Binns, D., O'Donovan, C., and Apweiler, R. (2009) The GOA database in 2009an integrated Gene Ontology Annotation resource. *Nucleic Acids Res.* **37**, D396–D403.

131. Prince, J. T., Carlson, M. W., Wang, R., Lu, P., and Marcotte, E. M. (2004) The need for a public proteomics repository. *Nat. Biotechnol.* **22**, 471–472.

132. Mathivanan, S., Ahmed, M., Ahn, N. G., Alexandre, H., Amanchy, R., Andrews, P. C., Bader, J. S., Balgley, B. M., Bantscheff, M., Bennett, K. L., Bjorling, E., Blagoev, B., Bose, R., Brahmachari, S. K., Burlingame, A. S., Bustelo, X. R., Cagney, G., Cantin, G. T., Cardasis, H. L., Celis, J. E., Chaerkady, R., Chu, F., Cole, P. A., Costello, C. E., Cotter, R. J., Crockett, D., DeLany, J. P., De Marzo, A. M., DeSouza, L. V., Deutsch, E. W., Dransfield, E., Drewes, G., Droit, A., Dunn, M. J., Elenitoba-Johnson, K., Ewing, R. M., Van Eyk, J., Faca, V., Falkner, J., Fang, X., Fenselau, C., Figeys, D., Gagne, P., Gelfi, C., Gevaert, K., Gimble, J. M., Gnad, F., Goel, R., Gromov, P., Hanash, S. M., Hancock, W. S., Harsha, H. C., Hart, G., Hays, F., He, F., Hebbar, P., Helsens, K., Hermeking, H., Hide, W., Hjerno, K., Hochstrasser, D. F., Hofmann, O., Horn, D. M., Hruban, R. H., Ibarrola, N., James, P., Jensen, O. N., Jensen, P. H., Jung, P., Kandasamy, K., Kheterpal, I., Kikuno, R. F., Korf, U., Korner, R., Kuster, B., Kwon, M. S., Lee, H. J., Lee, Y. J., Lefevre, M., Lehvaslaiho, M., Lescuyer, P., Levander, F., Lim, M. S., Lobke, C., Loo, J. A., Mann, M., Martens, L., Martinez-Heredia, J., McComb, M., McRedmond, J., Mehrle, A., Menon, R., Miller, C. A., Mischak, H.,

Mohan, S. S., Mohmood, R., Molina, H., Moran, M. F., Morgan, J. D., Moritz, R., Morzel, M., Muddiman, D. C., Nalli, A., Navarro, J. D., Neubert, T. A., Ohara, O., Oliva, R., Omenn, G. S., Oyama, M., Paik, Y. K., Pennington, K., Pepperkok, R., Periaswamy, B., Petricoin, E. F., Poirier, G. G., Prasad, T. S., Purvine, S. O., Rahiman, B. A., Ramachandran, P., Ramachandra, Y. L., Rice, R. H., Rick, J., Ronnholm, R. H., Salonen, J., Sanchez, J. C., Sayd, T., Seshi, B., Shankari, K., Sheng, S. J., Shetty, V., Shivakumar, K., Simpson, R. J., Sirdeshmukh, R., Siu, K. W., Smith, J. C., Smith, R. D., States, D. J., Sugano, S., Sullivan, M., Superti-Furga, G., Takatalo, M., Thongboonkerd, V., Trinidad, J. C., Uhlen, M., Vandekerckhove, J., Vasilescu, J., Veenstra, T. D., Vidal-Taboada, J. M., Vihinen, M., Wait, R., Wang, X., Wiemann, S., Wu, B., Xu, T., Yates, J. R., Zhong, J., Zhou, M., Zhu, Y., Zurbig, P., and Pandey, A. (2008) Human Proteinpedia enables sharing of human protein data. *Nat. Biotechnol.* **26**, 164–167.

Chapter 5

Analysis of Post-translational Modifications by LC-MS/MS

Hannah Johnson and Claire E. Eyers

Abstract

Post-translational modifications are highly dynamic and known to regulate many cellular processes. Both the site and the stoichiometry of modification of a given protein sequence can have profound effects on the regulation of protein function. Thus, the identification of sites of post-translational modification is crucial for fully deciphering the biological roles of any given protein. The acute regulation and typically low stoichiometry of many post-translational modifications makes characterization of the sites of modification challenging. Thus, the development of analytical strategies to aid the selective enrichment and characterization of these species is paramount. Ongoing developments in mass spectrometry resulting in increased speed and sensitivity of analysis mean that mass spectrometry has become the ideal analytical tool for the qualitative and quantitative analysis of protein modifications. This chapter provides an overview of the most popular LC-MS/MS-based strategies for the enrichment of modified peptides/proteins and mass spectrometric workflows targeted toward the analysis of specific post-translationally modified analytes.

Key words: Post-translational modification, mass spectrometry, LC-MS/MS, enrichment.

1. Introduction

Post-translational modification provides a dynamic mechanism for regulating protein function and effectively expanding the chemical diversity of functional groups and regulatory potential beyond those defined by the standard 20 amino acid side chains. The term post-translational modification (PTM) can refer to the addition or the removal of a functional group from an amino acid side chain, the modification of protein termini, the cleavage of the synthesized polypeptide chain, or the covalent cross-linking between separate protein domains. For a more comprehensive overview of PTMs, readers are referred to a review by Walsh (1).

P.R. Cutillas, J.F. Timms (eds.), *LC-MS/MS in Proteomics*, Methods in Molecular Biology 658,
DOI 10.1007/978-1-60761-780-8_5, © Springer Science+Business Media, LLC 2010

The most commonly analyzed PTMs are covalent modifications, largely because of the close association between their dysregulation and a variety of disease states, and due to the biologically important changes in protein function that can occur as a result of their addition or removal. Covalent PTMs encompass both chemical moieties such as phosphate and carbohydrates, and functional polypeptides such as ubiquitin and SUMO. The reversible addition of a chemical moiety such as phosphate, catalyzed by specific protein kinases, plays an essential role in regulating signaling in both prokaryotic and eukaryotic cells. Incorporation of one or more phosphate groups on specific amino acid side chains within a protein, with serine, threonine, tyrosine, and histidine being the most commonly studied, often induces significant protein conformational change and consequently profound effects on protein activity and protein–protein interactions (2–4). Protein glycosylation is significantly more complex, with two main mechanisms for covalent binding of the glycan to the polypeptide: *N*-glycosylation, where the glycan is attached to an Asn residue within a tripeptide consensus sequence (Asn-X-Ser/Thr) (where X represents any amino acid except Pro) and *O*-glycosylation, in which the glycan is attached to a Ser or a Thr residue (5–7). Protein glycosylation is functionally important for modulating inter- and intracellular protein activities, cell adhesion, coordination of immune functions, and is critical for mitosis and cell division.

The functional significance of protein ubiquitination is dependent on the number of ubiquitin monomers attached at a single site, as well as the site of ubiquitin chain linkage. Ubiquitin is a 76-amino acid (~ 8 kDa) protein that is covalently attached via its C-terminal carboxyl group to the ε-amino groups of lysine residues on target proteins (8). Poly-ubiquitination via Lys48 of ubiquitin plays an essential role in proteasome-mediated induced protein degradation, while mono-ubiquitination regulates enzyme activity, protein–protein interactions, and protein trafficking (8, 9).

Protein function and/or stability can also be altered by specific structural changes to the polypeptide chain; disulfide bonds between cysteine residues separated in primary sequence often help to stabilize tertiary structure. Additionally, amino acid side chains can be altered. For example, the deamidation of asparagine generates aspartate, which fundamentally changes the amino acid composition and charge of the polypeptide post-translationally. Many of the PTM events that occur on a single protein are known to be synergistic; phosphorylation can promote protein ubiquitination, thereby coupling an extracellular stimuli-initiated signal transduction event with subsequent protein degradation (8).

Over 300 PTMs have been reported in the literature (1) and since they regulate all aspects of cellular homeostasis, determination of the site and extent of modification is crucial for deciphering cellular signaling networks. Crucially, both the type and location of modification within the tertiary protein structure may influence its function, with the stoichiometry of modification potentially determining the extent of any effect. This chapter therefore discusses, in brief, the current LC-MS/MS methodologies that can be used to address these questions. For a more comprehensive review, readers are directed to Eyers and Gaskell (4).

2. Mass Spectrometry in the Analysis of PTMs

Due to its sensitivity and specificity, mass spectrometry (MS) is an analytical strategy widely used for the identification and characterization of gene products within biological systems. A variety of MS instruments can be used for the identification and quantification of proteins present in complex biological mixtures. Typically, this requires the separation of proteins and/or their constituent peptides, often using some form of chromatography, prior to MS analysis. Mass analysis of a single purified protein can theoretically permit the identification of the type of PTM present. However, more detailed analysis, usually at the peptide level, is usually required to pinpoint the site(s) of modification. A modified peptide derived from such a sample can be detected readily by virtue of a mass increment or deficit relative to the unmodified form of the same peptide. Certain modifications induce a consistent characteristic mass change (4). For example, observation of mass differences representative of phosphorylation (+80 Da), methylation (+14 Da), or *S*-nitrosylation (+29 Da) is a good indicator of modified peptide species. However, presence of a particular mass difference does not conclusively permit identification of the site of modification (for which tandem MS is required), although the site of modification can sometimes be directly inferred if only a single potential site is present in the putative peptide sequence. It should also be noted that observation of a defined mass increment may be indicative of a number of different modifications. For example, the mass change that arises as a result of trimethylation (+42.05 Da) and acetylation (+42.01 Da) may be indistinguishable when using mass spectrometers with limited mass accuracy. Similarly, phosphorylation (+79.9663 Da) and sulfation (+79.9568 Da) can only be distinguished unambiguously with instruments capable of high mass accuracy or by using modification-specific analytical strategies such as affinity enrichment or enzymatic (e.g., phosphatase) treatment.

Characterization of PTM sites by LC-MS/MS is currently achievable using a number of complementary approaches that can be broadly separated into four categories: (i) tandem MS using one or more of a number of available fragmentation mechanisms; (ii) removal of the modification between consecutive mass spectrometric analyses; (iii) selective enrichment of modified proteins or peptides based on the modified functional group prior to MS/MS, and (iv) PTM-specific multistage MS strategies. Tandem MS (MS/MS) as used in a typical product ion analysis (**Fig. 5.1**) refers to the mass spectrometric analysis of all ions in a sample, prior to the subsequent selection and fragmentation of an ion of a specific *m/z*, and the subsequent MS analysis of the resultant fragment ions. When applied to peptides, both the amino acid sequence and the type and site of modification can be determined. Large-scale characterization of the proteins present in a complex sample, such as a cell lysate, is typically performed using a standard "bottom-up" proteomics workflow, where proteins are digested using a suitable protease (most commonly trypsin) generating smaller analytes better suited to current MS analysis workflows. These peptides are then fractionated using one or more forms of liquid chromatography [e.g., strong cation exchange (SCX) and reverse phase (RP)] prior to

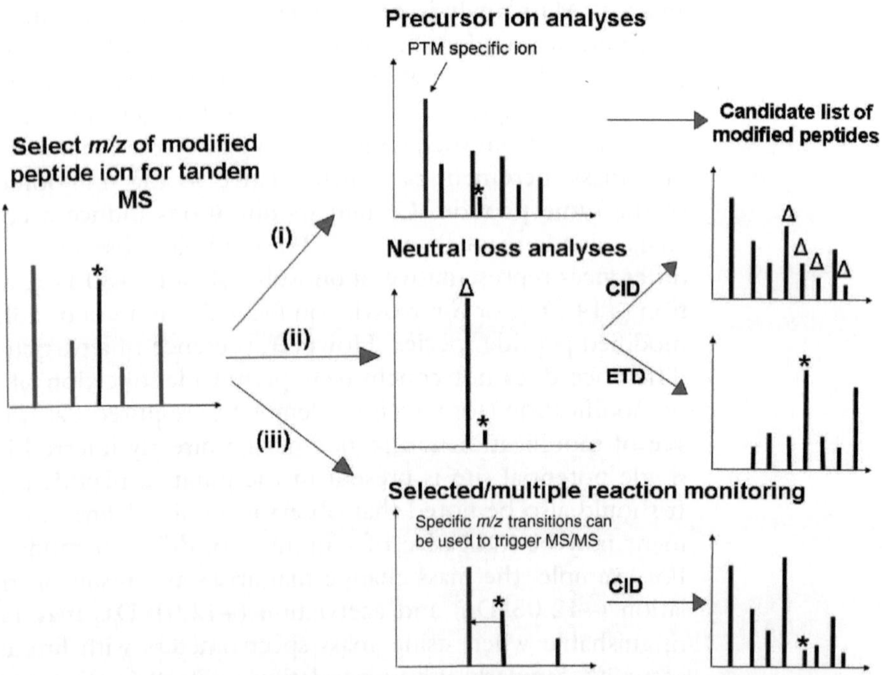

Fig. 5.1. PTM-targeted MS/MS strategies. (**i**) Precursor ion scanning: spectra are recorded only for those modified peptide ion precursors (indicated by *) that generate a characteristic product ion following CID at an *m/z* specific for the modification of interest; (**ii**) neutral loss-triggered MS/MS/MS: CID of a modified peptide ion precursor (indicated by *) which results in predominant loss of a neutral mass in the product ion spectrum (indicated as Δ) can be used to trigger either CID of the molecular ion minus the modification (MS/MS/MS) or MS/MS of the modified peptide ion using ETD; (**iii**) selected/multiple reaction monitoring tandem MS analyses can be used to trigger CID of a modified peptide ion precursor of defined *m/z* (indicated by *) following the detection of a positive transition of interest demonstrating modification.

MS analysis (10). Collision-induced dissociation (CID) remains the fragmentation method of choice for most proteomic applications. However, CID results in preferential fragmentation of the most facile bonds, and in the case of modified peptides, this is often the covalent bond linking the modification to the peptide backbone. Consequently, when peptides containing a PTM undergo CID, neutral loss of the modification is often the most abundant fragmentation product, leading to ambiguity in pinpointing the site of modification. Preferential loss of the PTM is thought to be due to the large amount of energy transferred to the analyte during collisional activation, which is randomized over many vibrational degrees of freedom before dissociation takes place. Vibrational dissipation of this excess translational energy results in dissociation of the weakest bonds, often at the bulky side chains containing the modification. Phosphopeptides are notoriously problematic in this regard. While CID has been invaluable in characterizing sites of phosphorylation on specific proteins (11) and also in large-scale proteomic studies (12, 13), CID results primarily in neutral loss of the phosphate group, observed as a loss either of 80 Da (HPO_3) or of 98 Da (H_3PO_4, β-elimination) (14). As the primary fragmentation pathway is loss of this phosphate moiety, the production of sequence-determining b- and y-fragment ions may be limited, making sequence analysis of the peptides difficult (**Fig. 5.2**). Unlike phosphorylated serine and threonine residues, phosphorylated tyrosine residues are much more stable to the neutral loss of H_3PO_4, simplifying identification. While glycosylated and sulfated peptides also characteristically exhibit this loss of functional group following CID, it would be incorrect to state that this is a universal issue for PTM analysis. Peptides containing many PTMs, including acetylation and methylation, can also be readily characterized using CID (15).

Alternate fragmentation mechanisms such as electron capture and electron transfer dissociation (ECD and ETD, respectively) do not promote analyte fragmentation by transfer of vibrational energy, rather they enable radical-initiated bond cleavage following the transfer of electrons (**Fig. 5.2**). Labile PTMs are therefore typically maintained on the peptide backbone ions during fragmentation. Specific sites of modification, particularly glycosylation and phosphorylation, can thus be more accurately identified using these alternate tandem MS strategies (16–19). However, the utility of ETD decreases with a decrease in charge state, with a shift toward non-dissociative electron transfer for peptides with a precursor ion charge of less than 3, as is most often generated by tryptic peptides. This results in an intact electron transfer product species, $[M+2H]^{+\bullet}$, and limited informative fragment ion generation (20). The application of a short burst of collisional dissociation after the ETD reaction, termed ETcaD (20) (supplemental activation with CID post-ETD) or

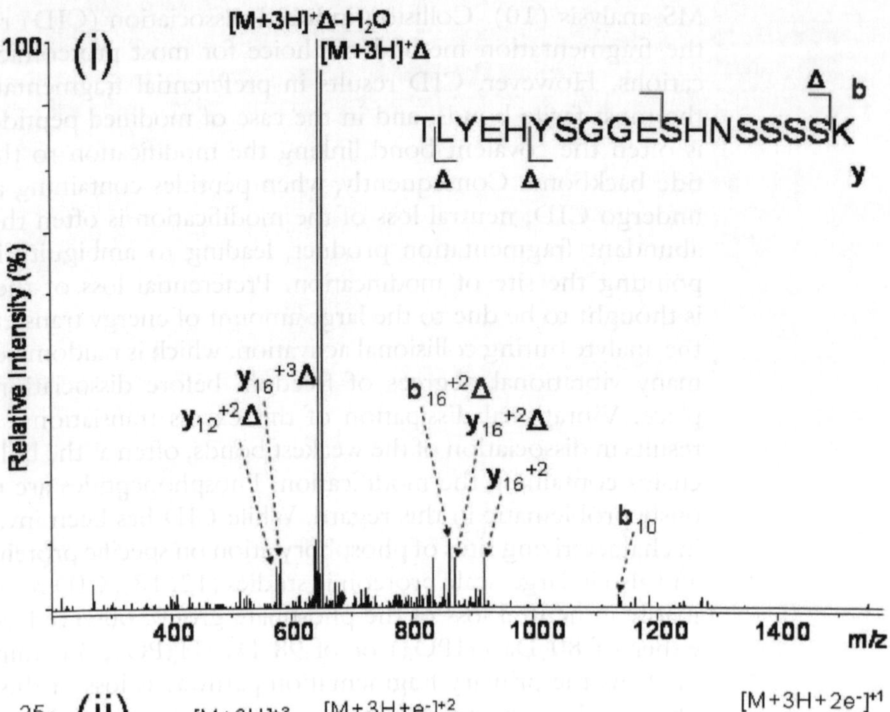

Fig. 5.2. CID and ETD tandem mass spectra of the singly phosphorylated triply protonated peptide TLYEHYSGGESHNSpSSSK. (i) CID – loss of H_3PO_4 from the peptide is indicated as Δ. (ii) ETD – the phosphate is maintained on the fragment ion during dissociation. N-terminal ions (c and b ions) and C-terminal ions (z and y ions) are indicated.

CID "tickling," dissociates the hydrogen-bonded, dissociated, charge-reduced species, significantly improving the sequence coverage by production of c^+ and $z^{+\bullet}$ fragment ions (21). The observation of improved sequence coverage when highly charged, generally larger peptides are subjected to ETD has encouraged proteolytic digestion with enzymes other than trypsin. Lys-C, which cleaves C-terminal to lysine resides, typically generates larger peptides than trypsin (which cleaves C-terminal to both lysine and arginine). The application of Lys-C digestion in conjunction with ETD has enabled better fragmentation, higher sequence coverage, and therefore increased PTM site discovery when compared to procedures using tryptic peptides (22, 23). Although ETD and ECD have proved extremely useful at overcoming some of the PTM-associated neutral loss issues observed during CID, these alternative fragmentation mechanisms do not completely overcome these problems. Sulfopeptides are more susceptible to neutral loss than are phosphopeptides of identical amino acid sequence, both during ECD and the marginally "gentler" process of ETD (24). However, it appears that the amount of neutral sulfate loss can be reduced (and therefore site identification improved) by fragmenting the peptide–alkali metal adduct as opposed to the protonated peptide ions (24).

3. Selective Enrichment

The highly dynamic nature of many PTMs often means that the stoichiometry of modification is relatively low, further complicating their characterization. Additionally, the efficiency of ionization and/or detection of modified peptide species (notably phosphopeptides) (25), when compared to their non-modified counterparts, can be compromised. Strategies have therefore been developed that target the explicit identification of peptides or proteins carrying specific post-translational modifications (**Fig. 5.3**). Selective enrichment techniques fall within three general categories: (i) incorporation of a tagged modification for selective enrichment; (ii) affinity purification using specific biological agents, and (iii) affinity enrichment based on chemical functionality of the modification.

3.1. Tagging

Large-scale analysis of sites of ubiquitination has been achieved using a strain of *Saccharomyces cerevisiae* expressing only a His_6-tagged form of ubiquitin. Ubiquitin-conjugated proteins were then affinity purified using nickel chromatography, prior to proteolysis with trypsin, SCX peptide separation, and LC-MS/MS. This approach was successful, identifying over 1,000 ubiquitin-modified proteins and characterizing 110 sites of ubiquitination in a single study (26). Whilst approaches such as these that rely on using specifically designed modified strains of an organism of interest undoubtedly provide large quantities of data for further

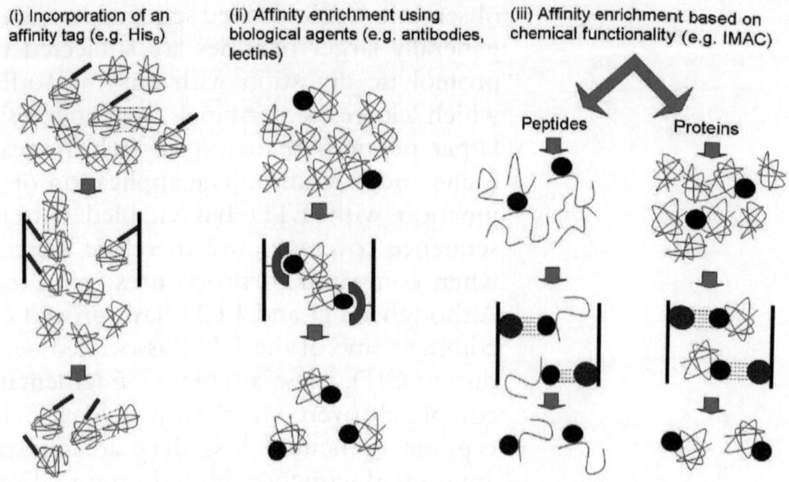

Fig. 5.3. Selective enrichment techniques. (**i**) Incorporation of a PTM-specific affinity tag (e.g., His$_6$ ubiquitin) permits selective enrichment of those proteins that are covalently modified and their associated proteins (indicated by dashed lines). (**ii**) Affinity enrichment using biological agents such as antibodies permits selective enrichment of those PTM-containing proteins which have affinity for that biological agent. (**iii**) Proteins and peptides can be enriched based on chemical functionality such as charge and polarity.

PTM analysis, such strategies do not lend themselves to generic applications. Specifically, they are unsuitable for the analysis of primary cells (e.g., clinical samples).

3.2. Affinity Purification with Biological Agents

Biological molecules that bind with high specificity to the modification or the modified residue of interest, such as antibodies and lectins, are ideal for the selective isolation of modified polypeptides. Phosphorylated proteins can be enriched either using generic phosphospecific antibodies, the most successful example of this being the anti-phosphotyrosine antibodies, which are used to enrich at both the peptide and the protein level (27, 28), or with phosphospecific antibodies designed around known consensus sequences. Antibodies raised against pSQ (29) and RxpS/pT (30, 31) are good examples of immunoaffinity reagents that have been used to successfully identify substrates of protein kinases whose consensus requirements for phosphorylation are known. Recent developments in the generation of ubiquitin-specific antibodies have also aided the selective enrichment of ubiquitin-conjugated proteins for mass spectrometric analysis (32). Along similar lines, lectins bind to carbohydrates, and they have therefore been used extensively for the selective enrichment of glycoproteins, glycopeptides, and oligosaccharides (33–35). Bioaffinity molecules such as lectins (and consensus specific antibodies) have an additional advantage due to the distinct carbohydrate-binding specificity of different families of lectins.

Specific lectins can therefore be exploited to affinity purify defined oligosaccharide structures on glycoconjugates (33).

3.3. Affinity Capture

In addition to biological affinity tools, there are a number of solid-phase materials that can be used for the enrichment of modified peptide or protein species based on chemical functionality. Such enrichment of phosphorylated species is largely based on the principle of immobilized metal affinity chromatography (IMAC) which adsorbs phosphopeptides to chelated metal ions (e.g., Fe^{3+}, Ga^{3+}) through metal–phosphate ion-pair interactions (36, 37). As an extension of this, TiO_2- and ZrO_2-based columns also adsorb phosphorylated peptide species well (38, 39). During IMAC or TiO_2 enrichment, phosphopeptides bound to the immobilized metal ions are eluted either by increasing the pH or by competitively eluting with phosphate ions. However, these methods are compromised by the (unwanted) concurrent enrichment of highly acidic peptides, which bind to the IMAC columns by the same chemical principles as the negatively charged phosphopeptides. Methyl esterification of acidic groups has been demonstrated to overcome some of these issues (40), although secondary modifications that occur during this reaction may influence subsequent peptide characterization by MS (11). Binding of acidic peptides to TiO_2 can also be reduced by sample binding in the presence of 2,5-dihydroxybenzoic acid (DHB) or glutamate (41, 42). Due to the differing affinity of TiO_2 and IMAC for the preferential enrichment of singly and multiply phosphorylated peptides respectively (40, 43, 44), many current phosphoproteomic studies are using a combination of solid-phase media for optimal phosphopeptide enrichment (43). Ion exchange chromatography [both strong cation exchange (45) and strong anion exchange (46)] have also been used successfully for the enrichment and partial fractionation of phosphopeptides prior to LC-MS/MS analysis, although these strategies are less specific.

A different method has also demonstrated utility for the enrichment of phosphorylated peptides from mixtures. Here, phosphopeptides are precipitated with an excess of calcium phosphate at alkaline pH, enabling them to be separated from the non-phosphorylated peptides by centrifugation (47). This technique is extremely rapid and could become an important step in phosphopeptide analysis workflows.

Hydrophilic interaction chromatography (HILIC) is now routinely used for the semi-selective enrichment of glycopeptides (48–50). Enrichment of glycans has been demonstrated to occur via a number of mechanisms including hydrogen bonding and ionic interactions and it can therefore often be used in the separation of isobaric species, an extremely useful feature that complements MS/MS analysis (51).

4. PTM-Targeted MS/MS Strategies

As indicated earlier, modified peptides often behave characteristically during MS/MS analysis, generating unique PTM signatures due to either loss of functional groups or the formation of diagnostic fragment ions. Mass spectrometric methods can often therefore be tailored to allow the specific selection and analysis of defined populations of modified peptides. Typically, peptide identification is performed using product ion analysis, where all of the product ions generated following decomposition of a precursor ion of defined *m/z* are recorded. To utilize the specific fragmentation patterns of modified peptide ions, analysis can be targeted toward precursor ions that generate a defined product or diagnostic fragment ion, termed precursor ion scanning (**Fig. 5.1**). Analysis can also be directed toward ions exhibiting a characteristic difference between their precursor and product ions (demonstrating loss of functional group); this is neutral loss scanning (**Fig. 5.1**).

4.1. Precursor Ion Scanning

A precursor ion scan determines the *m/z* of all of the precursor ions (in this case, peptide ions) that generate a specific product ion following fragmentation, in a single experiment (52). During precursor ion analysis, the entire mass range is scanned in the first mass analyzer, including the precursor ions whose fragmentation results in the selected product ion. The second mass analyzer is meanwhile fixed to monitor and transmit only those product ions of a specific *m/z* ratio. The resulting mass spectrum then details those precursor ions whose fragmentation resulted in the formation of the product ion of interest. This procedure increases the sensitivity of analysis by effectively filtering out chemical 'noise' from the mass spectrum. When analyzing phosphorylated peptides, a precursor ion scan of *m/z* 79 (PO_3^-) or 63 (PO_2^-) in negative mode produces a mass spectrum containing only the phosphorylated precursor ions present within the sample (53). Similarly, phosphotyrosine generates a diagnostic signal at *m/z* +216.042 (54). In a separate experiment, these identified precursor ions can then be targeted for MS/MS to elucidate their amino acid sequence and identify the site(s) of phosphorylation (53).

Due to the chemical complexity of glycosylated species, glycopeptides generate a number of diagnostic sugar oxonium fragment ions following CID, which can be used to help identify glycopeptide precursor ions. The most characteristic of these is *m/z* 204 for $HexNAc^+$ (55) since both *N*- and *O*-linked glycopeptides have an *N*-hexosamine sugar residue attached to the peptide backbone. Precursor ion analysis of peptides containing acetylated lysine residues is also achievable due to the characteristic production of an acetylated lysine immonium ion at *m/z* 143.1 and the related ion (demonstrating loss of NH_3) at *m/z* 126.1 (56, 57).

4.2. Neutral Loss Scanning

A neutral loss scan determines the *m/z* of parent ions which fragment to produce a product ion demonstrating loss of a specified (neutral) mass. The resultant product ion generated following this neutral loss can then be selected for further fragmentation in an MS/MS/MS experiment. Neutral loss experiments of this sort can be achieved only on those instruments capable of multiple stages of tandem MS, namely ion-trapping instruments. Modified peptides which demonstrate such a PTM-specific loss during CID can easily be characterized using this technique. However, as is the case with phosphorylated peptides, direct mapping of the site of modification may be difficult. In addition to performing an MS/MS/MS CID experiment, the observation of a neutral loss during CID can also be used to trigger ECD or ETD of the precursor of interest; this is particularly beneficial for those labile precursors for which sequence-determining fragment ions are difficult to generate (58). Combining ETD data generated in this manner with CID information could prove particularly useful for improving confidence in the identification of sites of phosphorylation, glycosylation, and methylated arginine amongst others (59).

4.3. Selected Reaction Monitoring/Multiple Reaction Monitoring

Selected reaction monitoring (SRM) and multiple reaction monitoring (MRM) rely on the fragmentation of a specific peptide ion to yield one or more defined product ions which are diagnostic for a specific series of amino acids. In SRM, a product ion of specified *m/z* derived from a particular precursor ion *m/z* window is used to confirm the presence of a specific analyte. This is extended with MRM, where a number of different fragment ion *m/z* values are used to confirm the presence of peptide ions of interest, providing greater confidence in the assignment. Due to the increase in sensitivity of identification and quantification achieved, SRM and MRM can be used in the identification of modified peptides derived from a known post-translationally modified protein. This technique, adapted under the acronym "multiple reaction monitoring-initiated detection and sequencing" (MIDAS), has been used successfully for the identification of sites of phosphorylation and ubiquitination from purified proteins containing known modifications (60, 61).

5. Quantification of PTMs

Elucidation of the stoichiometry of a given modification is important for correlation with the proportion of a protein exhibiting altered PTM-regulated functionality, be that an enzyme activity or a binding affinity. However, mass spectrometric signal intensity is dependent not only on the amount of analyte but also on its chemical properties. A comparison of the signal intensities of a peptide and its modified counterpart is therefore not an accurate representation of the proportion of peptide that is modified. Circumventing the potential issues of changes in ionization efficiency of differentially modified peptides of the same primary sequence, relative quantification of stoichiometry between two or more systems can be achieved using stable isotope labeling. Internal reference peptides that are chemically identical to the analyte of interest, but can be differentiated by mass due to the incorporation of heavy-isotope forms of particular elements (^{13}C, ^{15}N), are therefore used for the quantification of different peptide states. Absolute quantification of phosphorylated and ubiquitinated proteins has been achieved using isotope dilution methods (62–65). In the case of phosphorylation stoichiometry determination, phosphorylated isotope-labeled surrogate peptides are synthesized, quantified, and added to the sample of interest in known amounts, permitting calculation of the stoichiometry of modification of the native peptide based on comparable signal ion intensity. Stable isotope labeling with amino acids in cell culture (SILAC) permits calculation of the relative amounts of modified peptides present between multiple cell populations. This strategy has the advantage of permitting the combination of samples undergoing analysis at an early stage in the sample preparation workflow, minimizing artifactual differences, and is capable of determining the relative stoichiometry of a large number of modified peptides simultaneously (66–68). However, to specifically assess the stoichiometry of modified peptides, SILAC is best used in conjunction with one (or more) of the enrichment or MS-targeted strategies described previously (34).

As an alternative to stable isotope labeling, normalized ion currents can also be used for the relative quantification of modified peptides between LC-MS/MS runs. The relative amounts of modified peptides in two or more samples are then calculated using variations in the signal ion intensities of peptide species of interest (69), although this strategy is generally not as precise.

The continued interest of the scientific community in improving our ability to determine the presence and stoichiometry of sites of modification ensures that LC-MS/MS-based methodologies are constantly under development. However, it is only by characterizing these modified proteins as intact polymers, rather than as tryptic peptides that the community currently favors, that it will eventually be possible to assess the true extent of post-translational modifications and their influence on cellular function.

References

1. Walsh, C. (2005) Posttranslational modification of proteins: expanding nature's inventory. B. Roberts, Colorado.
2. Hunter, T. (2000) Signaling2000 and beyond. *Cell* **100**, 113–127.
3. Seet, B. T., Dikic, I., Zhou, M. M., and Pawson, T. (2006) Reading protein modifications with interaction domains. *Nat. Rev.* 7, 473–483.
4. Eyers, C. E., and Gaskell, S. J. (2008) Mass spectrometry to identify post-translational modifications. Wiley Encyclopedia of Chemical Biology. doi:10.1002/9780470048672.wecb469.
5. Morelle, W., and Michalski, J. C. (2007) Analysis of protein glycosylation by mass spectrometry. *Nat. Protoc* **2**, 1585–1602.
6. Dwek, R. A. (1996) Glycobiology: toward understanding the function of sugars. *Chem. Rev.* **96**, 683–720.
7. Helenius, A., and Aebi, M. (2004) Roles of N-linked glycans in the endoplasmic reticulum. *Annu. Rev. Biochem.* **73**, 1019–1049.
8. Weissman, A. M. (2001) Themes and variations on ubiquitylation. *Nat. Rev. Mol. Cell. Biol.* **2**, 169–178.
9. Drews, O., Zong, C., and Ping, P. (2007) Exploring proteasome complexes by proteomic approaches. *Proteomics* 7, 1047–1058.
10. Washburn, M. P., Wolters, D., and Yates, J. R., 3rd (2001) Large-scale analysis of the yeast proteome by multidimensional protein identification technology. *Nat. Biotechnol.* **19**, 242–247.
11. Haydon, C. E., Eyers, P. A., Aveline-Wolf, L. D., Resing, K. A., Maller, J. L., and Ahn, N. G. (2003) Identification of novel phosphorylation sites on *Xenopus laevis* Aurora A and analysis of phosphopeptide enrichment by immobilized metal-affinity chromatography. *Mol. Cell. Proteomics* **2**, 1055–1067.
12. Nuhse, T. S., Stensballe, A., Jensen, O. N., and Peck, S. C. (2003) Large-scale analysis of in vivo phosphorylated membrane proteins by immobilized metal ion affinity chromatography and mass spectrometry. *Mol. Cell. Proteomics* **2**, 1234–1243.
13. Reinders, J., and Sickmann, A. (2005) State-of-the-art in phosphoproteomics. *Proteomics* **5**, 4052–4061.
14. Schweppe, R. E., Haydon, C. E., Lewis, T. S., Resing, K. A., and Ahn, N. G. (2003) The characterization of protein post-translational modifications by mass spectrometry. *Acc. Chem, Res.* **36**, 453–461.
15. Fraga, M. F., Ballestar, E., Villar-Garea, A., Boix-Chornet, M., Espada, J., Schotta, G., Bonaldi, T., Haydon, C., Ropero, S., Petrie, K., Iyer, N. G., Perez-Rosado, A., Calvo, E., Lopez, J. A., Cano, A., Calasanz, M. J., Colomer, D., Piris, M. A., Ahn, N., Imhof, A., Caldas, C., Jenuwein, T., and Esteller, M. (2005) Loss of acetylation at Lys16 and trimethylation at Lys20 of histone H4 is a common hallmark of human cancer. *Nat. Genet.* **37**, 391–400.
16. Tyler, R. K., Chu, M. L., Johnson, H., McKenzie, E. A., Gaskell, S. J., and Eyers, P. A. (2009) Phosphoregulation of human Mps1 kinase. *Biochem. J.* **417**, 173–181.
17. Mirgorodskaya, E., Roepstorff, P., and Zubarev, R. A. (1999) Localization of O-glycosylation sites in peptides by electron capture dissociation in a Fourier transform mass spectrometer. *Anal. Chem.* **71**, 4431–4436.
18. Zubarev, R. A. (2004) Electron-capture dissociation tandem mass spectrometry. *Curr. Opin. Biotechnol.* **15**, 12–16.
19. Tsybin, Y. O., Ramstrom, M., Witt, M., Baykut, G., and Hakansson, P. (2004) Peptide and protein characterization by high-rate electron capture dissociation Fourier transform ion cyclotron resonance mass spectrometry. *J. Mass Spectrom.* **39**, 719–729.
20. Swaney, D. L., McAlister, G. C., Wirtala, M., Schwartz, J. C., Syka, J. E., and Coon, J. J. (2007) Supplemental activation method for high-efficiency electron-transfer dissociation of doubly protonated peptide precursors. *Anal. Chem.* **79**, 477–485.
21. Swaney, D. L., McAlister, G. C., Wirtala, M., Schwartz, J. C., Syka, J. E. P., and Coon, J. J. (2007) Supplemental activation method for high-efficiency electron-transfer dissociation of doubly protonated peptide precursors. *Anal. Chem.* **79**, 477–485.
22. Chi, A., Huttenhower, C., Geer, L. Y., Coon, J. J., Syka, J. E. P., Bai, D. L., Shabanowitz, J., Burke, D. J., Troyanskaya, O. G., and Hunt, D. F. (2007) Analysis of phosphorylation sites on proteins from *Saccharomyces cerevisiae* by electron transfer dissociation (ETD) mass spectrometry. 10.1073/pnas.0607084104. *Proc. Natl. Acad. Sci. USA* **104**, 2193–2198.
23. Zhang, Q., Schepmoes, A. A., Brock, J. W. C., Wu, S., Moore, R. J., Purvine, S. O., Baynes, J. W., Smith, R. D., and Metz, T. O. (2008) Improved methods for the enrichment and analysis of glycated peptides.

doi:10.1021/ac801704j. *Anal. Chem.* **80**, 9822–9829.

24. Medzihradszky, K. F., Guan, S., Maltby, D. A., and Burlingame, A. L. (2007) Sulfopeptide fragmentation in electron-capture and electron-transfer dissociation. *J. Am. Soc. Mass Spectrom.* **18**, 1617–1624.
25. Steen, H., Jebanathirajah, J. A., Rush, J., Morrice, N., and Kirschner, M. W. (2006) Phosphorylation analysis by mass spectrometry: myths, facts, and the consequences for qualitative and quantitative measurements. *Mol. Cell. Proteomics* **5**, 172–181.
26. Peng, J., Schwartz, D., Elias, J. E., Thoreen, C. C., Cheng, D., Marsischky, G., Roelofs, J., Finley, D., and Gygi, S. P. (2003) A proteomics approach to understanding protein ubiquitination. *Nat. Biotechnol.* **21**, 921–926.
27. Blagoev, B., Ong, S.-E., Kratchmarova, I., and Mann, M. (2004) Temporal analysis of phosphotyrosine-dependent signaling networks by quantitative proteomics. *Nat. Biotechnol.* **22**, 1139–1145.
28. Rush, J., Moritz, A., Lee, K. A., Guo, A., Goss, V. L., Spek, E. J., Zhang, H., Zha, X. M., Polakiewicz, R. D., and Comb, M. J. (2005) Immunoaffinity profiling of tyrosine phosphorylation in cancer cells. *Nat. Biotechnol.* **23**, 94–101.
29. Cortez, D., Glick, G., and Elledge, S. J. (2004) Minichromosome maintenance proteins are direct targets of the ATM and ATR checkpoint kinases. *Proc. Natl. Acad. Sci. USA* **101**, 10078–10083.
30. Gronborg, M., Kristiansen, T. Z., Stensballe, A., Andersen, J. S., Ohara, O., Mann, M., Jensen, O. N., and Pandey, A. (2002) A mass spectrometry-based proteomic approach for identification of serine/threonine-phosphorylated proteins by enrichment with phospho-specific antibodies: identification of a novel protein, Frigg, as a protein kinase A substrate. *Mol. Cell. Proteomics* **1**, 517–527.
31. Kane, S., Sano, H., Liu, S. C., Asara, J. M., Lane, W. S., Garner, C. C., and Lienhard, G. E. (2002) A method to identify serine kinase substrates. Akt phosphorylates a novel adipocyte protein with a Rab GTPase-activating protein (GAP) domain. *J. Biol. Chem.* **277**, 22115–22118.
32. Vasilescu, J., Smith, J. C., Ethier, M., and Figeys, D. (2005) Proteomic analysis of ubiquitinated proteins from human MCF-7 breast cancer cells by immunoaffinity purification and mass spectrometry. doi:10.1021/pr050265i. *J. Proteome Res.* **4**, 2192–2200.
33. Nawarak, J., Phutrakul, S., and Chen, S.-T. (2004) Analysis of lectin-bound glycoproteins in snake venom from the Elapidae and Viperidae families. *J. Proteome Res.* **3**, 383–392.
34. Qiu, R., and Regnier, F. E. (2005) Use of Multidimensional lectin affinity chromatography in differential glycoproteomics. doi:10.1021/ac048751x. *Anal. Chem.* **77**, 2802–2809.
35. Drake, R. R., Schwegler, E. E., Malik, G., Diaz, J., Block, T., Mehta, A., and Semmes, O. J. (2006) Lectin capture strategies combined with mass spectrometry for the discovery of serum glycoprotein biomarkers. 10.1074/mcp.M600176-MCP200. *Mol. Cell. Proteomics* **5**, 1957–1967.
36. Andersson, L., and Porath, J. (1986) Isolation of phosphoproteins by immobilized metal (Fe^{3+}) affinity chromatography. *Anal. Biochem.* **154**, 250–254.
37. Posewitz, M. C., and Tempst, P. (1999) Immobilized gallium(III) affinity chromatography of phosphopeptides. doi:10.1021/ac981409y. *Anal. Chem.* **71**, 2883–2892.
38. Larsen, M. R., Thingholm, T. E., Jensen, O. N., Roepstorff, P., and Jorgensen, T. J. D. (2005) Highly selective enrichment of phosphorylated peptides from peptide mixtures using titanium dioxide microcolumns. 10.1074/mcp.T500007-MCP200. *Mol. Cell. Proteomics* **4**, 873–886.
39. Kweon, H. K., and Hakansson, K. (2006) Selective zirconium dioxide-based enrichment of phosphorylated peptides for mass spectrometric analysis. doi:10.1021/ac0522355. *Anal. Chem.* **78**, 1743–1749.
40. Ficarro, S. B., McCleland, M. L., Stukenberg, P. T., Burke, D. J., Ross, M. M., Shabanowitz, J., Hunt, D. F., and White, F. M. (2002) Phosphoproteome analysis by mass spectrometry and its application to *Saccharomyces cerevisiae*. *Nat. Biotechnol.* **20**, 301–305.
41. Larsen, M. R., Thingholm, T. E., Jensen, O. N., Roepstorff, P., and Jorgensen, T. J. (2005) Highly selective enrichment of phosphorylated peptides from peptide mixtures using titanium dioxide microcolumns. *Mol. Cell. Proteomics* **4**, 873–886.
42. Wu, J., Shakey, Q., Liu, W., Schuller, A., and Follettie, M. T. (2007) Global profiling of phosphopeptides by titania affinity enrichment. *J. Proteome Res.* **6**, 4684–4689.
43. Thingholm, T. E., Jensen, O. N., Robinson, P. J., and Larsen, M. R. (2008) SIMAC (sequential elution from IMAC), a

phosphoproteomics strategy for the rapid separation of monophosphorylated from multiply phosphorylated peptides. *Mol. Cell. Proteomics* 7, 661–671.

44. Jensen, S. S., and Larsen, M. R. (2007) Evaluation of the impact of some experimental procedures on different phosphopeptide enrichment techniques. *Rapid Commun. Mass Spectrom.* **21**, 3635–3645.
45. Beausoleil, S. A., Jedrychowski, M., Schwartz, D., Elias, J. E., Villen, J., Li, J., Cohn, M. A., Cantley, L. C., and Gygi, S. P. (2004) Large-scale characterization of HeLa cell nuclear phosphoproteins. *Proc. Natl. Acad. Sci. USA* **101**, 12130–12135.
46. Han, G., Ye, M., Zhou, H., Jiang, X., Feng, S., Tian, R., Wan, D., Zou, H., and Gu, J. (2008) Large-scale phosphoproteome analysis of human liver tissue by enrichment and fractionation of phosphopeptides with strong anion exchange chromatography. *Proteomics* **8**, 1346–1361.
47. Zhang, X., Ye, J., Jensen, O. N., and Roepstorff, P. (2007) Highly efficient phosphopeptide enrichment by calcium phosphate precipitation combined with subsequent IMAC enrichment. *Mol. Cell. Proteomics* **6**, 2032–2042.
48. Hagglund, P., Bunkenborg, J., Elortza, F., Jensen, O. N., and Roepstorff, P. (2004) A new strategy for identification of N-glycosylated proteins and unambiguous assignment of their glycosylation sites using HILIC enrichment and partial deglycosylation. *J. Proteome Res.* **3**, 556–566.
49. Calvano, C. D., Zambonin, C. G., and Jensen, O. N. (2008) Assessment of lectin and HILIC based enrichment protocols for characterization of serum glycoproteins by mass spectrometry. *J. Proteomics* **71**, 304–317.
50. Picariello, G., Ferranti, P., Mamone, G., Roepstorff, P., and Addeo, F. (2008) Identification of N-linked glycoproteins in human milk by hydrophilic interaction liquid chromatography and mass spectrometry. *Proteomics* **8**, 3833–3847.
51. Wuhrer, M., de Boer, A. R., and Deelder, A. M. (2009) Structural glycomics using hydrophilic interaction chromatography (HILIC) with mass spectrometry. *Mass Spectrom. Rev.* **28**, 192–206.
52. IUPAC (1997) Compendium of chemical terminology, 2nd (The "Gold Book") ed, Blackwell Scientific Publications, Oxford.
53. Carr, S. A., Huddleston, M. J., and Annan, R. S. (1996) Selective detection and sequencing of phosphopeptides at the femtomole level by mass spectrometry. *Anal. Biochem.* **239**, 180–192.
54. Steen, H., Pandey, A., Andersen, J. S., and Mann, M. (2002) Analysis of tyrosine phosphorylation sites in signaling molecules by a phosphotyrosine-specific immonium ion scanning method. 10.1126/stke.2002.154.pl16. *Sci. STKE* **2002**, pl16.
55. Huddleston, M. J., Bean, M. F., and Carr, S. A. (1993) Collisional fragmentation of glycopeptides by electrospray ionization LC/MS and LC/MS/MS: methods for selective detection of glycopeptides in protein digests. *Anal. Chem.* **65**, 877–884.
56. Bean, M. F., Annan, R. S., Hemling, M. E., Mentzer, M., Huddleston, M. J., and Carr, S. A. (1995) LC-MS methods for selective detection of posttranslational modifications in proteins: glycosylation, phosphorylation, sulfation, and acylation techniques in protein chemistry. In Crabb, J. W. (Ed.), Vol. 6, pp. 107–116, Academic Press.
57. Kim, J. Y., Kim, K. W., Kwon, H. J., Lee, D. W., and Yoo, J. S. (2002) Probing lysine acetylation with a modification-specific marker ion using high-performance liquid chromatography/electrospray-mass spectrometry with collision-induced dissociation. doi:10.1021/ac0256080. *Anal. Chem.* **74**, 5443–5449.
58. Sweet, S. M., Mardakheh, F. K., Ryan, K. J., Langton, A. J., Heath, J. K., and Cooper, H. J. (2008) Targeted online liquid chromatography electron capture dissociation mass spectrometry for the localization of sites of in vivo phosphorylation in human Sprouty2. *Anal. Chem.* **80**, 6650–6657.
59. Zubarev, R. A., Zubarev, A. R., and Savitski, M. M. (2008) Electron capture/transfer versus collisionally activated/induced dissociations: solo or duet? *J. Am. Soc. Mass Spectrom.* **19**, 753–761.
60. Unwin, R. D., Griffiths, J. R., Leverentz, M. K., Grallert, A., Hagan, I. M., and Whetton, A. D. (2005) Multiple reaction monitoring to identify sites of protein phosphorylation with high sensitivity. 10.1074/mcp.M500113-MCP200. *Mol. Cell. Proteomics* **4**, 1134–1144.
61. Sahana Mollah, I. E. W., Phung, Q., Arnott, D., Dixit, V. M., Lill, J. R. (2007) Targeted mass spectrometric strategy for global mapping of ubiquitination on proteins. *Rapid Commun. Mass Spectrom.* **21**, 3357–3364.
62. Hegeman, A. D., Harms, A. C., Sussman, M. R., Bunner, A. E., and Harper, J. F. (2004)

An isotope labeling strategy for quantifying the degree of phosphorylation at multiple sites in proteins. *J. Am. Soc. Mass Spectrom.* **15**, 647–653.

63. Zhang, X., Jin, Q. K., Carr, S. A., and Annan, R. S. (2002) N-terminal peptide labeling strategy for incorporation of isotopic tags: a method for the determination of site-specific absolute phosphorylation stoichiometry. *Rapid Commun. Mass Spectrom.* **16**, 2325–2332.
64. Mayor, T., Graumann, J., Bryan, J., MacCoss, M. J., and Deshaies, R. J. (2007) Quantitative profiling of ubiquitylated proteins reveals proteasome substrates and the substrate repertoire influenced by the Rpn10 receptor pathway. *Mol. Cell. Proteomics* **6**, 1885–1895.
65. Mayya, V., Rezual, K., Wu, L., Fong, M. B., and Han, D. K. (2006) Absolute quantification of multisite phosphorylation by selective reaction monitoring mass spectrometry: determination of inhibitory phosphorylation status of cyclin-dependent kinases. *Mol. Cell. Proteomics* **5**, 1146–1157.
66. Bonenfant, D., Towbin, H., Coulot, M., Schindler, P., Mueller, D. R., and van Oostrum, J. (2007) Analysis of dynamic changes in post-translational modifications of human histones during cell cycle by mass spectrometry. *Mol. Cell. Proteomics* **6**, 1917–1932.
67. Pan, C., Gnad, F., Olsen, J. V., and Mann, M. (2008) Quantitative phosphoproteome analysis of a mouse liver cell line reveals specificity of phosphatase inhibitors. *Proteomics* **8**, 4534–4546.
68. Amanchy, R., Kalume, D. E., and Pandey, A. (2005) Stable isotope labeling with amino acids in cell culture (SILAC) for studying dynamics of protein abundance and posttranslational modifications. *Sci STKE* **2005**, pl2.
69. Steen, H., Jebanathirajah, J. A., Springer, M., and Kirschner, M. W. (2005) Stable isotope-free relative and absolute quantitation of protein phosphorylation stoichiometry by MS. *Proc. Natl. Acad. Sci. USA* **102**, 3948–3953.

Part II

Structural Analysis of Post-Translational Modifications

Chapter 6

In-Depth Analysis of Protein Phosphorylation by Multidimensional Ion Exchange Chromatography and Mass Spectrometry

Maria P. Alcolea and Pedro R. Cutillas

Abstract

Protein phosphorylation controls fundamental biological functions that are often deregulated in disease. Therefore, system-level understanding of complex pathophysiological processes requires methods that can be used to profile and quantify protein phosphorylation as comprehensively as possible. Here we present a detailed protocol to enrich phosphopeptides from total cell lysates in a form amenable to downstream analysis by mass spectrometry. Using these techniques, we have detected several thousands of phosphorylation sites in the NIH-3T3 cell line.

Key words: Phosphoproteomics, cell signaling, cancer, quantification.

1. Introduction

Phosphorylation is one of the most common posttranslational modifications known to modulate the biophysical properties of proteins. Protein kinases and phosphatases are the enzymes responsible for catalysing the transfer and the removal of phosphate groups on proteins, respectively. Intense research in this field has shown the relevance of these phosphorylation events for the regulation of many different biochemical pathways in eukaryotic cells. As for their biological function, proliferation, differentiation, apoptosis, transcription, metabolism and migration represent just a short list of all the processes in which protein kinases

P.R. Cutillas, J.F. Timms (eds.), *LC-MS/MS in Proteomics*, Methods in Molecular Biology 658,
DOI 10.1007/978-1-60761-780-8_6, © Springer Science+Business Media, LLC 2010

and phosphatases play central regulatory roles (1). Indeed, the ubiquitous role of protein phosphorylation in the regulation of cell biology is reflected by the notion that one-third of all proteins in higher eukaryotes are under protein kinase and/or phosphatase control at a given time (2).

Thus, the ability to quantify kinase and phosphatase activities in a comprehensive manner is being instrumental for unravelling many biological processes such as those involved in the development of complex and poorly understood diseases such as cancer. It is therefore well accepted that further advances in our understanding of pathways controlled by phosphorylation at the system level, perhaps the ultimate aim of fields such as signal transduction research, require methods that can be used for comprehensive analysis of phosphorylation. In this regard, recent developments in mass spectrometry-based proteomic approaches have been successfully applied to study protein phosphorylation, allowing identification and quantification of thousands of phosphorylation sites per experiment (3).

These large-scale phosphoproteomic studies are possible because of the recent availability of highly selective phosphopeptide isolation and separation techniques, such as those based on cation exchange chromatography (SCX), immunoprecipitation, immobilised metal ion affinity chromatography (IMAC), titanium dioxide (TiO_2), calcium precipitation and chemical modifications, among others (4). These enrichment techniques are an essential requirement for successful phosphoproteomic experiments because of the low stoichiometry of phosphorylation on phosphoproteins relative to their non-phosphorylated counterparts. In order to reduce the complexity of biological samples and hence increase the fraction of the phosphoproteome analysed, it is also important to perform a separation step prior to the enrichment steps. Thus workflows in which SCX are coupled to IMAC or TiO_2 are common (5–7).

Our laboratory has been implementing and optimising techniques for comprehensive analysis of phosphorylation. The protocol presented here involves the digestion of cellular proteins with trypsin, reducing the peptide mixture complexity by an initial fractionation step using SCX HPLC and phosphopeptide enrichment of the obtained fractions by a combination of orthogonal ion exchange techniques, namely IMAC and TiO_2 chromatography. This combination of enrichment steps was introduced by Thingholm and colleagues (8) as an improved phosphopeptide enrichment strategy that consists of sequentially eluting phosphopeptides from the IMAC material in combination with the different selectivity of TiO_2, which ultimately allows obtaining a better coverage of the whole phosphoproteome. Here, we present an adaptation of this simple protocol for in-depth analysis of protein phosphorylation.

2. Materials

2.1. Cell Culture and Lysis

1. Murine NIH-3T3 fibroblasts (American Tissue Culture Collection, LGC Standards, Teddington, UK) as a cell model commonly used in cell-signalling studies (*see* **Note 1**).
2. Phosphate-buffered saline (PBS): Prepare 10× stock with 1.37 M NaCl, 27 mM KCl, 100 mM Na_2HPO_4, 18 mM KH_2PO_4 (adjust to pH 7.4 with HCl if necessary) and autoclave before storage at room temperature. Prepare working solution by dilution of one part with nine parts water (*see* **Note 2**).
3. Dulbecco's modified Eagle's medium (DMEM) (Gibco/BRL, Bethesda, MD) supplemented with 10% foetal bovine serum (FBS; HyClone, Ogden, UT), 100 units/ml penicillin and 100 μg/ml streptomycin (Invitrogen, Carlsbad, CA).
4. Urea-denaturing lysis buffer supplemented with protease and phosphatase inhibitors: 20 mM HEPES (pH 8.0), 8 M urea, 1 mM sodium vanadate, 1 mM sodium fluoride, 2.5 mM sodium pyrophosphate, 1 mM ß-glycerol phosphate. Prepare the stock lysis buffer (20 mM HEPES, pH 8.0, 8 M urea) and store in aliquots at −20°C. On the day of the experiment, supplement 1 ml of stock lysis buffer with the following: 2 μl of 0.5 N NaF, 10 μl of 100 mM Na_3VO_4, 10 μl of 250 mM sodium pyrophosphate and 1 μl of 1 M ß-glycerol phosphate (*see* **Notes 3** and **4**).
5. Internal standards: α-casein and ß-casein. Prepare stock solutions at 4 pmol/μl.
6. Reducing and alkylating solutions are 1 M dithiothreitol (DTT) and 415 mM iodoacetamide, respectively.
7. Bio-Rad protein assay solution (Bio-Rad, Hercules, CA) (*see* **Note 5**).

2.2. Cell Lysate Protein Digestion

1. Trypsinisation buffer: 20 mM HEPES (pH 8.0).
2. Trypsin digestion is performed using immobilised TLCK-trypsin (Thermo Fisher Scientific, Massachusetts, USA).
3. 10% Trifluoroacetic acid (TFA).

2.3. Solid-Phase Extraction (SPE) Desalting

1. SPE cartridges (Sep-Pack C_{18} columns; Waters UK Ltd, Manchester, UK).
2. SPE solutions: (A) Conditioning solution: 100% acetonitrile (ACN); (B) loading/washing solution: 0.1% TFA; (C) eluting solution: 50% ACN, 0.1% TFA.

2.4. SCX HPLC

1. SCX mobile phase A: 0.1% formic acid, 25% ACN.
2. SCX mobile phase B: 300 mM ammonium acetate, 25% ACN.
3. Polysulphoethyl A (Poly LC) SCX column 4.6 mm × 100 mm (PolyLC, Columbia, MD, USA).
4. HPLC system with fraction collector (1200 series; Agilent, Wokingham, UK).

2.5. Sequential Elution from Immobilised Metal Ion Affinity Chromatography (SIMAC)

2.5.1. Immobilised Metal Ion Affinity Chromatography

1. Ni(III)-coated Sepharose high-performance beads (GE Healthcare, Little Chalfont, UK).
2. IMAC chelating solution: 200 mM EDTA.
3. IMAC charging solution: 100 mM $Fe(III)Cl_3$.
4. IMAC loading/washing solution: 50% ACN, 0.1% TFA.
5. IMAC elution solutions: (A) 20% ACN, 1% TFA; (B) NH_4OH (pH 11.3) (prepare by mixing 940 μl of water and 60 μl of 25% NH_4OH; can be stored at −20°C); (C) NH_4OH (pH 11.3) prepared in 50% ACN.
6. 100% Formic acid (FA).

2.5.2. Titanium Dioxide (TiO_2) Chromatography

1. TiO_2 beads (Hichrom Ltd, Theale, UK).
2. 100% ACN.
3. 4 M Urea.
4. 1% Sodium dodecyl sulphate (SDS).
5. TiO_2 loading buffer: 80% ACN, 5% TFA, 1 M glycolic acid.
6. TiO_2 washing solutions: (A) 80% ACN, 5% TFA; (B) 10% ACN.
7. TiO_2 elution buffers: (A) NH_4OH (pH 11.3) (prepare as indicated in **2.5.1 step 5**); (B) 30% ACN.
8. 100% FA.

2.6. MS Identification

Peptide separation is performed by nanoflow reverse-phase chromatography using a C18 column in a high-pressure liquid chromatography (HPLC) system. In our example, we use a nanoflow ultrahigh-pressure liquid chromatograph (UPLC; Acquity, Waters/Micromass) connected on line to a quadrupole time-of-flight (Q-TOF) mass spectrometer (Waters/Micromass UK Ltd, Manchester, UK). In our example, we used a BEH C_{18} 100-μm × 100-mm column (Waters/Micromass UK Ltd, Manchester, UK).

1. LC-MS/MS mobile phases: solution A (0.1% FA in LC-MS grade water) and solution B (0.1% FA in LC-MS grade ACN).
2. Sample resuspension solution: 0.1% TFA.

3. Methods

Protein phosphorylation is a reversible and dynamic process that regulates essential biological functions in health and disease. The identification of phosphorylation sites can provide mechanistic information at the molecular level and serve as read-outs of pathway activities. In this sense, mass spectrometry combined with phosphopeptide enrichment methodologies has emerged as a powerful method for phosphoproteomic studies (9, 10).

Among the numerous phosphopeptide enrichment strategies available, IMAC and TiO_2 have become the most popular techniques due to their simplicity and great selectivity to isolate phosphopeptides. IMAC is based on electrostatic interactions between the positive charge of the immobilised metal ions (such as Fe(III) or Ga(III)) and the negative charge of phosphopeptides (11). Although non-specific recovery of peptides containing acidic amino acids represents a problem, pH/acid-controlled conditions raise the specificity of IMAC (12). TiO_2 instead has shown higher selectivity for phosphopeptides than IMAC, with lower unspecific binding, and a higher recovery of less acidic (mainly monophosphorylated) peptides (13, 14). These findings suggest that the complementarity of the two approaches may have potential for comprehensive phosphoproteomic profiling. This concept was applied by Thingholm and colleagues to develop a combined approach named 'sequential elution from IMAC' (SIMAC) (8), which is based on differential elution of phosphopeptides according to their acidity from the IMAC material using acidic and basic elution conditions. The less acidic fractions eluted from IMAC (containing a larger proportion of monophosphopeptides) are then further enriched using TiO_2 chromatography.

The SIMAC method produces three fractions per sample. This is often insufficient fractionation when the aim is to perform an in-depth analysis of the phosphoproteome. We propose therefore a combination of SCX sample fractionation followed by phosphopeptide enrichment by SIMAC in order to more com-

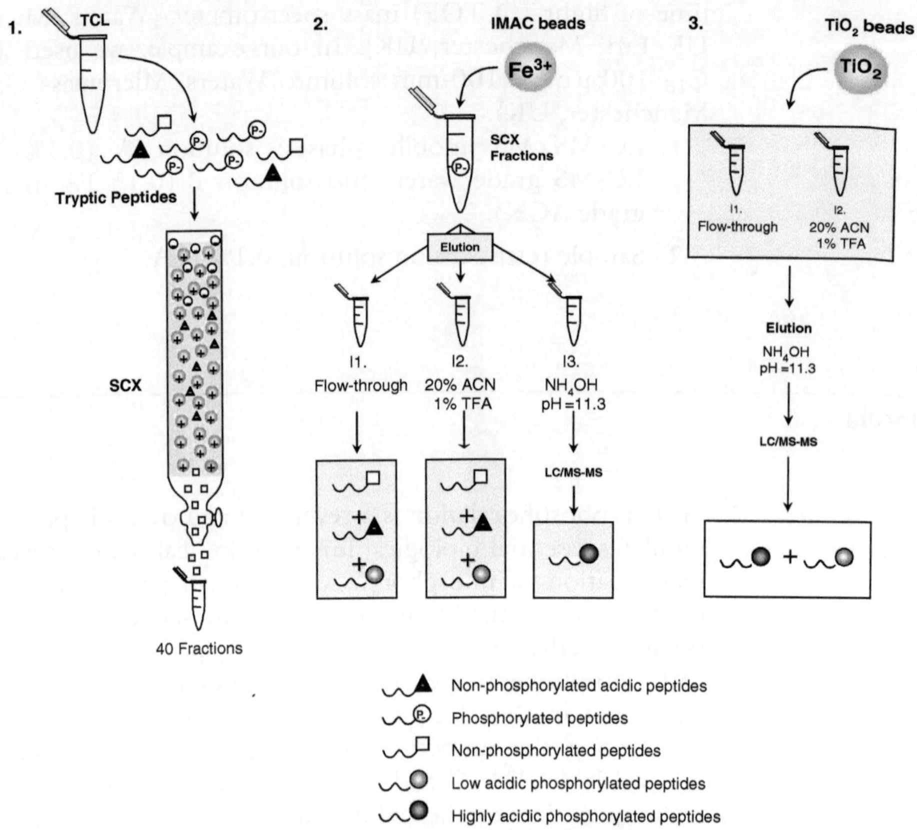

Fig. 6.1. Scheme for the enrichment of phosphopeptides using multidimensional chromatography.

prehensively identify phosphorylated peptides. Here we describe this protocol in detail (the analytical workflow is summarized in **Fig. 6.1**).

3.1. Sample Preparation for Large-Scale Phosphoproteomics

1. Murine NIH-3T3 fibroblasts are routinely cultured at 37°C in a humidified atmosphere at 5% CO_2 and grown in DMEM medium supplemented with 10% foetal bovine serum, 100 units/ml penicillin and 100 μg/ml streptomycin. Cells are maintained at 10–75% confluency (*see* **Note 6**).
2. Prior to the experiment, cells are seeded at about 20% confluency and cultured for 48 h when cells reach approx. 75% confluency (*see* **Note** 7). The use of at least 1×10^7 cells per sample is recommended for samples that are not going to undergo SCX fractionation and $\sim 5 \times 10^8$ cells when the aim is to perform an SCX separation step prior to further phosphopeptide enrichment. The current protocol is given for a sample containing 2×10^8 NIH-3T3.

3. The medium should be changed the day before the experiment.
4. On the day of the experiment, supplement the stock lysis buffer as indicated in **Section 2.1**, Step 4. PBS is also supplemented with 1 mM sodium vanadate and 1 mM sodium fluoride (1 ml of 100 mM sodium vanadate and 200 μl of 0.5 N sodium fluoride/100 ml of PBS). Both PBS and lysis buffer must be prechilled at 4°C before use and kept on ice (*see* **Note 8**).
5. Please note that cell harvesting should be performed on ice (*see* **Note 8**). For harvesting purposes, the DMEM is aspirated from the flask and cells are washed twice with the supplemented PBS. Scrape the cells off the flask at approx. 2×10^7 cells/ml of urea-denaturing lysis buffer. Stand on ice for 15 min to allow lysis to take place. Solubilise cellular proteins further by sonication (three pulses, half intensity) (*see* **Note 9**).
6. Clear the lysate from debris by centrifugation at 20,000×*g* for 10 min at 4°C. The protein concentration of the supernatants is determined using standard protein quantification procedures.
7. Optionally add 20 pmol of internal standards to each sample (*see* **Note 10**).
8. Reduce and alkylate cell lysate proteins with DTT at a final concentration of 10 mM and iodoacetamide at 120 mM, respectively (*see* **Note 11**). For 10 ml of cell lysate, add 100 μl of 1 M DTT and incubate for 45 min at 30°C. Proceed by adding 2,900 μl of 415 mM iodoacetamide and incubate for 45 min at room temperature in the dark.
9. Dilute protein extracts using 20 mM HEPES (pH 8.0) to a final concentration of 2 M urea (1:4 dilution) to allow protein digestion by trypsinisation. For this, add 30 ml of 20 mM HEPES (pH 8.0) to 10 ml lysate (*see* **Note 12**).
10. Perform trypsinisation using 300 TAME units of TLCK of trypsin/2×10^8 cells for 16 h at 37°C with vigorous shaking (*see* **Note 13**). Please note that trypsin beads must be washed three times with 20 mM HEPES before adding them to the samples.
11. Stop digestions by acidifying the samples with TFA at a final concentration of 1%.
12. Remove trypsin beads by spinning the samples at 2,000×*g* for 5 min.

13. At this stage, the samples can be frozen prior to desalting by solid-phase extraction (SPE).

3.2. Solid-Phase Extraction

1. The peptide solution resulting from the trypsinisation step is desalted using Sep-Pack C_{18} columns, or any other SPE cartridges (*see* **Note 14**).
2. Condition Sep-Pack C_{18} 1-ml columns using 10 ml SPE conditioning solution (100% ACN).
3. Equilibrate columns with 10 ml SPE loading/washing solution (0.1% TFA).
4. Load 40 ml sample onto four Sep-Pak C18 columns (*see* **Note 15**).
5. Wash the column with 10 ml SPE loading/washing solution (0.1% TFA).
6. Finally, bound peptides can be eluted with 5 ml SPE eluting solution (50% ACN, 0.1% TFA) (*see* **Note 16**).

3.3. Strong Cation Exchange High-Performance Liquid Chromatography (SCX HPLC)

Desalted peptides can be subjected to phosphopeptide enrichment as explained in **Section 3.4**. Alternatively, peptides can be separated into fractions by SCX HPLC prior to further phosphopeptide enrichment. This optional step reduces the sample complexity (at the expense of increasing LC-MS/MS running time) and hence results in a greater coverage of the phosphoproteome by providing an orthogonal separation step.

1. Connect the SCX column to a HPLC system and set the flow rate to 1 ml/min.
2. Condition the SCX column with SCX solvent B (300 mM ammonium acetate, 25% ACN) for 30 min.
3. Equilibrate SCX column with SCX solvent A (0.1% formic acid, 25% ACN) for at least 10 min and until a stable baseline on the UV trace is achieved.
4. Load the desalted peptides onto the equilibrated SCX column.
5. Apply gradient elution from 0 to 25% B for 20 min followed by 25–100% B for another 20 min. Collect 1-min fractions (40 fractions in total).
6. Combine fractions based on UV absorption to obtain 10 fractions.
7. Dry SCX fractions in speed-vac to remove organic solvents and redissolve in 5 ml SPE loading/washing solution (0.1% TFA).
8. Desalt SCX fractions by SPE as described in **Section 3.2**.

3.4. Sequential Elution from Immobilised Metal Ion Affinity Chromatography (SIMAC)

3.4.1. Immobilised Metal Ion Affinity Chromatography

3.4.1.1. Bead Conditioning

1. Prior to start of IMAC chromatography for phosphopeptide enrichment per se, the nickel from the Ni(III)-coated Sepharose beads must be replaced by Fe(III) (**Note 17**).
2. Take 300 μl of commercial beads (50% slurry) per sample to be enriched. The current protocol is given for one sample, i.e., 300 μl of beads (*see* **Note 18**). Spin them down at 1,000 × *g* for 2 min, discarding the supernatant that contains the preservative solution. Use these spinning conditions for all the following steps.
3. Wash the beads with 300 μl of water, spin down and keep the beads.
4. Incubate with 300 μl IMAC chelating solution (200 mM EDTA) for 5 min to release the nickel from the beads (*see* **Note 19**), spin down and keep the beads. Perform this step twice. The beads must change the colour, from blue to colourless.
5. Wash the beads with 300 μl IMAC loading/washing buffer (50% ACN, 0.1% TFA), spin down and discard the supernatant. Wash the beads three times.
6. Load the beads with 300 μl IMAC charging solution (100 mM Fe(III)Cl_3) for 5 min, centrifuge and keep the beads. Repeat the loading once again. The beads should become yellow (*see* **Note 20**).
7. Wash the beads with 300 μl IMAC loading/washing solution (50% ACN, 0.1% TFA), spin down and discard the supernatant. Wash the beads for six times.
8. The beads will be ready to use after resuspending them in 50% slurry with IMAC loading/washing buffer (50% ACN+ 0.1% TFA). For 300 μl of starting beads, one obtains 150 μl of pure beads, which are resuspended with 150 μl of loading buffer.

3.4.1.2. Phosphopeptide Enrichment

1. Incubate the desalted samples with 300 μl of conditioned IMAC beads for 30 min at room temperature (*see* **Note 21**).

2. Centrifuge the samples at 1,000×g for 5 min. Take aside the supernatant (unbound fraction) for further enrichment with TiO_2 beads.
3. Beads are now washed with 300 μl of IMAC loading/washing buffer (50% ACN+ 0.1% TFA).
4. Combine the unbound and washed fraction as the I1 fraction for later will TiO_2 enrichment (*see* **Note 22**).
5. For the elution of non-phosphorylated acidic peptides and less acidic phosphopeptides (mainly monophosphopeptides), treat beads with 300 μl IMAC elution solution A (20% ACN, 1% TFA) for 5 min. Spin down and keep the supernatant as the I2 fraction. This will also be further enriched with TiO_2 beads (*see* **Note 23**).
6. Highly acidic phosphopeptides, enriched in multiphosphopeptides (I3 fraction), are obtained by eluting with 300 μl IMAC elution solution B (ammonia water, pH 11.3, prepared as indicated in **Section 2.4**) for 5 min. The supernatant obtained after centrifugation constitutes the I3 fraction. Repeat this step twice and combine the two ammonia water elutions.
7. Wash the beads with 50 μl of IMAC elution buffer C (ammonia water, pH 11.3, containing 50% ACN). Combine the resulting supernatant with the previous ammonia elutions rendering the final I3 fraction (*see* **Note 24**). At this stage the beads can be discarded.
8. Ammonia fraction (I3) should be acidified with 10% FA final concentration. In our example, we will use 65 μl 100% FA for I3 acidification.
9. Lyophilise I2 and I3 fraction.
10. Resuspend the I3 fraction in 10 μl of 0.1% TFA for analysis by LC-MS.

3.4.2. Titanium Dioxide (TiO_2) Chromatography

1. Resuspend the TiO_2 beads with 100% ACN to 50% slurry. Use 10 μl of beads per sample (*see* **Note 25**).
2. Spin the beads down and remove the ACN supernatant.
3. Wash the beads first with 200 μl of water and then twice with 200 μl TiO_2 loading buffer (1 M glycolic acid in 80% ACN, 5% TFA) (*see* **Note 26**).
4. Resuspend I2 fraction with 2 μl 4 M urea, 3 μl 1% SDS and 200 μl TiO_2 loading buffer.
5. Incubate factions I1 and I2 with 10 μl TiO_2 beads for 30 min at room temperature. After this time, centrifuge the samples at 1,000×g for 5 min and discard the supernatant.

6. Wash beads sequentially with 200 μl TiO_2 loading buffer, then with 200 μl TiO_2 washing solution A (80% ACN/5% TFA) and finally with 200 μl TiO_2 washing solution B (10% ACN). Discard the resulting supernatant in each case.
7. Elution of bound phosphopeptides is achieved by incubating the beads with 50 μl TiO_2 elution solution A (ammonia water, pH 11) (*see* **Section 2.4**) for 5 min, spinning and collecting the supernatants (*see* **Note 27**).
8. Beads are subjected to a second elution round with TiO_2 elution solution B (*see* **Note 28**). Combine this elution with the previous one to obtain the final phosphopeptide-enriched sample.
9. Acidify eluates with 10% FA final concentration, lyophilise and resuspend in 10 μl 0.1% TFA for analysis by LC-MS/MS (*see* **Note 29**).

3.5. Identification of Phosphopeptides by LC-MS

1. The identification of phosphopeptides by mass spectrometry can be performed using, for example, a quadrupole time-of-flight (Q-TOF) mass spectrometer connected on line to a nanoflow ultrahigh-pressure liquid chromatograph.
2. HPLC separations can be done using a BEH reverse-phase column. The suggested mobile phases are solution A (0.1% FA in LC-MS grade water) and solution B (0.1% FA in LC-MS grade ACN).
3. The recommended gradient run is from 1% B to 35% B in 100 min followed by a 5-min wash at 85% B and a 7-min equilibration step at 1% B (*see* **Note 30**).
4. Perform a mass spectrometry data-dependent analysis (DDA) in which the three most abundant multiply charged ions present in the survey spectrum are automatically mass selected and fragmented by collision-induced dissociation in each cycle. MS scans of 500 ms, followed by three MS/MS scans of 1 s each are suggested to be acquired within a mass range of 50–2,000 m/z (*see* **Note 31**).
5. The phosphopeptide sequences can be identified by loading the MS/MS data files in a protein search engine, such as Mascot, Seaquest and Phoenix, among others.

4. Notes

1. This assay can be performed with any cell line to be studied. Here we exemplify the approach by using the murine NIH-3T3 fibroblasts as a cell model commonly used in cell-signalling studies.

2. Routine solutions should be prepared in water that has a resistivity of 18 MΩ cm, unless otherwise stated. The solutions used for both solid-phase extraction and SIMAC protocols, along with all the solvents used in the LC-MS system, must be LC-MS grade. LC-MS grade solvents used here were sourced from LGC Promochem (Middlesex, UK), but any other LC-MS grade solvents should be adequate.
3. The lysis buffer used in this protocol is urea based. Urea acts as a denaturing agent at high concentrations such as that used in the current lysis buffer (8 M). When the lysis buffer is mixed with the cells, the urea denatures the proteins disrupting the cellular structures and acting as a lysing agent. Due to its denaturing nature, urea also protects phosphopeptides from protease degradation and phosphatases activities present in the biological samples. Please note that this compound is toxic; hence one must be extremely careful when handling it.
4. For this type of experiment, it is essential that the lysis buffer contains different protease and phosphatase inhibitors such as Na_3VO_4 and NaF (tyrosine phosphatase inhibitors), sodium pyrophosphate and ß-glycerol phosphate (phosphatase inhibitors of broad specificity). This will preserve the phosphosites of the cell lysate proteins.
5. Any other reagent for protein quantification may be suitable.
6. The murine NIH-3T3 fibroblasts used in this example should not be maintained in culture for a long number of passages (maximum 25–30 passages) and should be kept between 10 and 80% confluency. This is due to the high transformation potential of this cell line. When culturing NIH-3T3 at 100% confluency and/or for long passages, they stop behaving like normal fibroblast and become transformed.
7. Cells are harvested at 75% confluency when they are actively dividing and when proliferative signalling networks are activated. This favours the study of kinase activity differences between experimental conditions.
8. Phosphopeptides are very sensitive to degradation by proteases and phosphatases present in the sample. For this reason, when performing phosphoproteomic experiments, it is very important to use prechilled solutions and to keep the sample on ice unless otherwise stated. This will reduce any phosphatase and protease activity that may remain in the sample despite the treatment with urea and inhibitors.

9. In order to obtain a maximum coverage of the phosphoproteome, it is essential to make sure that the proteins located in cellular membranes and nuclei are released and solubilised. This is easily solved by sonicating cell lysates as indicated.
10. Internal standards are phosphoproteins that can be added to the sample at a known concentration. This will allow us to normalise and correct for differences introduced during sample handling.
11. By reducing the cell lysate proteins with DTT, the disulphide bonds are broken. However, in order to prevent disulphide bond reformation, they must also be alkylated with reagents such as iodoacetamide. This step is introduced to facilitate the enzymatic protein digestion, in our example by trypsin.
12. Trypsinisation cannot be performed at 8 M urea due to its denaturing effect. Thus, before trypsinisation, the concentration of urea must be diluted to 2 M or less to preserve trypsin structure and hence function.
13. The use of immobilised TLCK/trypsin has the advantage that it can be removed from the sample after trypsinisation. However, it is very important to ensure vigorous shaking during incubation with trypsin to maintain beads in suspension, otherwise trypsin efficiency is dramatically reduced.
14. Prior to phosphopeptide enrichment, it is essential to remove the added salts since these may interfere with subsequent ion exchange chromatographic steps. For this purpose, we need to desalt the samples by SPE.
15. In this step, the peptides from the sample are retained in the Sep-Pack reverse-phase column, while inorganic salts are not retained and washed away.
16. The Sep-Pack eluting solution is the loading solution for the downstream enrichment step. Therefore, the Sep-Pack eluent can be directly loaded onto the IMAC beads.
17. The principle of the IMAC phosphopeptide enrichment protocol is based on ionic interactions between the negatively charged phosphate groups of such peptides and the positively charged metal ions coating the IMAC beads. One of the most widely used IMAC methods for phosphopeptide enrichment uses Fe^{3+}-coated beads; however other cations such as Ga^{3+} and Al^{3+} have been reported as well (4).
18. In our experience, 300 μl IMAC beads (50% slurry) represent the optimal amount of beads for our specific experimental conditions. However, it is strongly recommended

to optimise the amount of beads needed for efficient binding of phosphopeptides when studying different cell lines.

19. Ethylenediaminetetraacetic acid (EDTA) is a chelating agent that acts by sequestering di- and tricationic metal ions. Thus, after EDTA treatment, the Ni^{3+} is released from the Sepharose beads, which can then be charged with Fe^{3+} ions.
20. Because we have not tested the stability of the Fe^{3+}-conditioned beads, we suggest using them fresh after loading the iron.
21. Fe(III)-IMAC chromatography has been shown to have higher affinity for multiply phosphorylated peptides than for monophosphorylated peptides, which actually represent the vast majority of phosphorylated peptides (14). In contrast, TiO_2 chromatography has proven to be more efficient at yielding monophosphopeptides. The SIMAC approach developed by Thingholm et al. takes advantage of both techniques in order to obtain a broader coverage of the whole phosphoproteome (8).
22. The first fraction obtained from the IMAC beads (I1) mainly contains non-phosphorylated peptides and some less acidic phosphopeptides (mainly monophosphorylated peptides) that did not bind to the IMAC beads during the incubation (8). This faction will later be further enriched in phosphopeptides using TiO_2 beads.
23. Since acidic conditions have been shown to preferentially elute less acidic phosphorylated peptides from the IMAC beads, by using 20% ACN, 1% TFA, one recovers a fraction (I2) enriched in monophosphorylated peptides along with most of the acidic peptides that also bind to the IMAC beads in an unspecific fashion (8). This fraction will later be further enriched in phosphopeptides using TiO_2 beads.
24. This last elution with an organic solvent allows recovery of any phosphorylated peptides that may remain bound to the IMAC beads due to hydrophobic interactions.
25. TiO_2 chromatography has been shown to have higher selectivity and capacity for phosphopeptides than does IMAC (9). Thus, the volume of beads needed to perform the phosphopeptide enrichment is low in comparison.
26. Besides the high selectivity of TiO_2 chromatography for phosphopeptides, these beads allow the use of higher concentrations of TFA and glycolic acid, both of which further reduce the unspecific binding of non-phosphorylated acidic peptides (1.17 and 1.18). The increased concentration of TFA reduces the pH of the solution, neutralising the charge of non-phosphorylated acidic peptides and

hence avoiding their interaction with the Titanium. Due to the acidic properties of the glycolic acid, this compound competes with the non-phosphorylated acidic peptides for the Titanium binding but it does not compete with the phosphopeptides owing to their higher acidity.

27. The retained phosphorylated peptides are released from the TiO_2 beads by using alkaline buffers such as ammonia water (pH 11.3).
28. This last organic elution allows recovery of any phosphorylated peptides that may remain bound to the TiO_2 beads due to hydrophobic interactions.
29. It is recommended to include in the run one blank sample just containing the digested internal standards (20 pmol α- and β-casein). This will enable computing the percentage of phosphopeptide recovery.
30. Due to the complexity of the sample, we suggest to use a 100-min gradient. This will result in a good separation of the peptides, minimising the problem of MS undersampling at least to some extent.
31. There are different fragmentation techniques that can be used for phosphopeptide identification. In this sense, electron capture/transfer dissociation (ECD/ETD) has proven to be an efficient alternative to collision-induced dissociation (CID). Phosphopeptide fragmentation by CID often results in phosphate neutral loss, which sometimes causes a poor backbone fragmentation and hence low probability of phosphopeptide identification. This problem can be addressed by acquiring MS^3 spectra of the neutral loss product ions in cases where ion trap instruments are available. As an alternative, ECD/ETD fragmentation primarily fragments the peptide backbone chain with no phosphate loss, which in some cases results in an increased probability of detecting phosphopeptides and determining their precise phosphorylation site (10). However, CID may produce better fragmentation spectra than may ECD/ETD for low charge phosphopeptides; therefore, these two fragmentation techniques are complementary.

References

1. Manning, G., Whyte, D. B., Martinez, R., Hunter, T., and Sudarsanam, S. (2002) The protein kinase complement of the human genome. *Science* **298**, 1912–1934.
2. Bodenmiller, B., Malmstrom, J., Gerrits, B., Campbell, D., Lam, H., Schmidt, A., Rinner, O., Mueller, L. N., Shannon, P. T., Pedrioli, P. G., Panse, C., Lee, H. K., Schlapbach, R., and Aebersold, R. (2007) PhosphoPepa phosphoproteome resource for systems biology research in *Drosophila* Kc167 cells. *Mol. Syst. Biol.* **3**, 139.
3. de la Fuente van Bentem, S., Mentzen, W. I., de la Fuente, A., and Hirt, H. (2008)

Towards functional phosphoproteomics by mapping differential phosphorylation events in signaling networks. *Proteomics* **8**, 4453–4465.
4. Paradela, A., and Albar, J. P. (2008) Advances in the analysis of protein phosphorylation. *J. Proteome Res.* 7, 1809–1818.
5. Olsen, J. V., Blagoev, B., Gnad, F., Macek, B., Kumar, C., Mortensen, P., and Mann, M. (2006) Global, in vivo, and site-specific phosphorylation dynamics in signaling networks. *Cell* **127**, 635–648.
6. Trinidad, J. C., Specht, C. G., Thalhammer, A., Schoepfer, R., and Burlingame, A. L. (2006) Comprehensive identification of phosphorylation sites in postsynaptic density preparations. *Mol. Cell. Proteomics* **5**, 914–922.
7. Villen, J., and Gygi, S. P. (2008) The SCX/IMAC enrichment approach for global phosphorylation analysis by mass spectrometry. *Nat. Protoc.* **3**, 1630–1638.
8. Thingholm, T. E., Jensen, O. N., Robinson, P. J., and Larsen, M. R. (2008) SIMAC (sequential elution from IMAC), a phosphoproteomics strategy for the rapid separation of monophosphorylated from multiply phosphorylated peptides. *Mol. Cell. Proteomics* 7, 661–671.
9. Smith, J. C., and Figeys, D. (2008) Recent developments in mass spectrometry-based quantitative phosphoproteomics. *Biochem. Cell Biol.* **86**, 137–148.
10. Schreiber, T. B., Mausbacher, N., Breitkopf, S. B., Grundner-Culemann, K., and Daub, H. (2008) Quantitative phosphoproteomicsan emerging key technology in signal-transduction research. *Proteomics* **8**, 4416–4432.
11. Andersson, L., and Porath, J. (1986) Isolation of phosphoproteins by immobilized metal (Fe^{3+}) affinity chromatography. *Anal. Biochem.* **154**, 250–254.
12. Tsai, C. F., Wang, Y. T., Chen, Y. R., Lai, C. Y., Lin, P. Y., Pan, K. T., Chen, J. Y., Khoo, K. H., and Chen, Y. J. (2008) Immobilized metal affinity chromatography revisited: pH/acid control toward high selectivity in phosphoproteomics. *J. Proteome Res.* 7, 4058–4069.
13. Pinkse, M. W., Uitto, P. M., Hilhorst, M. J., Ooms, B., and Heck, A. J. (2004) Selective isolation at the femtomole level of phosphopeptides from proteolytic digests using 2D-NanoLC–ESI–MS/MS and titanium oxide precolumns. *Anal. Chem.* 76, 3935–3943.
14. Bodenmiller, B., Mueller, L. N., Mueller, M., Domon, B., and Aebersold, R. (2007) Reproducible isolation of distinct, overlapping segments of the phosphoproteome. *Nat. Methods* **4**, 231–237.

Chapter 7

Mapping of Phosphorylation Sites by LC-MS/MS

Bertran Gerrits and Bernd Bodenmiller

Abstract

Reversible protein phosphorylation ranks among the most important post-translational modifications that occurs in the cell. It is therefore highly relevant to elucidate the phosphorylation states of a given biological system, albeit challenging. Most notably the often low stoichiometry of phosphorylation is inherently incompatible with standard LC-MS analysis of a complex protein digest mixture, primarily due to the relative low dynamic range of current mass analyzers. Therefore a need for specific enrichment of phosphorylated peptides or proteins exists. Significant progress surrounding the biochemical analysis of reversible protein phosphorylation in the past years has led to the development of several new techniques to isolate or enrich phosphopeptides, particularly in large-scale analyses. This chapter deals with three such examples.

Key words: Phosphoprotein, phosphopeptide, phosphoproteomics, IMAC, TiO_2, PAC, phosphopeptide isolation, mass spectrometry.

1. Introduction

The procedure of identifying and characterizing proteinphosphorylations can be roughly divided into six steps, namely (i) sample preparation and protein extraction, (ii) preservation of the phosphorylation state of the biological sample (*see* **Note 1**), (iii) generation of MS-amenable molecules, (iv) isolation of phosphopeptides, (v) acquisition of mass spectrometric data, and (vi) identification of phosphopeptides and phosphosites using database searching. There are three common phosphopeptide isolation methods, namely affinity based methods either using a titanium dioxide (TiO_2) resin or immobilized metal affinity chromatography (IMAC) (1), titanium dioxide (TiO_2) (2, 3), and phosphoramidite chemistry (PAC) (4–6). These methods and

P.R. Cutillas, J.F. Timms (eds.), *LC-MS/MS in Proteomics*, Methods in Molecular Biology 658,
DOI 10.1007/978-1-60761-780-8_7, © Springer Science+Business Media, LLC 2010

variants thereof will detect different, partially overlapping parts of the phosphoproteome (7), meaning that no method can give a complete overview, but rather a distinct subset of the biological state. The last section of the chapter deals with some general aspects of the mass spectrometric acquisition of phosphopeptide samples plus a few directional hints to the evaluation and data analysis.

2. Materials

2.1. IMAC

1. Sample/wash/equilibration solution: 30% acetonitrile 250 mM acetic acid solution.
2. Optional: methanolic HCl: Prepare methanolic HCl by adding 160 μL of acetyl chloride to 1 mL of anhydrous methanol.
3. Phosphate buffer A: Prepare 50 mM phosphate buffer (pH 8.9) by dissolving 0.0063 g monosodium phosphate monohydrate and 1.3 g disodium phosphate heptahydrate in 100 mL water (*see* **Note 2**).
4. Phosphate buffer B: Prepare 100 mM phosphate buffer (pH 8.9) by dissolving 0.0126 g monosodium phosphate monohydrate and 2.66 g disodium phosphate heptahydrate in 100 mL water.
5. 1 M Hydrochloric acid (HCl) solution.
6. PHOS-Select^{TM} gel (Sigma-Aldrich).
7. Mobicol spin column (MoBiTec, Göttingen, Germany).
8. C18 (ultra) microspin columns (Harvard Apparatus Ltd, Edenbridge, United Kingdom).

2.2. TiO_2

1. Sample-/resin-/wash solution: saturated phthalic acid solution: Dissolve 125 mg of phthalic acid in 1 mL of 80% acetonitrile and 3.5% trifluoroacetic acid (TFA). Some phthalic acid will not dissolve.
2. Wash solution 1: 0.1% TFA solution in water.
3. Wash solution 2: 80% acetonitrile (ACN), 0.1% TFA in water.
4. Ammonium hydroxide solution: 0.3 M ammonium hydroxide (NH_4OH) solution.
5. Titanium dioxide resin (GL Science, Saitama, Japan).
6. Mobicol spin column (MoBiTec, Göttingen, Germany).

2.3. PAC

For the isolation of phosphopeptides using the phosphoramidite chemistry (PAC) method, a number of chemicals are required.

Please note the subdivision between the different packing materials, glass beads, and dendrimers.

1. Bio-Spin column (Bio-Rad).
2. Anhydrous *N*,*N*-dimethylformamide, 99.8% (Sigma-Aldrich).
3. Aminopropyl CPG beads (Proligo).
4. *N*-Hydroxybenzotriazole (HoBT) (Nova Biochem).
5. 3-Maleimidopropionic acid, 97% (Sigma-Aldrich).
6. *N*,*N*-Diisopropylcarbodiimide, 99% (Sigma-Aldrich).
7. Anhydrous dichloromethane, ≥99.8% (Sigma-Aldrich).
8. Ninhydrin reagent solution (Sigma-Aldrich).
9. Acetyl chloride, ≥99% (Sigma-Aldrich).
10. Anhydrous methanol, 99.8% (Sigma-Aldrich).
11. Imidazole (Sigma-Aldrich).
12. Anhydrous acetonitrile, 99.8% (Sigma-Aldrich).
13. *N*-(3-Dimethylaminopropyl)-*N′*-ethylcarbodiimide hydrochloride (EDC) (Sigma-Aldrich).
14. Cystamine (Sigma-Aldrich).
15. MES, low moisture content, ≥99% (titration) (Sigma-Aldrich).
16. Empty spin columns (MoBiTec).
17. Filter for spin columns (MoBiTec).
18. Glass tubes (Waters).
19. Trifluoroacetic acid (TFA) (Pierce).
20. Tris(2-carboxyethyl) phosphine hydrochloride (TCEP) (Pierce).
21. For 1 mg of peptide: Sep-Pak® Vac C18 cartridge (Waters).
22. For micrograms of peptide: MicroSpin™ Columns (Nest Group).

3. Methods

3.1. Isolation of Phosphopeptides Using IMAC

The isolation of phosphopeptides using IMAC can be done on peptides with and without methyl esterification. Methyl esterification reduces the amount of non-specific binding of non-phosphopeptides. However, as a first step, choose the phosphopeptide enrichment without the methyl esterification. Please note that the ratio of peptide to resin (7) as well as the correct pH is critical (*see* **Note 3**) in the binding step.

3.1.1. Preparation of the Sample Solution

Without methyl esterification prior to the isolation: Reconstitute 1.0 mg of lyophilized peptides in 30% ACN, 250 mM acetic acid at pH 2.7 and mix with 30 μL equilibrated PHOS-Select™ gel in a blocked Mobicol spin column as described in **Section 3.1.2**.

With methyl esterification (see **Note 4***) prior to the isolation*: Reconstitute 1.0 mg of lyophilized peptide in 500 μL of methanolic HCl. Allow the methyl esterification to proceed for 120 min at 12°C. Remove the solvent immediately under vacuum and dry conditions (*see* **Note 5**). Reconstitute the dry, methylated peptides in 30% ACN, 250 mM acetic acid at pH 2.7 and mix with 30 μL equilibrated PHOS-Select™ gel in a blocked Mobicol spin column as described in **Section 3.1.2**.

3.1.2. Detailed Protocol: IMAC

1. Preparation of samples obtained from a protein tryptic digest (*see* **Note 6**).
2. Wash/equilibration of affinity gel: Carefully mix the PHOS-Select Iron Affinity Gel beads until they are completely and uniformly suspended.
3. Immediately add 60 μL of the 50% slurry (30 μL of gel) to the Mobicol spin column. To dispense beads, use a wide orifice pipette tip or cut approximately 5 mm off the end of a regular pipette tip to allow unrestricted flow of the bead suspension.
4. Add 300 μL of wash/equilibration solution (250 mM acetic acid with 30% ACN) to the tube, vortex, and centrifuge in a microcentrifuge for 30 s at 2,200×*g*.
5. Discard the flow-through liquid. Repeat Step 4 twice.
6. Sample loading: Place an end cap onto the column outlet and place the column in a new collection tube. Add up to 350 μL of the sample solution to the equilibrated gel (*see* **Section 3.1.1** and **Note 7**).
7. Incubate for 120 min with mixing (end-over-end rotation recommended).
8. After incubation, remove the end cap and centrifuge as described in Step 4. The flow-through liquid in the collection tube contains unbound peptides, which may be pooled with the washes from Step 9 and analyzed further.
9. Affinity gel wash: Place the column in the collection tube saved from Step 8.
10. Add 300 μL of wash/equilibration solution (250 mM acetic acid with 30% ACN) to the tube, vortex, and centrifuge in a microcentrifuge for 30 s at 2,200×*g*.
11. The flow-through liquid in the collection tube contains unbound peptides, which may be discarded or pooled with the flow-through liquid from Step 8 and analyzed further. Repeat the washing step once.

12. Affinity gel wash: Wash the gel once with 300 μL of water to remove any residual wash/equilibration solution prior to elution.
13. Sample elution: Place an end cap onto the column outlet and place the column in a new collection tube.
14. Add 150 μL of 50 mM phosphate buffer A (pH 8.9) and incubate for 3 min with shaking.
15. Elute by centrifugation and retain the flow-through. Repeat the elution step once with 150 μL of 100 mM phosphate buffer B (pH 8.9).
16. Purify phosphopeptides using Harvard (Ultra) microspin columns according to manufacturer's instructions and analyze as soon as possible by LC-MS.

3.2. Isolation of Phosphopeptides Using TiO_2

Please note that the ratio of peptide to resin (7) as well as the correct pH is critical (*see* **Note 3**) in the binding step.

1. Place 1.25 mg of TiO_2 resin in a Mobicol spin column.
2. Wash the 1.25 mg TiO_2 resin with 200 μL H2O.
3. Wash the 1.25 mg TiO_2 resin with 200 μL methanol.
4. Equilibrate 1.25 mg TiO_2 resin with 300 μL phthalic acid solution.
5. Load 1–2 mg peptide dissolved in 300 μL phthalic acid solution onto the resin and incubate for 30 min with end-over-end rotation.
6. Wash the resin twice with 300 μL phthalic acid solution.
7. Wash the resin twice with 300 μL of 80% ACN and 0.1–0.5% TFA solution.
8. Wash the resin twice with 300 μL of 0.1–0.5% TFA solution.
9. Elute phosphopeptides twice with 150 μL of 0.3 M ammonium hydroxide solution.
10. Adjust pH of eluent to 2.7 and clean phosphopeptides using reverse-phase column (microspin columns; Nest Group).
11. Dry sample under vacuum and analyze as soon as possible by LC-MS.

3.3. Isolation of Phosphopeptides Using PAC: Maleimide-Containing Solid Phase

3.3.1. Synthesizing Beads

1. Take 120 μmol HoBT and 120 μmol 3-maleimidopropionic acid and dissolve in 0.8 mL dry DMF.

2. Add 120 μmol diisopropylcarbodiimide to make the link solution.
3. Place 100 mg amino glass beads (5 mg good for 1 phosphopeptide isolation) into washing columns.
4. Wash with two volumes of DMF.
5. Add the link solution to the beads, close the column, and incubate for 90 min at RT with end-over-end rotation.
6. Wash beads three times with DMF and then two times with CH_4Cl_2.
7. Resuspend in CH_4Cl_2, transfer to tube, and remove the supernatant.
8. Dry under vacuum.

3.3.2. Phosphopeptide Isolation

The following is for 1–3 mg peptide starting material. For smaller starting amounts, the columns and volumes must be adjusted accordingly.

1. Add 160 μL dry acetyl chloride to 1 mL dry anhydrous methanol (methanolic HCl).
2. Reconstitute 1 mg dry peptides in 500 μL–1 mL methanolic HCl (for less peptide use, corresponding amount of methanolic HCl).
3. Incubate for 120 min at 12°C under dry conditions (methyl esterification).
4. Dry the sample quickly under vacuum in a cold Speedvac.
5. Dissolve the sample in 40 μL methanol, 40 μL water, and 80 μL ACN prior to derivatization.
6. Place 5 mg EDC (final concentration of 50 mM) and 225 mg cystamine (final concentration of 2 M) into an Eppendorf tube.
7. Add imidazole to 50 mM and MES to 100 mM.
8. Adjust pH to 5.8 with micro pH meter and fill up with water to 500 μL.
9. Add 250 μL reaction solution to the peptides, verify that pH is between 5.5 and 5.8.
10. Shake vigorously for 8 h at room temperature. Do not exceed this time as otherwise the formation of side products will significantly increase.
11. Wash Sep-Packs columns (use appropriate column) with one volume of methanol, then 80% ACN, then 0.1% TFA.
12. Acidify derivatized phosphopeptides to pH 2.7.
13. Load derivatized peptides onto spin column.
14. Load the flow-through once more.

15. Wash once with 0.1% TFA.
16. Add 2 mL of 10 mM TCEP in 100 mM phosphate buffer (pH 6.0); the pH is crucial!
17. Press halfway through, incubate for 8 min at room temperature.
18. Wash with 0.1% TFA.
19. Elute with 750 μL of 80% ACN, 0.1% TFA.
20. Add 100 μL phosphate buffer, check and adjust pH to 6.3 and partly remove ACN to a final of 30% in a speedvac device.
21. Take 5–10 mg of prepared maleimide beads, place in blocked Mobicol column.
22. Add solution with the derivatized peptides (pH 6.3) to the beads.
23. Incubate for 1 h at room temperature.
24. Wash the beads under shaking two to three times with 280 μL each of 3 M NaCl, water, methanol, and 80% ACN.
25. Resuspend the beads in 80% ACN and transfer to glass tube.
26. Remove excess ACN (supernatant).
27. Incubate for 1 h in 10% TFA and 30% ACN (phosphopeptide recovery).
28. Spin tube down and transfer supernatant to new glass tube.
29. Add 40% ACN to beads, vortex, spin down, and take out the supernatant.
30. Add 40% ACN to beads, vortex, spin down, and take out the supernatant.
31. Pool all supernatants.
32. Dry under vacuum.
33. Resuspend in 0.1% TFA and 0.005% phosphoric acid for MS analysis.

3.3.3. Isolation of Phosphopeptides Using PAC: Dendrimer

This section describes the isolation of phosphopeptides also using phosphoramidite chemistry. Instead of derivatized glass beads, a dendrimer is used to capture the phosphopeptides. Methylation of peptides can be performed as described in **Section 3.1.1**. Before use, adjust pH of dendrimer to 6. Dry under vacuum and redissolve in reaction solution.

1. The methylated peptide sample is dissolved in a solution of 40 μL of methanol, 40 μL of water, and 80 μL of acetonitrile.
2. Dissolve peptide methyl esters in 250 μL of a reaction solution containing 50 mM EDC, 50 mM imidazole (pH 6.5),

100 mM MES (pH 6.0), and 1 M polyamine such as polyallylamine or PAMAM dendrimer generation 5.

3. Adjust pH to 5.5, if necessary, with 1 M NaOH.
4. Shake vigorously for 12–18 h at room temperature.
5. Transfer the solution to a Biomax filter device with molecular weight of 10,000 Da.
6. In the following steps, mix the polymer/dendrimer well to avoid unspecific binding of non-phosphopeptides.
7. Wash three times with 500 μL of 3 M NaCl and discard the flow-through.
8. Wash two times with 10% methanol and discard the flow-through.
9. Wash two times with water and discard the flow-through.
10. Add 10% TFA for 30 min to recover phosphopeptides.
11 Wash the polymer with 10% methanol several times and collect the flow-through.
12. Dry the collected flow-through under vacuum.
13. Resuspend for LC-MS analysis.

3.4. Identification of Phosphopeptides from LC-MS Data by Database Searching

There are many workflows possible based on different search engines like Mascot (8) or Sequest (9) and different mass spectrometric platforms to identify phosphopeptides, but most have the following in common. After phosphopeptide enrichment, peptide fractions are separated prior to MS analysis on a reverse-phase-based column. Typical characteristics of this setup are packing material of C18, inner diameter of the column 75–125 μm, particle size 2–5 μm, and an ACN/water/formic acid solvent system. Many large-scale phosphoproteomic projects (10–12) have acquired their data on Fourier transform-based mass spectrometers such as the LTQ-Orbitrap (13).

Both phosphopeptide identification and site determination are affected by frequent losses of phosphoric acid from the peptide, particularly in ion trap CID fragmentation spectra resulting in a dominant pseudo-neutral loss peptide ion. Often this results in low intensity of the peptide fragment ions. To combat this phenomenon, several mass spectrometric methods have been developed, such as MS^3 scans for neutral loss-containing MS^2 ions and multi-stage activation to further fragment these same ions. However, it has been recently shown that with high mass accuracy mass spectrometers, this is no longer beneficial (14).

After acquisition of the mass spectrometric data, the raw data are converted to an appropriate format for the specific search engine. It has to be noted here that data filtering can have a dramatic influence on the search output. For a phosphopeptide search, variable (or dynamic) modifications have to be set for the residues researched. Typically, these modifications are set only for serine (S), threonine (T), and tyrosine (Y), but phosphorylations have also been reported on aspartic acid (D) (15), histidine (H) (16), cysteine (C) (17), and arginine (R) (18), albeit less commonly.

Although identifying phosphopeptides using database search engines is rather straightforward with high accuracy mass spectrometers, the assignment of the phosphorylation site is in most cases not trivial. This problem occurs when there are multiple residues that can be phosphorylated in the peptides sequence assigned. For this ambiguity to be resolved, fragment ions unique to that phosphosite must be detected. Manual interpretation of tandem mass spectra is often biased and not feasible for large data sets per definition. An eloquent way to remove this bias is to automatically calculate the so-called ambiguity score (Ascore) (19). This score is a measure for the likelihood of matching a difference of at least the number of matched phosphopeptide-specific fragment ions by chance from the top two candidate sites as determined by the search algorithm. An Ascore of 20 ($P=0.01$) should then result in the site being localized with a 99% certainty.

4. Notes

1. In order to preserve the endogenous phosphorylation state of the biological system, it is recommended to add phosphatase inhibitors to the lysis buffer (17).
2. Water in this text is referred to as HPLC-grade water.
3. Underloading of the gel bead results in the co-purification of non-phosphorylated peptides, therefore use the correct peptide to resin ratio.
4. If you want to perform quantification, use D_0–D_3-methanol for the methyl esterification.
5. Drying under vacuum, lyophilizing under centrifugation, vacuum concentration, and Speedvac are used throughout the text and refer to the same technique.
6. Phosphate, phospholipids, DNA, RNA, high salt, and EDTA disturb the phosphopeptide isolation by IMAC and TiO_2.
7. *Important*: Make sure that the pH is between 2.5 and 3. If necessary, adjust the pH.

References

1. Andersson, L., and Porath, J. (1986) Isolation of phosphoproteins by immobilized metal (Fe-3+) affinity-chromatography. *Anal. Biochem.* **154**, 250–254.
2. Larsen, M. R., Thingholm, T. E., Jensen, O. N., Roepstorff, P., and Jorgensen, T. J. (2005) Highly selective enrichment of phosphorylated peptides from peptide mixtures using titanium dioxide microcolumns. *Mol. Cell. Proteomics* **4**, 873–886.
3. Pinkse, M. W. H., Uitto, P. M., Hilhorst, M. J., Ooms, B., and Heck, A. J. R. (2004) Selective isolation at the femtomole level of phosphopeptides from proteolytic digests using 2D-nanoLC–ESI–MS/MS and titanium oxide precolumns. *Anal. Chem.* **76**, 3935–3943.
4. Zhou, H., Watts, J. D., and Aebersold, R. A (2001) Systematic approach to the analysis of protein phosphorylation. *Nat. Biotechnol.* **19**, 375–378.
5. Tao, W. A. et al. (2005) Quantitative phosphoproteome analysis using a dendrimer conjugation chemistry and tandem mass spectrometry. *Nat. Methods* **2**, 591–598.
6. Bodenmiller, B., Mueller, L. N., Pedrioli, P. G., Pflieger, D., Jünger, M. A., Eng, J. K., Aebersold, R., Tao, W. A. (2007) An integrated chemical, mass spectrometric and computational strategy for (quantitative) phosphoproteomics: application to *Drosophila melanogaster* Kc167 cells. *Mol. Biosyst.* **4**, 275–286.
7. Bodenmiller, B., Mueller, L. N., Mueller, M., Domon, B., and Aebersold, R. (2007) Reproducible isolation of distinct, overlapping segments of the phosphoproteome. *Nat Methods.* **4**(3), 231–237.
8. Perkins, D. N., Pappin, D. J., Creasy, D. M., and Cottrell, J. S. (1999) Probability-based protein identification by searching sequence databases using mass spectrometry data. *Electrophoresis* **20**(18), 3551–3567.
9. Yates, J. R., 3rd, Eng, J. K., McCormack, A. L., and Schieltz, D. (1995) Method to correlate tandem mass spectra of modified peptides to amino acid sequences in the protein database. *Anal. Chem.* **67**(8), 1426–1436.
10. Pan, C., Gnad, F., Olsen, J. V., and Mann, M. (2008) Quantitative phosphoproteome analysis of a mouse liver cell line reveals specificity of phosphatase inhibitors. *Proteomics* **8**(21), 4534–4546.
11. Villén, J., and Gygi, S. P. (2008) The SCX/IMAC enrichment approach for global phosphorylation analysis by mass spectrometry. *Nat. Protoc.* (10), 1630–1638.
12. Bodenmiller, B., Campbell, D., Gerrits, B., Lam, H., Jovanovic, M., Picotti, P., Schlapbach, R., and Aebersold, R. (2008) PhosphoPepa database of protein phosphorylation sites in model organisms. *Nat. Biotechnol.* **26**(12), 1339–1340.
13. Makarov, A., Denisov, E., Kholomeev, A., Balschun, W., Lange, O., Strupat, K., and Horning, S. (2006) Performance evaluation of a hybrid linear ion trap/orbitrap mass spectrometer. *Anal. Chem.* **78**(7), 2113–2120.
14. Villén, J., Beausoleil, S. A., and Gygi, S. P. (2008) Evaluation of the utility of neutral-loss-dependent MS3 strategies in large-scale phosphorylation analysis. *Proteomics* **8**(21), 4444–4452.
15. Bastide, F., Meissner, G., Fleischer, S., and Post, R. L. (1973) Similarity of the active site of phosphorylation of the adenosine triphosphatase for transport of sodium and potassium ions in kidney to that for transport of calcium ions in the sarcoplasmic reticulum of muscle. *J. Biol. Chem.* **248**, 8385–8391.
16. Hultquist, D. E. (1968) The preparation and characterization of phosphorylated derivatives of histidine. *Biochim. Biophys. Acta* **153**, 329–340.
17. Pas, H. H., and Robillard, G. T. (1988) *S*-Phosphocysteine and phosphohistidine are intermediates in the phosphoenolpyruvate-dependent mannitol transport catalyzed by *Escherichia coli* EII(Mtl). *Biochemistry* **27**, 5835–5839.
18. Wakim, B. T., and Aswad, G. D. (1994) Ca(2+)–calmodulin-dependent phosphorylation of arginine in histone 3 by a nuclear kinase from mouse leukemia cells. *J. Biol. Chem.* **269**, 2722–2727.
19. Gordon, J. A. (1991) Use of vanadate as protein–phosphotyrosine phosphatase inhibitor. *Methods Enzymol.* **201**, 477–482.

Chapter 8

Analysis of Carbohydrates on Proteins by Offline Normal-Phase Liquid Chromatography MALDI-TOF/TOF-MS/MS

Theodora Tryfona and Elaine Stephens

Abstract

Glycoproteins typically exist as a diverse population of glycoforms, which can consist of a great number of different glycan structures (including structural isomers) that vary in their degree of occupancy at given glycosylation sites. Hence, the detailed characterization of such glycans can be a very demanding task. Liquid chromatography in combination with mass spectrometric techniques can provide a fast and sensitive tool for the analysis of these structurally complex molecules. Here we describe a sensitive method that employs capillary normal-phase HPLC coupled offline to MALDI-TOF/TOF-MS/MS for the detailed characterization of enzymatically released *N*-glycans. The normal-phase chromatography allows the separation of some isobaric glycan structures and analysis of the glycans by MALDI high-energy CID gives sequence, linkage and branching information.

Key words: Protein glycosylation, *N*-glycans, MALDI-TOF-MS, MALDI-CID, normal-phase chromatography.

1. Introduction

Glycan chains on proteins play key roles in several biological processes (1), such as cell adhesion, protein folding, solubility and aggregation, and therefore are of great biological interest. Nearly all glycoproteins consist of glycoforms in which a single protein is diversified by a heterogeneous array of glycans at each glycosylation site. Detailed characterization of the glycans attached to proteins therefore depends on the availability of sensitive methods that are capable of successfully analysing mixtures. Mass spectrometry is one such method and has been extremely

P.R. Cutillas, J.F. Timms (eds.), *LC-MS/MS in Proteomics*, Methods in Molecular Biology 658,
DOI 10.1007/978-1-60761-780-8_8, © Springer Science+Business Media, LLC 2010

successful in the detailed structural elucidation of many different types of carbohydrates (2–5). The use of chromatographic methods, in combination with mass spectrometric techniques is very advantageous for extremely complex samples and important for the detection of minor oligosaccharide isomers in mixtures (5). Normal-phase HPLC (NP-HPLC) using amide-based columns is a well-established, robust separation technique used by many laboratories to obtain high-resolution separation of *N*-linked glycans released from glycoproteins. We have combined this method with MALDI-TOF/TOF-MS/MS for the detailed characterization of carbohydrate mixtures. It is successfully employed in our laboratory for the detailed analysis of a wide range of carbohydrate-based polymers including *N*- and *O*-linked glycans and plant polysaccharides. The following protocol describes the preparation and NP-HPLC-MALDI-TOF-MS of *N*-linked glycans released from plant, insect and mammalian glycoproteins.

2. Materials

2.1. Protein Precipitation

1. Ice-cold (–20°C) acetone.
2. 60% Trichloroacetic acid in H_2O.

2.2. Release of N-Glycans from Plant and Insect Glycoproteins

1. Pepsin digestion buffer: 5% formic acid (vol:vol) pH 3 (adjust pH with 1 N NaOH).
2. Pepsin (Sigma; *see* **Note 1**) solution: 1 mg of enzyme in 1 mL of pepsin digestion buffer.
3. PNGase A digestion buffer: 50 mM ammonium acetate (pH 5; adjust pH with 5% acetic acid).
4. Peptide *N*-glycosidase A (PNGase A) (5 mU in 100 μL; Roche; *see* **Note 2**).
5. Pall NanoSep centrifugal devices with 3 K Omega Membrane (Sigma).
6. Vacuum centrifuge (Speed Vac; Savant).

2.3. Release of N-Glycans from Mammalian Glycoproteins

1. 100 mM Ammonium bicarbonate (NH_4HCO_3).
2. HPLC-grade acetonitrile (Rathburn Chemicals, Walkerburn, Scotland).
3. 200 mM Dithiothreitol (DTT) in 100 mM NH_4HCO_3.
4. 1 M Iodoacetamide (IAA) in 100 mM NH_4HCO_3 (*see* **Note 3**).
5. Bovine trypsin (protein sequencing grade; Roche) (*see* **Note 4**) solution in 50 mM NH_4HCO_3 (*see* **Note 5**).

6. PNGase F digestion buffer: 100 mM NH_4HCO_3 (pH 8.4).
7. Lyophilized peptide *N*-glycosidase F (PNGase F; Roche) (*see* **Note 6**).
8. Vacuum centrifuge (Speed Vac; Savant).

2.4. N-Glycan Purification

1. Classic Sep-Pak C_{18} 1-cm^3 column (Waters) (*see* **Note 7**).
2. HPLC-grade methanol (Fisher) (*see* **Note 8**).
3. 5-mL Borosilicate glass Luer-Lip syringe (Sigma).
4. 5% (vol/vol) Acetic acid solution.
5. Dowex cation exchange resin (50WX8-100).
6. 4 M HCl.
7. HPLC-grade water (*see* **Note 9**).
8. Disposable polystyrene columns (Pierce).
9. pH indicator strips.

2.5. Removal of Sialic Acids from Mammalian N-Glycans

1. 2 M Acetic acid solution.

2.6. Reductive Amination of Glycans

1. Vacuum centrifuge (Speed Vac; Savant).
2. Dry (anhydrous) DMSO (Romil; *see* **Note 10**).
3. Glacial acetic acid.
4. Anthranilic acid: 2-aminobenzoic acid (2-AA; Fluka) or 2-aminobenzamide acid (2-AB; Fluka).
5. Sodium cyanoborohydride (Aldrich) (*see* **Note 11**).
6. GlykoClean S cartridge (Glyko) (*see* **Note 12**).
7. HPLC-grade water.
8. HPLC-grade acetonitrile (Rathburn Chemicals, Walkerburn, Scotland; *see* **Note 13**).

2.7. Capillary Normal-Phase Chromatography

1. 80% Acetonitrile in water for sample loading.
2. Capillary HPLC system with UV detector set at 254 nm (e.g., Ultimate HPLC system with capillary calibrator; Dionex, UK).
3. MALDI plate spotter for automated fraction collection (Probot; Dionex, UK).
4. Capillary amide-80 normal-phase column (300 μm × 25 cm; 3 μm or 5 μm particle size) (Dionex).
5. Solvent A: 50 mM ammonium formate made up with HPLC-grade H_2O and pH adjusted to 4.4 with formic acid (*see* **Note 14**).

6. Solvent B: 80% acetonitrile containing 20% of solvent A.
7. MALDI target.

2.8. Acquisition of MS Spectra

1. 2,5-Dihydroxybenzoic acid (DHB; Fluka) solution: 10 mg/mL DHB in 50% methanol (*see* **Note 15**).
2. Desiccator with vacuum pump and solvent trap assembly.
3. 4700 Proteomics Analyser (Applied Biosystems, Foster City, CA, USA).

3. Methods

3.1. Protein Precipitation

This is carried out to concentrate and remove contaminants such as salts and detergents from the glycoprotein samples. If the protein solution is very dilute (0.1–0.4 mg/mL), the TCA method gives the best protein recovery.

3.1.1. Acetone Precipitation

1. Suspend the glycoprotein sample into cold (–20°C) acetone volume five times that of the protein samples to be precipitated, mix well and incubate for 24 h at –20°C.
2. Centrifuge at 10,000×*g* for 30 min at 4°C , decant and properly dispose of the supernatant, being careful not to dislodge the protein pellet.
3. Allow the acetone to evaporate from the uncapped tube at room temperature for 30 min but make sure not to over-dry the pellet as it may not dissolve properly.

3.1.2. TCA Precipitation

1. To a dilute solution of glycoprotein (e.g., 200 μL at 40–200 μg/mL), add 40 μL of 60% TCA.
2. Mix well and incubate mixture overnight on ice.
3. Centrifuge at 10,000×*g* for 30 min at 4°C.
4. Carefully remove the supernatant and add 100 μL of 90% ice-cold acetone to wash the pellet.
5. Incubate on ice for 15 min and centrifuge as above.
6. Carefully remove the acetone-containing supernatant and allow acetone to evaporate from the uncapped tube at room temperature for 30 min but make sure not to over-dry the pellet as it may not dissolve properly.

3.2. N-Glycan Release

3.2.1. Plant and Insect Glycoproteins

1. Dissolve glycoprotein sample in 100 μL of pepsin digestion buffer and add 28 μL of pepsin solution.

2. Incubate at 37°C for 24 h.
3. Inactivate pepsin by incubation at 100°C for 5 min and immediately cool the digest mixture on ice (*see* **Note 16**).
4. Freeze and lyophilize sample using a freeze dryer.
5. Redissolve 2.5 mU of commercially available peptide-*N*-glycosidase A (*see* **Note 2**) in 250 μL of HPLC-grade water and apply it to dialysis by threefold centrifugation in Nanosep centrifugal devices (3K) at 5,000×*g* for about 30 min at 4°C.
6. For a complete exchange of buffer, dilute the concentrated enzyme in the upper layer (volume ~ 50 μL) each time with ammonium acetate buffer to final volume of 500 μL. After three cycles of centrifugation, dilute the concentrated enzyme with ammonium acetate buffer (pH 5) to an enzyme concentration of 5 mU in 100 μL.
7. Dissolve the lyophilized samples in 100 μL of PNGase A digestion buffer and add 5 μL of the buffer-exchanged PNGase A enzyme.
8. Incubate samples at 37°C for 16–18 h or overnight.
9. Stop digestion by drying the samples.

3.2.2. Mammalian Glycoproteins

Mammalian glycans often contain sialic acid residues, which are extremely susceptible to cleavage at elevated temperatures under acidic conditions. Therefore, pepsin digestion is not recommended and the following, more lengthy procedure should be followed to efficiently release *N*-glycans:

1. Redissolve the protein pellet in 90 μL of 50 mM NH_4HCO_3 under agitation (*see* **Note 17**).
2. Supplement with 10% acetonitrile (add 10 μL).
3. Add 5 μL of DTT solution and heat it up to 95°C for 5 min.
4. Alkylate the protein by adding 4 μL of IAA solution. The reaction will take place at 25°C for 45 min in the darkness.
5. Stop the alkylation by adding 20 μL of DTT solution and by incubating the solution at 25°C for 45 min.
6. TCA precipitate according to **Section 3.1.2**.
7. Redissolve in 90 μL of 50 mM NH_4HCO_3 and add 10 μL of acetonitrile (*see* **Note 17**).
8. Prepare a trypsin solution in NH_4HCO_3 and add it to the protein solution in a trypsin to protein ratio of 1:50 to 1:20.
9. Incubate at 37°C for 16–18 h or overnight.
10. Inactivate the trypsin by incubation at 95°C in a water bath and cool the digest for 30 min on ice.

11. Let the trypsinized solution come to room temperature.
12. Add 2.5 μL (2.5 U) of PNGase F and incubate for 4 h at 37°C. Add another 2.5 μL of PNGase F and continue incubating overnight at a final enzyme concentration of 50 U/mL.
13. Stop digestion by lyophilizing samples.

3.3. Purification of N-Glycans

3.3.1. RP C_{18} Spin Cartridge Clean-up

This treatment removes all hydrophobic (non-polar) peptides and enzymes from the mixture and the hydrophilic glycans will be found in the flow-through. A protocol using classic C_{18} Sep-Pak SPE cartridges is described here, which can bind and remove large quantities of peptides. However, for smaller total protein amounts (up to 30 μg), smaller spin cartridges (PepClean C_{18} Spin Columns; Pierce) can be used to keep sample volumes down (follow the manufacturer's instructions carefully).

1. After PNGase A or PNGase F digestion, redissolve the lyophilized samples in 200 μL of 5% acetic acid.
2. Clamp 5-mL glass syringe onto the retort stand, attach the C_{18} Sep-Pak cartridge onto the tip and wash with 5 mL of methanol, making sure that no air bubbles appear in the cartridge.
3. Wash with 5 mL of 5% acetic acid.
4. Load the sample onto the column, making sure to wash out the sample tube.
5. Elute glycans with 3 mL of 5% acetic acid and collect this fraction into an appropriate container as it contains free *N*-glycans (*see* **Note 18**).

3.3.2. Dowex Cation Exchange

This step removes any remaining hydrophilic peptides and all salts are converted to their corresponding acids.

1. Prepare Dowex beads by washing three times with 4 M HCl.
2. Wash repeatedly with HPLC-grade water until the pH of the water is the same as the water supply and follow by washing three times with 5% acetic acid (*see* **Note 19**).
3. Plug a disposable polystyrene column with glass wool and pack with 500 μL of prepared Dowex beads. Wash with two volumes of 5% acetic acid (approx. 1 mL) (*see* **Note 20**).
4. Re-suspend dry sample in 50 μL of 5% acetic acid and load directly onto the column.
5. Elute with two volumes (approx. 1 mL) of 5% acetic acid, collect fractions and lyophilize using a freeze dryer (*see* **Note 18**).

3.4. Removal of Sialic Acids from Mammalian N-Glycans

Good fragmentation of sialylated *N*-glycans by MALDI-TOF/TOF-MS/MS is made difficult because of the instability of sialic acid on the glycan species in positive ion mode. In order to stabilize the sialic acid residues, methyl esterification of the carboxyl group (6) or permethylation of complete glycan (7) can be carried out prior to MALDI-TOF-MS of the mixture. However, for the NP-LC-MALDI-TOF/TOF-MS/MS experiments described here, we recommend that sialic acid residues are removed by mild acid treatment prior to reductive amination as described below:

1. Redissolve purified glycans in 100 μL of 2 M acetic acid.
2. Incubate at 80°C for 2 h.
3. Lyophilize using a freeze dryer.

3.5. Reductive Amination of Enzymatically Released Glycans

1. Redissolve purified glycans from cation exchange purification (or the desialylated mammalian *N*-glycans) in 20 μL of H_2O and re-dry using a Speed Vac (*see* **Note 21**).
2. To 350 μL of DMSO (anhydrous) add 150 μL of glacial acetic acid.
3. Add 100 μL of the above solution to 6 mg of anthranilic acid (2-AA) or 5 mg of 2-aminobenzoic acid (2-AB) and mix well.
4. In a separate tube, weigh out 6 mg of sodium cyanoborohydride and add the DMSO/acetic acid/2-AA or 2-AB mixture (*see* **Note 11**).
5. Mix well by pipetting at 65°C.
6. Add 10 μL of this solution to dry native oligosaccharides and vortex. Briefly centrifuge so that reagent is at the bottom of the tube.
7. Incubate at 65°C for 3 h with intermittent mixing/centrifuging.
8. Directly purify glycans from reductive amination reagents using a GlykoClean S cartridge.

3.6. Purification of Reductively Aminated Glycans

1. Condition GlykoClean S cartridge by washing with 1 mL HPLC-grade water, then with 5 mL of 30% acetic acid solution, followed by 3 mL of 100% acetonitrile and finally an additional 1 mL of 100% acetonitrile (*see* **Note 22**).
2. Make sure glycan samples are at or below room temperature.
3. Spot each sample onto a freshly washed cartridge membrane, spreading the sample over the entire membrane surface (*see* **Note 23**).
4. Leave for 15 min to allow the glycans to absorb on the membrane. For a maximum recovery, rinse the sample vial

with 100 μL of 100% acetonitrile and apply to the cartridge allowing time for it to penetrate into the membrane.

5. Wash the cartridge with 1 mL of 100% acetonitrile and allow to drain.
6. Wash five times with 1 mL 96% acetonitrile solution, allowing each aliquot to drain before the next step is applied. Discard the flow-through.
7. Place the cartridge over a collection vessel suitable for drying 1.5 mL of water and elute glycans with three washes of 0.5 mL water, allowing each wash to drain before the next one is applied.
8. Lyophilize samples (*see* **Note 24**).

3.7. Normal-Phase Chromatography

1. Dissolve *N*-glycan oligosaccharides in 80% acetonitrile and load onto an amide-80 column (300 μm × 25 cm; 5 μm bead size; LC Packings/Dionex) and with the capillary HPLC system, generate a gradient that flows at 3 μL/min.
2. Elute oligosaccharides with increasing aqueous concentration: equilibrate the NP column with 5% A. For elution of *N*-glycans, the gradient is initiated 5 min after injection and increased linearly to 52% A over 96 min. The column is then equilibrated for 10 min with 5% solvent A prior to the next injection.
3. If using a UV detector, monitor at 254 nm for 2-AA- and 2-AB-labelled glycans.
4. Pass column effluent directly to a Probot sample fractionation system and spot onto a MALDI target plate at 20 s intervals. Leave spots to air-dry.

3.8. Acquisition of MALDI-TOF-MS

1. Overlay all sample spots with 0.6 μL of 2,5 DHB solution (10 mg/mL in 50% methanol) and rapidly dry in a vacuum desiccator. This drying method produces small crystals which facilitate easy spectral acquisition (*see* **Note 25**).
2. Fire the laser on a few spots over the plate in order to determine the correct laser intensity for the automatic acquisition of the MS data.
3. Acquire spectra automatically by exposing each spot to 1,500–2,500 shots over a range between *m/z* 850 and 5,000.

3.9. Analysis of MS Results

1. Generate a base peak chromatogram (*see* **Fig. 8.1**a) and match observed signals in the MALDI spectra to putative glycan structures (*see* **Table 8.1**).

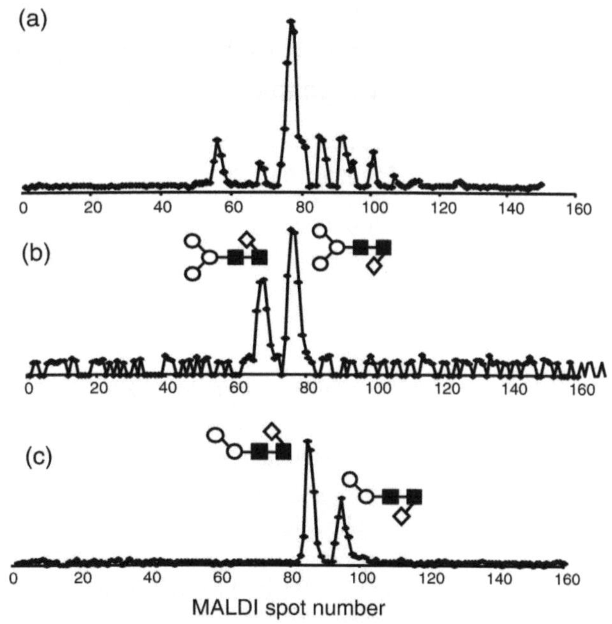

Fig. 8.1. Normal-phase HPLC-MALDI-TOF-MS of PNGase A-released *N*-glycans from honey bee PLA_2 labelled with 2-aminobenzoic acid (2-AA). (**a**) Base peak chromatogram of released *N*-glycans, (**b**) extracted ion chromatogram for $Man_2GlcNAc_2Fuc$ and (**c**) extracted ion chromatogram for $Man_3GlcNAc_2Fuc$. The elution positions of the structural isomers of $Man_{2-3}GlcNAc_2Fuc$ having α1–6 and α1–3 linked core fucose are shown. Symbol representation of glycan structure: ○, Man; ■, GlcNAc; ◇, Fuc; /, 1-3 linkage; \ 1-6 linkage; –, 1–4 linkage.

2. For 2-AA- or 2-AB-labelled *N*-glycans, $[M + H]^+$ and $[M + Na]^+$ ions are detected, which can be easily identified in the mass spectra due to their 22 mass unit difference.
3. Structural isomers may be separated by the chromatography; so it is important to carry out extracted ion chromatograms in order to detect these. For example, in the glycan mixture shown in **Fig. 8.1b** and **c**, structural isomers of $Man_{2-3}GlcNAc_2Fuc$-containing α1–6- or α1–3-linked core fucose are clearly separated by the normal-phase column. This separation is extremely advantageous for the identification of such core fucosylated isomers.

3.10. Acquisition of MALDI High-Energy CID Spectra

1. Use a 4700 DHB MS/MS acquisition method with the collision energy set at 1 or 2 kV (*see* **Note 26**).
2. Fill the collision cell with argon to a pressure of around 2×10^6 Torr.
3. Tune the instrument using an oligosaccharide standard to ensure best fragment ion sensitivity.

Table 8.1
Masses of most common monosaccharide residues found in *N*-glycans

Monosaccharide	Monosaccharide residue masses (monoisotopic)
Pentose (xylose)	132.0423
Deoxyhexose (fucose)	146.0579
Hexose (mannose, galactose)	162.0528
N-Acetylhexosamine (*N*-acetylglucosamine)	203.0794
N-Acetylneuraminic acid (Sialic acid)	291.0954
Sum of terminal group masses (including 2-AB label) for calculation of $[M + Na]^+$	161.0685

4. Collect the high-energy CID spectra on selected $[M + Na]^+$ ions (*see* **Note 27**). It is best to do this manually, stopping the acquisition when the desired signal-to-noise ratios for the fragments are achieved.

3.11. Interpretation of MALDI High-Energy CID Spectra

The high-energy CID spectra are characterized by signals for cross-ring fragments and "elimination" ions. The fragments are assigned according to the Domon and Costello (8) nomenclature (*see* **Fig. 8.2**). All of these ions shown (Z_n, Y_n, C_n, X_n and A_n) provide sequence information. However, the cross-rings (A_n and X_n), as well as the elimination ions (D_n and E_n) (9), give more detailed information concerning antenna substitutions and linkage positions. The following section describes how to identify the most abundant and structurally informative fragments observed.

1. The spectrum will contain Y_n and C_n ions, which occur from cleavage of the glycosidic bond between the sugar residues. In high-energy CID spectra, these normally occur concomitantly with their Y_n–2 and C_n–2 counterparts. Therefore, these ions can be easily identified in the spectra as doublets that differ by 2 Da (*see* **Fig. 8.3b** at *m/z* 365 and 363 for C_2 and C_2-2, respectively).
2. A predominant C-type cleavage always occurs between the distal *N*-acetyl glucosamine in the chitobiose unit and core β-mannose (e.g., *see* C_2 and C_2-2 in both spectra). These ions are accompanied by a D_n ion, which is derived from

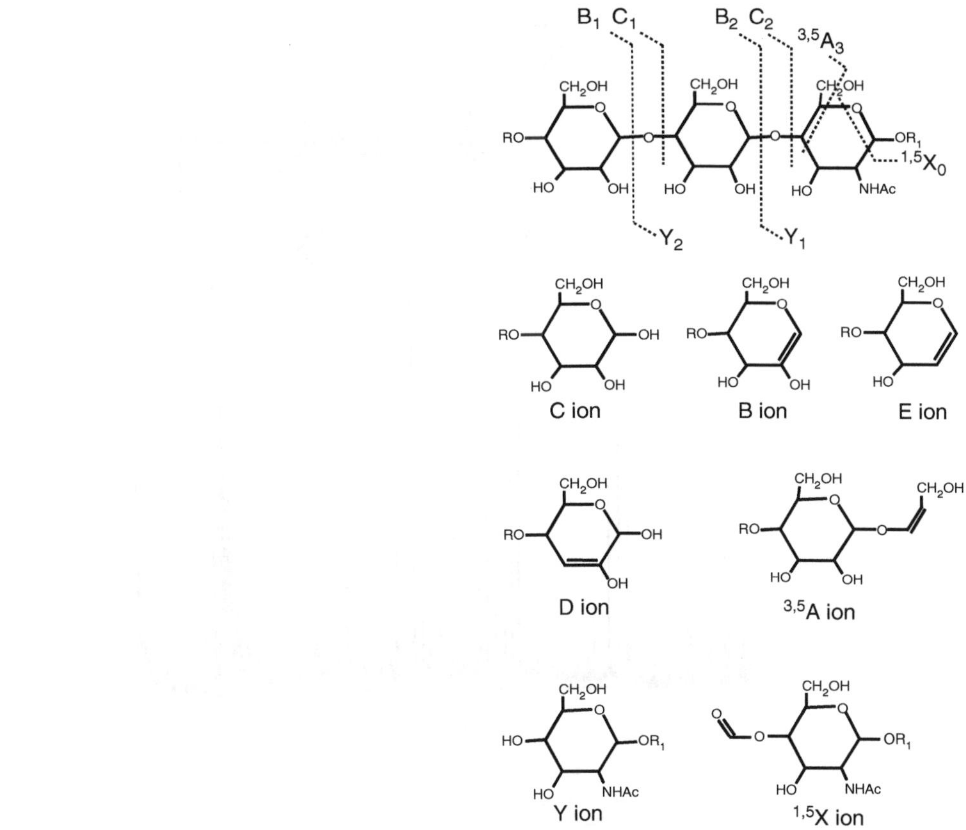

Fig. 8.2. Domon and Costello nomenclature for describing the fragmentation of carbohydrates (8, 9).

the elimination of the lower antenna that is linked to the 3-position of core β-mannose. For example, the D_2 ion (at *m/z* 347 Da) in **Fig. 8.3a** is 180 mass unit less than its corresponding C_2 ion because it has lost mannose (162 Da) and H_2O (18 Da) due to the elimination process. The structure shown in **Fig. 8.3b** has no lower antenna and so loses water (18 Da) only from the C_2 fragment to give the D_2 ion at *m/z* 347. These characteristic D ions are therefore particularly useful for determining antenna positions and their substitutions.

3. Another abundant ion occurs by fragmentation of the glycosidic linkage between the two *N*-acetylglucosamine residues in the chitobiose core to give a B_n ion. Its corresponding C ion is normally minor by comparison (*see* B_3 and C_3 in **Fig. 8.3a** and **b**).
4. These B_n ions are often accompanied by E_n ions, which occur by elimination of the substituent from the C-2 position.

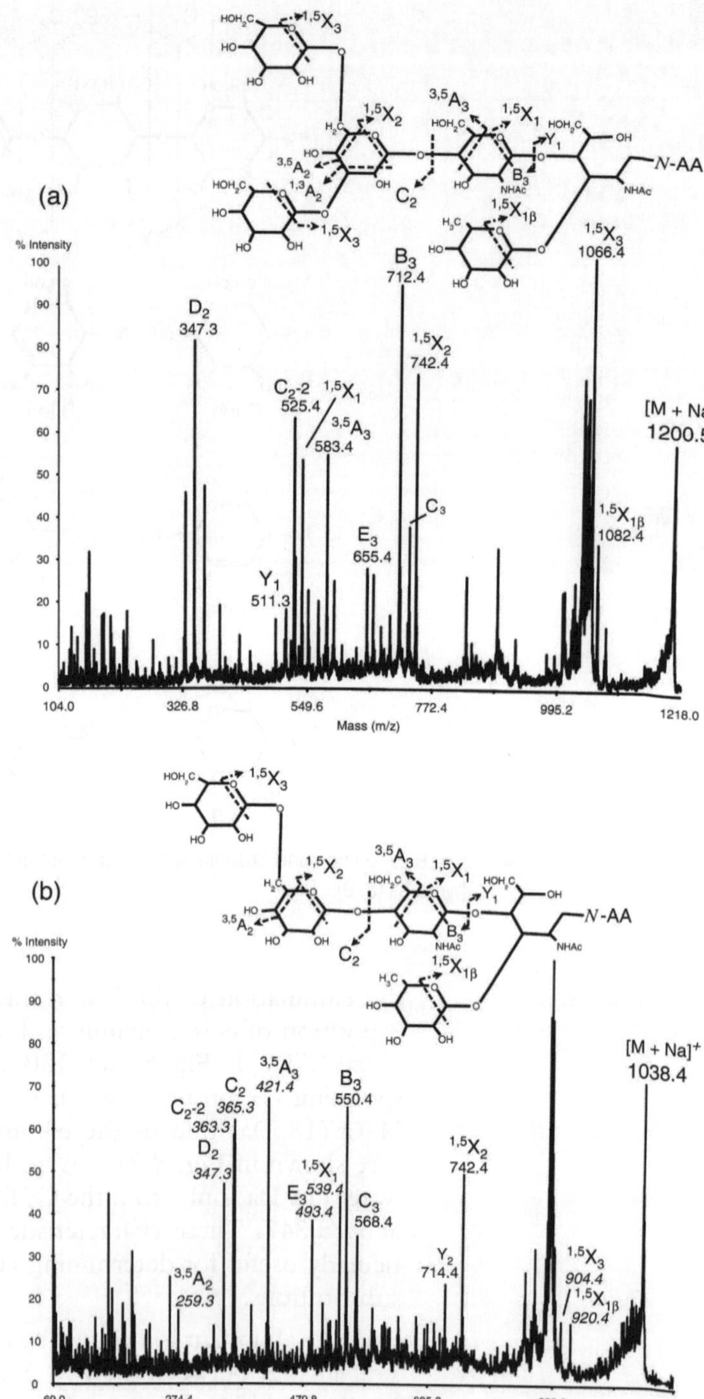

Fig. 8.3. MALDI-CID spectrum of (a) $Man_3GlcNAc_2Fuc$ and (b) $Man_2GlcNAc_2Fuc$ labelled with 2-aminobenzoic acid (2-AA).

5. Other dominant ions include the cross-ring fragments (X_n and A_n). The X_n ions provide only sequence information. However, the A_n ions, which incorporate the non-reducing end, often give linkage information. The most abundant A_n ions observed in *N*-glycan high-energy CID are normally the $^{3,5}A_n$ fragments that occur around the central β-mannose residue and its neighbouring *N*-acetylglucosamine in the chitobiose core (*see* **Fig. 8.3a** and **b**).

4. Notes

1. Pepsin is stable at 60°C. The pH optimum for the enzyme is 2–4 and the activity is irreversibly lost at pH>6. Pepsin cleaves preferentially C-terminal to Phe, Leu and Glu; it does not cleave Val, Ala and Gly. Other residues may also be cleaved, with very variable rates.
2. PNGase A preparations should not contain glycerol, which is present as a stabilizing reagent in commercially obtainable preparations, since this can interfere with subsequent reductive amination reaction efficiencies. The enzyme is therefore dialyzed into 50 mM ammonium acetate (pH 5.0), which can be added directly to the glycopeptide mixture.
3. Solution is light sensitive, so wrap the container in aluminium foil and use immediately.
4. Trypsin is autolytic and is not very stable above 40°C. The pH optimum for the enzyme is 7–9, and its activity is irreversibly lost at pH>11. Trypsin cleaves very specifically at Arg-X and Lys-X bonds, but if X is Pro, then no cleavage occurs.
5. Prepare freshly before use and keep it on ice; adjust trypsin concentration in order to achieve a 1:100 to 1:20 trypsin to protein ratio (between 200 and 500 ng/mL).
6. It is recommended that freeze-dried PNGase F be used and not enzyme solutions in glycerol. If the enzyme contains glycerol, make sure it is buffer exchanged into a volatile buffer such as ammonium bicarbonate (pH 8) prior to use.
7. Classic Sep-Pak C_{18} cartridges (Waters) should be kept at room temperature, away from moisture and exposure to air should be kept to a minimum.
8. Methanol is toxic by inhalation or if swallowed, thus it must be handled in the fume hood.

9. All solutions should be prepared in HPLC-grade water, unless otherwise stated. In this text, we refer to HPLC-grade water as 'water'.

10. Anhydrous DMSO should be kept in a desiccator at room temperature.

11. Sodium cyanoborohydride is flammable, very toxic and dangerous to the environment, thus it is highly recommended that it should be handled in the fume hood and that all suitable protective clothing, eye protection and gloves be used. Avoid release to the environment.

12. Store in a refrigerator.

13. Acetonitrile should be kept in a cool dry place. Avoid breathing the fumes.

14. Dedicated glassware specifically for making up eluents should be kept separate from other general use glassware. It is recommended that no pipettes be inserted into solvent bottles; instead pour solvents directly into graduated cylinder and avoid using plastic containers.

15. Always prepare fresh DHB matrix solution. Store DHB container in a cool area making sure that the temperature will never reach above 20°C.

16. It is important to keep the heating time to a minimum to prevent the hydrolysis of fucosylated residues.

17. It is important to be sure that the precipitated proteins are thoroughly re-solubilized and that the final pH is adjusted to pH 8.0.

18. Released *N*-glycans are stable for several months when stored at −20°C.

19. Equilibrated Dowex beads can be stored in 5% acetic acid for several weeks.

20. Slurry-pack the column taking care to remove air bubbles. Do not allow the column to run dry.

21. To avoid an incomplete derivatization of glycans, a careful concentration of the sample to the bottom of the microvial is necessary. Direct analysis of the labelled glycans by MALDI-TOF mass spectrometry will show incomplete derivatization by the occurrence of additional peaks 120 (2-AA) or 119 Da (2-AB) lower than expected in the mass spectra.

22. Allow the cartridge to drain completely after each wash.

23. Make sure that the membrane is still wet with acetonitrile. If the membrane has dried, it must be re-wetted by washing with 0.5 mL acetonitrile prior to loading the sample.

24. Labelled glycans can be stored at –20°C in the dark.
25. DHB is the most commonly used matrix for MALDI-TOF analysis of carbohydrates. Make sure to exclude high-intensity matrix ions at low mass (*m/z* <850), which are particularly abundant in NP-LC-MALDI.
26. MS/MS methods set up for the CID of peptides in α-cyano-4-hydroxycinnamic acid (CHCA) will not give good fragmentation.
27. Fragmentation of $[M + H]^+$ will give predominantly glycosidic cleavages (B_n and Y_n ions) and little cross-ring fragments.

Acknowledgements

This work was funded by the Food Standards Agency (for TT).

References

1. Varki, A. (1993) Biological roles of oligosaccharides: all of the theories are correct. *Glycobiology* **3**, 97–130.
2. Dell, A. and Morris, H. R. (2001) Glycoprotein structure determination by mass spectrometry. *Science* **291**, 2351–2356.
3. Stephens, E., Sugars, J., Maslen, S. L., Williams, D. H., Packman, L. C., and Ellar, D. J. (2004) The *N*-linked oligosaccharides of aminopeptidase-*N* from *Manduca sexta* – site localization and identification of novel *N*-glycan structures. *Eur. J, Biochem.* **271**, 4241–4258.
4. Harvey, D. J. (1999) Matrix-assisted laser desorption/ionization mass spectrometry of carbohydrates. *Mass Spectrom. Rev.* **18**, 349–450.
5. Maslen, S. L., Goubet, F., Adam, A., Dupree, P., and Stephens, E. (2006) Arabinoxylan oligosaccharide structure elucidation by normal phase HPLC-MALDI-TOF/TOF-MS/MS. *Carbohydrate Res.* **342**, 724–735
6. Powell, A. K. and Harvey, D. J. (1996) Stabilization of sialic acids in N-linked oligosaccharides and gangliosides for analysis by positive ion matrix-assisted laser desorption/ionization mass spectrometry. *Rapid Commun. Mass Spectrom.* **10**, 1027–1032.
7. Ciucanu, I. and Kerek, F. (1984) A simple and rapid method for the permethylation of carbohydrates. *Carbohydrate Res.* **131**, 209–217.
8. Domon, B. and Costello, C. E. (1988) A systematic nomenclature for carbohydrate fragmentations in FAB-MS/MS spectra of glycoconjugates. *Glycoconjugate J.* **5**, 397–409.
9. Spina, E., Sturiale, L., Romeo, D., Impallomeni, G., Garozzo, D., Waidelich, D., and Glueckmann, M. (2004) New fragmentation mechanisms in matrix-assisted laser desorption/ionization time-of-flight/time-of-flight tandem mass spectrometry of carbohydrates. *Rapid Commun. Mass Spectrom.* **18**, 392–398.

Part III

Techniques for Quantitative LC-MS

Chapter 9

Selected Reaction Monitoring Applied to Quantitative Proteomics

Reiko Kiyonami and Bruno Domon

Abstract

Proteomics is gradually shifting from pure qualitative studies (protein identification) to large-scale quantitative experiments, prompted by the growing need to analyze consistently and precisely a large set of proteins in biological samples. The selected reaction monitoring (SRM) technique is increasingly applied to quantitative proteomics because of its selectivity (two levels of mass selection), its sensitivity (non-scanning mode), and its wide dynamic range. This account describes the different steps in the design and the experimental setup of SRM experiments.

Key words: Selective reaction monitoring (SRM), quantitative proteomics, triple quadrupole, isotopically labeled internal standards.

1. Introduction

Proteomics is gradually shifting from pure qualitative studies (protein identification) to large-scale quantitative experiments, prompted by the growing demand for protein biomarker qualification and verification in larger cohorts on one hand and the need for consistent quantitative data sets in systems biology to perform modeling on the other. Due to the difficulties inherent in analyzing mixtures of intact proteins, most quantitative proteomic strategies rely on the use of peptides generated by enzymatic digestion as surrogates for targeted proteins. Liquid chromatography hyphenated to mass spectrometry (LC-MS and LC-MS/MS) has become the standard platform for quantitative proteomics as it has the capability of analyzing a large number of peptides in complex biological samples. LC-MS also offers

P.R. Cutillas, J.F. Timms (eds.), *LC-MS/MS in Proteomics*, Methods in Molecular Biology 658,
DOI 10.1007/978-1-60761-780-8_9, © Springer Science+Business Media, LLC 2010

robustness and the ease of automation enabling high-throughput studies.

The development of tandem mass spectrometers, in particular triple quadrupole spectrometers, has revolutionized analytical chemistry. The selected reaction monitoring (SRM) technique is increasingly applied to quantitative proteomics because of its selectivity (two levels of mass selection), its sensitivity (non-scanning mode), and its wide dynamic range. In fact, LC-MS analysis in electrospray ionization mode has been used routinely for two decades to quantify small molecules and has become a benchmark for quantifying drug metabolites in complex biological samples.

The concomitant development of nano-HPLC systems using narrow bore columns (75 or 150 μm ID) directly interfaced with a mass spectrometer provides a further sensitivity increase, which is highly desired in proteomics where samples are very complex and often the components of interest are in very low concentrations. By definition, a quantitative study requires the systematic determination of the amount of each component of interest present in a series of samples, including the replication of the analyses. Precise LC-MS quantification is routinely carried out using stable isotope dilution, in which internal standards (isotopically labeled synthetic peptides) are added to the samples prior to the LC-MS analysis (1). Often proteomic studies are focused on measuring changes in the level of expression of targeted proteins in different samples (e.g., control and disease samples). Such experiments are based on relative quantification, i.e., comparing the signal intensities observed in different samples (after proper normalization to correct for instrument drifts and the sample amount injected). This method is referred to as a label-free quantification.

Regardless of the approach used, there is a need to reduce the sample complexity, as it is the actual limiting factor to achieve high-sensitivity measurements in biological samples. Hundreds of thousands of components present in a tryptic digest overwhelm the peak capacity of any state-of-the-art LC-MS system. In order to achieve acceptable limits of detection and quantification, reduction of the biochemical background is essential, and thus the sample preparation is critical. In addition, there is a need for a robust protocol that provides a high recovery at all stages of analysis to ensure reliable quantification. Furthermore, the concentrations of proteins to be analyzed in a biological sample span over a wide dynamic range: six orders of magnitude for a yeast cell lysate and (2) up to eleven for blood plasma (3). Simultaneous and systematic quantification of a large number of analytes in a complex sample remains challenging. At present, the best performing LC-MS systems cover a dynamic range of four to five orders of magnitude.

Although the SRM technology has been established for decades for small molecules, there are specifics when applied to proteomics. First, the multiply charged precursor ions result in fragments at m/z values significantly above the precursor ions (in particular for triply charged precursors), which requires instruments with an extended mass range to enable analysis of large y ions, which are preferred because of their higher selectivity and the better signal-to-noise ratio in that region of the spectrum. Second, the experimental design and the corresponding bioinformatic effort are performed upfront. This requires the selection of peptides as surrogates for the proteins of interest, and these peptides have to be uniquely associated with one protein. However, the uniqueness is a relative concept specific to one proteome (i.e., one species) as the transitions are dependent on the biochemical environment, i.e., non-specific proteolytic background. Last, in contrast to classical small molecule studies, which are focused on a few analytes, proteomic assays aim at analyzing large sets of peptides (hundreds) and thus intend to have very large numbers of transitions in one single experiment. Ongoing developments of the SRM technology to increase the selectivity and density of measurements are crucial to expand the capability of the method.

In this account, we describe a protocol to design and carry out SRM experiments, the selection of the peptides, and the associated transitions, as well as the optimization and validation of the SRM assays. Precise quantification of targeted proteins using internal standards and differential analysis using the so-called label-free approach is discussed.

2. Materials

1) Nano-HPLC system capable of delivering small flow rates reliably, with minimal dead volumes, and equipped with an auto-sampler with a reproducible microliter injection capability.
2) Reversed-phase C18 column (3–5 μm particles).
3) HPLC buffer A: HPLC-grade water containing 0.1% formic acid.
4) HPLC buffer B: HPLC-grade acetonitrile containing 0.1% formic acid.
5) A triple quadrupole mass spectrometer operated in SRM mode (*see* **Note 1**).
6) Isotope-labeled internal standard peptides for quantification.

3. Methods

3.1. Instrument Settings

This section describes typical instrument settings for both the HPLC system and the mass spectrometer.

1) Analyses are typically carried out in the nano-flow mode using 75–150 μm ID columns, at a flow rates ranging from 0.3 to 1.0 μL/min. The nano-HPLC system must be capable of delivering small flow rates reliably, have minimal dead volumes, and be equipped with an auto-sampler with a reproducible microliter injection capability.
2) HPLC separations are performed on a reversed-phase column (C18 column; 3–5 μm particles) using a gradient, for instance. 5% B to 45% B in 40 min. The gradient is generated by mixing two mobile phases: water containing 0.1% formic acid (A) and acetonitrile containing 0.1% formic acid (B).
3) LC-MS measurements are performed on a calibrated TSQ Vantage triple quadrupole mass spectrometer (*see* **Note 2**) operated in SRM mode using time-based SRM (t-SRM, see below). For ionization, spray voltage and capillary temperature were set at 1,800 V and 200°C, respectively. The selectivity for both Q1 and Q3 were set at 0.7 Da (FWHM). Argon was used as collision gas, and the nominal gas pressure was set at 1.2 mTorr. A cycle time of 2 s and window size of 4-min are used for *t-SRM* experiments.
4) The instrument methods used to perform multiplexed SRM experiments have to be carefully designed prior to actual experiment to ensure success (*see* **Section 3.4**).

3.2. Sample Preparation

As mentioned above, the sample preparation is a critical part of a quantitative experiment. Protocols ensuring a high recovery and the reduction of sample complexity are a prerequisite to achieve low limits of detection and quantification.

1) The biological samples to be analyzed (e.g., cell lysates, bodily fluids) are processed using established proteomic procedures to isolate the protein fraction. After a (optional) crude fractionation, the protein samples are reduced, alkylated, and digested with trypsin prior to LC-MS analysis. It is advisable to perform a sample cleanup using reversed-phase cartridges to improve the analysis consistency by removal of contaminants.
2) The digest is directly used for the LC-MS quantitative analysis. In a typical experiment, up to a 1 μg total protein digest can be injected.

3) To determine the absolute amounts of targeted proteins, the isotope-labeled peptides, which are the counterparts of endogenous analytes, are spiked into the digest mixture at well-defined concentrations prior to the LC-MS analysis (typically several dilutions are performed).

3.3. Design of an SRM Experiment

SRM experiments have the specific objective to quantify a predefined set of peptides in complex mixtures, exploiting the high sensitivity and selectivity due to the two-stage mass selection of the triple quadrupole and the narrow mass selection windows.

1) Dedicated software packages are available to expedite the generation of the method, through the automated selection of peptides and their transitions. Several such tools have been proposed, for instance, TIQAM (4) and Pinpoint (5).
2) From a practical point of view, several workflows can be envisioned depending on the scope of the experiment and the existing background information. Two typical proteomic applications are biomarker validation studies to precisely determine the concentration of candidates in a larger number of samples (e.g., serum), and large screens to detect peptides and estimate their abundances in complex biological samples. Both approaches have different objectives and specific expectations in terms of the quality of the data to be generated and thus adequate experimental design (*see* **Fig. 9.1**).

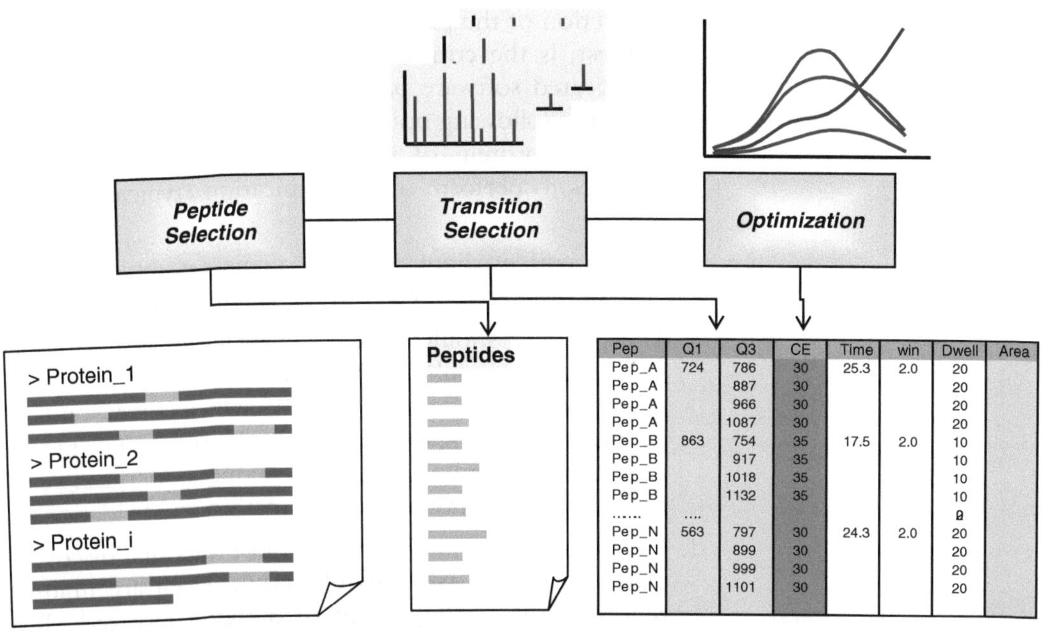

Fig. 9.1. Workflow to design an SRM experiment.

3) For precise quantification, the number of peptides to be quantified is limited (usually to a few dozen) and references, i.e., isotopically labeled peptides, must be available (*see* **Note 3**). The limits of detection and quantification associated with the targeted peptides and the linear dynamic range are established by spiking isotope-labeled peptides into the digest. The absolute amount of each targeted endogenous peptide is determined by computing the peak area of the signals of the endogenous and isotope-labeled peptides.
4) For large peptide screens in which hundreds of peptides are analyzed, a primary (initial) screen is performed without reference peptides. The information necessary to design the experiment is assembled from previous proteomic studies [i.e., query existing proteomic databases (6)] or predictive bioinformatics tools [i.e. calculation of retention times (7) or fragment ions of peptides (8)]. Without internal standards, it is only possible to estimate the amount of peptides or to perform relative quantification based only on the actual ion counts, as the response factors remain unknown. The aim of such studies is mainly qualitative, i.e. detection of the peptides in the sample of interest and estimation of their amounts (through comparison of multiple samples).

3.4. Generation of an SRM Method

SRM studies are by definition hypothesis driven, i.e., focused on a specific set of proteins corresponding to a specific biological or clinical question, e.g., a set of biomarkers specific to a disease, a biochemical pathway, or a protein network.

1) The selection of the peptides, as surrogates for the proteins of interest, is the critical part of the experimental design, and dedicated software packages facilitate this step, which includes the following points:
   a. For each protein of interest, select a minimal set of peptides to perform the quantification (typically two or three).
   b. These peptides should be proteotypic, i.e., unique to the protein of interest (or at least a protein family).
   c. The peptides should ionize well in electrospray mode, i.e., produce doubly or triply charged ions in the practical mass range (m/z 400–1,500), which corresponds to peptides containing 8–20 amino acid residues. For practical reasons, peptides containing residues undergoing easy modification are excluded, e.g., methionine oxidation.
   d. For proteins previously analyzed in proteomic discovery experiments, the peptides observed, which are unique, represent the starting point to design a targeted SRM

method. Information can be extracted from either public repositories such as Peptide-Atlas or MRM-Atlas (9) or in-house discovery results. They are directly imported in the software package used to design the method (e.g., Pinpoint). Existing MS/MS spectral libraries are queried to extract fragment ions to generate the SRM transitions, usually the most intense assigned fragment ions.

e. If the proteins have not been reported before, the peptide selection is performed in silico using the protein sequences. Precursor and fragment ion m/z are calculated, and the elution time is predicted using hydrophobicity indexes (7).

2) For each peptide, a series of transitions are defined, to ensure a high degree of selectivity. If the MS/MS spectra of the peptides were previously reported, five to eight of the most intense y ions are selected from the reference spectrum (*see* **Note 4**). If no MS/MS information is available, the m/z values of the fragments are calculated, and y fragments with m/z values above the precursor ion are favored.

3) The collision energy (CE) in an initial study is calculated for each peptide based on a generic formula of CE (V) = 2 + 0.034 × precursor ion m/z.

4) This initial transition list is exported as a csv file and uploaded into the instrument method.

3.5. LC-MS Analysis in SRM Mode

The instrument settings for an SRM experiment are described in **Section 3.1**.

1) Import the SRM method, which includes the Q1 and Q3 m/z values and the collision energy of each transition.

2) By default, the start time and the stop time of each transition are set to 0 and the end of the gradient time, respectively. Specific elution time windows are active in *t-SRM* experiments (see below).

3) The SRM is performed using the HPLC, MS settings, and the specific SRM method.

3.6. Data Analysis and Evaluation

1) The raw data files are processed with the instrument software (e.g., Pinpoint and Xcalibur) and the results are exported in a tabular format (e.g., Excel sheet).

2) The peptide identity is confirmed by verifying the co-elution of multiple fragment ions of each analyte and their intensities are compared with a reference MS/MS spectrum.

3) If isotopically labeled reference peptides are used, the co-elution of the reference and the endogenous peptides is a strong evidence of the identity, together with their

fragmentation patterns. **Figure 9.2** shows an example of co-eluting reference and endogenous peptides confidently identified and presenting well-resolved HPLC traces which will be retained for systematic quantitative assays. By spiking the reference peptides at different concentration levels, the LOD/LOQ and dynamic range of the peptides is established by the addition of reference peptides to generate dilution curves, as illustrated in **Fig. 9.3**.

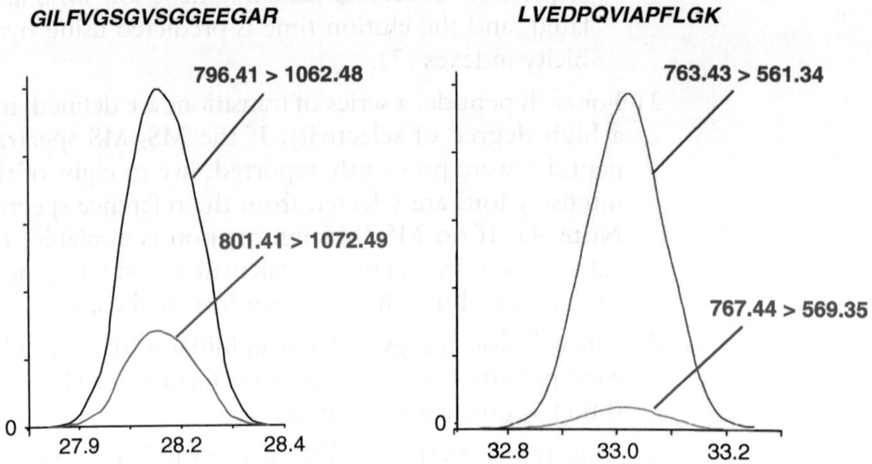

Fig. 9.2. SRM traces of the endogenous peptide and the internal standard.

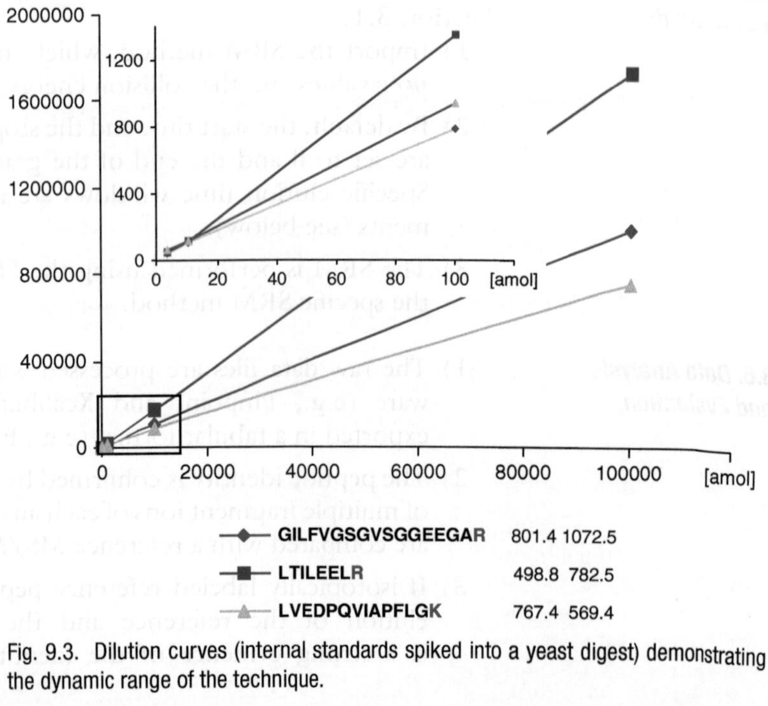

Fig. 9.3. Dilution curves (internal standards spiked into a yeast digest) demonstrating the dynamic range of the technique.

4) Typically, for each peptide to be analyzed, three to five transitions are selected for an optimized SRM assay. In addition, the elution time of each peptide is included in the method, and the SRM analysis is performed in the *t-SRM* mode. The time window for each peptide is user defined.

3.7. Optimization of SRM Experiments

In high-sensitivity experiments, it is important to tune the instrument parameters, in particular, the collision energy to maximize the signal of each transition. In our hands, the information derived from an MS/MS spectrum is usually sufficient to establish an initial method to screen for the presence of peptides and estimate their abundance. However, for the quantification of low-abundance components, optimization of the collision energy is often desirable to increase the signal-to-noise ratio.

1) To optimize the method, the experiment is repeated with different parameters; the initial analysis is performed with default collision energy setting (calculated using an empirical formula: CE = $+2.0 \times 0.034 \times m/z$ value of the precursor).
2) In a second phase, the experiment is repeated with lower and higher settings (e.g., 3 and 6 V below and above the default value); the best conditions are retained for the final method. The optimization often increases the sensitivity by a factor of 2–3, which is not negligible for low-intensity signals (*see* **Note 5**).

3.8. Large-Scale Experiment: Time-Based SRM (T-SRM)

As mentioned earlier, typical proteomic experiments aim at measuring a larger number of peptides in one single experiment. Recent developments of instrument data acquisition software enable us to use the observed or predicted elution time of a peptide and to monitor the signals associated with that peptide only during a small time window. This allows a more effective use of the instrument time and will significantly increase the number of transitions measured. The *t-SRM* method includes the instrument parameters and the specific parameters associated with each transition.

1) Import the optimized SRM method, which includes information of Q1 m/z, Q3 m/z, collision energy, and retention time for each transition.
2) Set the time window for the *t-SRM* experiment. The triple quadrupole instrument is operated with a cycle time of 2 s, the collision gas set at 1.2 mTorr, and a resolving power of Q1 and Q3 of 0.7 unit (FWHM).
3) Set up the HPLC method, including the gradient.
4) Triplicate the t-SRM experiment using the optimized SRM method.

3.9. Quantification Using Stable Isotope Dilution

1) The peptides used as internal standards are selected according to the criteria discussed above, i.e., they are unique to the targeted proteins, and present good mass spectrometric properties. In addition, these peptides may fulfill additional criteria regarding their composition, i.e., exclusion of amino acids or sequences undergoing easy chemical modification, i.e., oxidation of methionine residues or facile deamidation.
2) The ^{13}C stable isotopes are commonly incorporated into the C-terminal arginine or lysine residues.
3) The data acquisition method is modified to include the transitions associated with the internal standards; this step can be performed automatically by the method building software.
4) Add well-defined amounts of isotopically labeled peptide (e.g., 1 fmol/μL) to the peptide digest to be analyzed.
5) The LC-MS analysis is performed in the *t-SRM* mode using a method of targeting the endogenous and isotope-labeled peptide pairs.
6) During the data analysis, the absolute amount of each targeted peptide present in the sample is computed based on the peak area ratio of the endogenous and isotope-labeled peptide, and the amount of internal standard added to the sample.
7) For rigorous quantification analysis, one might consider to repeat the analysis with the internal standard at several concentrations in order to establish the linear response range.

3.10. Evaluating the Robustness of the Method

To evaluate the robustness of an optimized SRM method, replicate analyses are performed.

1) The results from replicated runs (minimally triplicates) allow determination of the precision of the method, both at the individual transition level and at the peptide level.
2) Measurements that exhibit large variability should be excluded; the ultimate assay will retain the transitions which showed lowest coefficient of variance (CV).
3) Whenever possible, at least two peptides for each protein of interest (covering different regions of the sequence) should be analyzed to ensure reliability of the analyses.

4. Notes

1) Although, any triple quadrupole instrument is capable of SRM measurements, a time-scheduling capability is essential for monitoring a large number of SRM transitions during a

single LC-MS run. The TSQ Vantage (Thermo Scientific) has been used for the SRM experiments described in this account.

2) The instrument needs to be properly calibrated and tuned for peptide quantification. As quality control, a mixture of peptide standards with well-defined concentrations (such as 1 fmol/μL) is used to assess the HPLC separation and instrument performances before analyzing biological samples.

3) The isotope-labeled peptides are also used to optimize the experimental conditions to maximize the overall sensitivity, by fine-tuning specific instrument parameters such as the collision energy and the source potentials. Synthetic peptides in which stable isotopes are incorporated into the C-terminal amino acid are commonly used as they have become commercially available from several sources at reasonable pricing. Alternative isotope dilution approaches are emerging, including the addition of concatenated recombinant protein, referred to as QconCAT composed of the peptides to be quantified (10).

4) If a good correlation between MS/MS spectra measured on an ion trap instrument and triple quadrupole instrument has been demonstrated (11), this allows the use of library spectra from a different platform to design SRM experiments.

5) For larger scale studies, i.e., analysis of large number of samples, the optimization of LC-MS will have a large impact on the overall time. The complexity of the sample is often the limiting factor. In essence, a relatively short gradient would be preferred if the sample complexity accommodates it. Most experiments in our hands were performed using 30- or 40-min gradients.

Acknowledgments

Dr. Andreas Huhmer and David Fischer are gratefully acknowledged for helpful discussion.

References

1. Gerber, S. A., Rush, J., Stemman, O., Kirschner, M. W., and Gygi, S. P. (2003) Absolute quantification of proteins and phosphoproteins from cell lysates by tandem MS. *Proc. Natl. Acad. Sci. USA* **100**, 6940–6945.

2. Picotti, P., Bodenmiller, B., Mueller, L. N., Domon, B., and Aebersold, R. (2009) Full dynamic range proteome analysis of *S. cerevisiae* by targeted proteomics. *Cell* **138**, 795–806.

3. Anderson, N. L., Polanski, M., Pieper, R., Gatlin, T., Tirumalai, R. S., Conrads, T. P., Veenstra, T. D., Adkins, J. N., Pounds, J. G., Fagan, R., and Lobley, A. (2004) The human plasma proteome: a nonredundant list developed by combination of four separate sources. *Mol. Cell. Proteomics* **3**, 311–326.
4. Stahl-Zeng, J., Lange, V., Ossola, R., Eckhardt, K., Krek, W., Aebersold, R., and Domon, B. (2007) High sensitivity detection of plasma proteins by multiple reaction monitoring of *N*-glycosites. *Mol. Cell. Proteomics* **6** 1809–1817.
5. Prakash, A., Kiyonami, R., Shoen, A., Nguyen, H., Peterman, S., Huhmer, A., Lopez, M., and Domon, B. (2009) Integrated workflow to design methods and analyze data in large-to-extremely-large scale SRM experiments. *Proc. ASMS 2009 Poster*, TH695.
6. Wojcik, J., and Schächter, V. (2000) Proteomic databases and software on the web. *Oxford J. Brief Bioinform.* **1**, 250–259.
7. Krokhin, O. V., Craig, R., Spicer, V., Ens, W., Standing, K. G., Beavis, R. C., and Wilkins, J. A. (2004) An improved model for prediction of retention times of tryptic peptides in ion-pair reverse-phase HPLC: its application to protein peptide mapping by off-line HPLC–MALDI MS. *Mol. Cell. Proteomics* **3**, 908–319.
8. Chalkley, R.J., Baker, P. R., Medzihradszky, K. F., Lynn, A. J., and Burlingame, A. L. (2008) In-depth analysis of tandem mass spectrometry data from disparate instrument types. *Mol. Cell. Proteomics* **7**, 2386–2398.
9. Picotti, P., Lam, H., Campbell, D., Deutsch, E. W., Mirzaei, H., Ranish, J., Domon, B., and Aebersold, R. (2008) A database of mass spectrometric assays for the yeast proteome. *Nat. Methods* **5**, 913–914.
10. Pratt, J. M., Simpson, D. M., Doherty, M. K., Rivers, J., Gaskell, S. J., and Beynon, R. J. (2006) Multiplexed absolute quantification for proteomics using concatenated signature peptides encoded by QconCAT genes. *Nat. Protoc.* **1**, 1029–1043.
11. Prakash, A., Tomazela, D. M., Frewen, B., Maclean, B., Merrihew, G., Peterman, S., Maccoss, M. J. (2008) Expediting the development of targeted SRM assays: using data from shotgun proteomics to automate method development. *J. Proteome Res.* **8**, 2733–2739.

Chapter 10

Basic Design of MRM Assays for Peptide Quantification

Andrew James and Claus Jorgensen

Abstract

With the recent availability and accessibility of mass spectrometry for basic and clinical research, the requirement for stable, sensitive, and reproducible assays to specifically detect proteins of interest has increased. Multiple reaction monitoring (MRM) or selective reaction monitoring (SRM) is a highly selective, sensitive, and robust assay to monitor the presence and amount of biomolecules. Until recently, MRM was typically used for the detection of drugs and other biomolecules from body fluids. With increased focus on biomarkers and systems biology approaches, researchers in the proteomics field have taken advantage of this approach. In this chapter, we will introduce the reader to the basic principle of designing and optimizing an MRM workflow. We provide examples of MRM workflows for standard proteomic samples and provide suggestions for the reader who is interested in using MRM for quantification.

Key words: Multiple reaction monitoring (MRM), precursor and fragment ion selection, optimization and validation of transitions, liquid chromatography, triple quadrupole mass spectrometer.

Abbreviations

QqQ	triple quadrupole mass spectrometer
SRM	selective reaction monitoring
MRM	multiple reaction monitoring
sMRM	scheduled MRM
CID	collision-induced dissociation
Amu	atomic mass unit
LC	liquid chromatography
LC-MS/MS	liquid chromatography mass spectrometry

1. Introduction

Mass spectrometry is commonly used for identification and quantification of a wide range of analytes such as lipids, drugs, proteins, and peptides. The analysis of proteins is, in most

P.R. Cutillas, J.F. Timms (eds.), *LC-MS/MS in Proteomics*, Methods in Molecular Biology 658, DOI 10.1007/978-1-60761-780-8_10, © Springer Science+Business Media, LLC 2010

instances, conducted on the peptide level, where proteins are digested by a protease such as trypsin, followed by the analysis of the resulting peptides.

Workflows for proteomics experiments can crudely be divided into two categories: discovery-based proteomics and targeted proteomics. In discovery-based experiments, samples are analyzed in an unbiased manner that does not require any prior knowledge of the sample composition. This approach is exemplified by shot-gun proteomics experiments where the mass spectrometer is operated to select only the most intense precursor ions for fragmentation. In contrast, in targeted proteomics approaches, only peptides that have been pre-selected are analyzed. In these cases, prior knowledge of the protein(s) of interest and sample composition is required. In a targeted proteomics assay, peptides of interest are selected and the mass spectrometer is set to ignore all other peptides present in the sample. This setup provides high analytical reproducibility, a better signal-to-noise, and increased dynamic range (1). Targeted approaches are typically conducted using multiple reaction monitoring (MRM), which is a non-scanning quantitative mass spectrometric workflow typically run on a triple quadrupole mass spectrometer (QqQ) (**Fig. 10.1A**).

MRM is typically used to detect and quantify between ten and a few hundred peptides, but can in principle be used for thousands of peptides in one single assay. Typical applications of MRM assays include detection and quantification of serum proteins present at low nanogram per milliliter concentrations (2, 3), which in combination with the high reproducibility (4) makes MRM a first choice for biomarker validation. In addition, MRM has also been applied to phospho-proteomics analysis (5, 6) and for systems analysis of the yeast metabolic enzymes (7).

Here we describe how to plan, design, and conduct MRM experiments exemplified by commonly available standards for mass spectrometry. We will first go through some of the basic concepts and considerations for designing MRM experiments. We will describe how to use discovery-based data to design the MRM assay, how to obtain selectivity, and how to use internal and external controls to attain a carefully designed MRM experiment.

2. Materials

(1) Triple quadrupole mass spectrometer

(2) Liquid Chromatography system

(3) Trypsin-digested BSA and alpha-casein

(4) Glu-fibrinogen

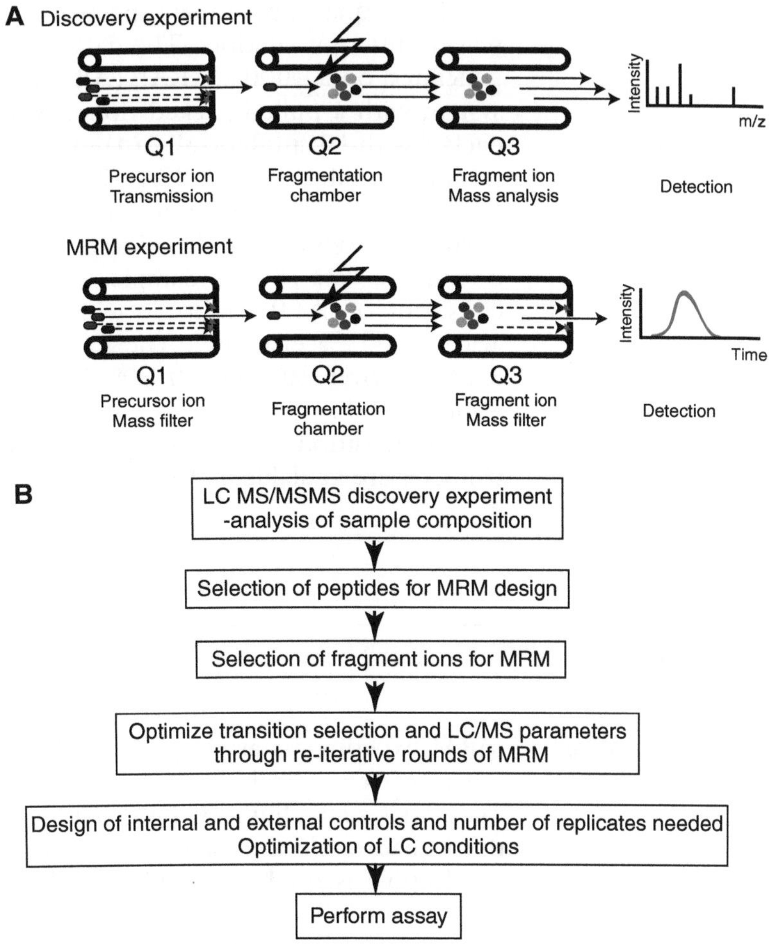

Fig. 10.1 (**A**) Outline of information-dependent analysis (IDA) mass spectrometry and MRM mass spectrometry. In IDA experiments, the most abundant precursor ion is selected (in Q1) and then fragmented (in Q2). This is followed by subsequent analysis of all fragment ions (in Q3). The resulting MS/MS spectrum is used for identification of the fragmented precursor ions. In an MRM experiment only pre-defined peptides (in Q1) are selected and fragmented (in Q2). Fragment ions, which have been pre-defined, are selectively passed through Q3 and detected. Thus in comparison, MRM is a selective workflow for mass spectrometry, which only detects a pre-defined combination of precursor and fragment ions. (**B**) Flow diagram for MRM design.

(5) Formic acid (HCOOH)
(6) Acetonitrile (CH_3CN)
(7) OmniSolv-grade water
(8) Trypsin
(9) Iodoacetamide (ICH_2CONH_2)
(10) Dithiothreitol ($HSCH_2CH(OH)CH(OH)CH_2SH$)
(11) C18 reverse phase, Reprosil AQ pur (Dr. Maisch)

Triple quadrupole mass spectrometers can be acquired from several different vendors. The data presented in this chapter was generated on a 4000 QTRAP (AB/Sciex) using an Eksigent LC system with a pulled packed tip sprayer ((360 OD/75 μm ID) packed with 10 cm Reprosil AQ pur). Tryptic digestions of bovine serum albumin (30 fmol, BSA with carboxymethyl-Cys, Michrom Bioresources, Inc.) and alpha-casein (60 fmol, reduced and alkylated with iodoacetamide, digested in-house) were injected to obtain the data shown. Samples were run using mobile phases consisting of (A) 0.1% formic acid in water and (B) 0.1% formic acid in 100% acetonitrile. The LC gradient was run as follows: 5–30% B over 30 min, 30–80% B over 5 min, and 80% B for 5 min at 400 nl/min.

All transitions as well as the specific settings of the mass spectrometer are available in **Table 10.1**.

3. Methods

3.1. Principle of MRM

MRM experiments are typically run on triple quadrupole mass spectrometers, which contain three serially placed quadrupole mass analyzers (**Fig. 10.1A**). The first quadrupole and the third quadrupole are used to scan for ions, whereas the second quadrupole is used as a collision cell and to transmit all fragment ions to Q3. In an MRM experiment, only the pre-selected precursor ion(s) of the specific *m/z* ratio(s) will be transmitted through the first quadrupole (Q1), while all other ions will be excluded. In the second quadrupole (Q2), the precursor ion(s) will be fragmented by collision-induced dissociation (CID), and all resulting fragment ions will subsequently be transmitted to the third quadrupole (Q3). In this quadrupole, only ions corresponding to the pre-set *m/z* ratios of selected fragment ion(s) will be transmitted to the detector, whereas all remaining ions will be excluded.

MRMs are highly selective since a signal is registered only when a pre-selected fragment ion is produced by a pre-selected precursor. Thus, in essence only two pieces of information are required to build an MRM for a peptide of interests: the precursor mass-to-charge ratio and the mass-to-charge ratio of its fragments following CID. The MRM workflow supports both relative and absolute quantification and is advantageous due to the higher dynamic range and selective use of the scan time.

Table 10.1
List of transitions used for control MRM (on a 4000 QTRAP). Included is the precursor ion *m/z*, the fragment ion *m/z*, the peptide sequence, and the collision energy

Q1	Q3	Dwell time (min)	Protein/sequence/mass/charge/fragmention	CE
409,71	646,34	20	BSA.ATEEQLK.410(2+)-646(y5)	23,027
409,71	747,39	20	BSA.ATEEQLK.410(2+)-747(y6)	23,027
417,21	746,35	20	BSA.FKDLGEEHFK.417(3+)-746(y6)	18,357
417,21	974,46	20	BSA.FKDLGEEHFK.417(3+)-974(y8)	18,357
443,72	656,36	20	BSA.DDSPDLPK.444(2+)-656(y6)	24,523
443,72	569,33	20	BSA.DDSPDLPK.444(2+)-569(y5)	24,523
450,24	625,35	20	BSA.LC[CM]VLHEK.450(2+)-625(y5)	24,789
450,24	786,35	20	BSA.LC[CM]VLHEK.450(2+)-785(y6)	24,789
461,74	722,41	20	BSA.AEFVEVTK.462(2+)-722(y6)	25,317
461,74	575,34	20	BSA.AEFVEVTK.462(2+)-575(y5)	25,317
464,25	651,35	20	BSA.YLYEIAR.464(2+)-651(y5)	25,427
464,25	488,28	20	BSA.YLYEIAR.464(2+)-488(y4)	25,427
474,23	722,3	20	BSA.SLHTLFGDELC[CM]K.474(3+)-721(y6)	20,852
474,23	869,33	20	BSA.SLHTLFGDELC[CM]K.474(3+)-868(y7)	20,852
487,73	746,35	20	BSA.DLGEEHFK.488(2+)-746(y6)	26,46
487,73	560,28	20	BSA.DLGEEHFK.488(2+)-560(y4)	26,46
489,19	918,28	20	BSA.TC[CM]VADESHAGC[CM]EK.489(3+)-1103(y10)	21,495
489,19	602,25	20	BSA.TC[CM]VADESHAGC[CM]EK.489(3+)-602(fragment)	21,495
499,3	601,39	20	BSA.[PGQ]-QTALVELLK.499(2+)-601(y5)	26,969
499,3	502,32	20	BSA.[PGQ]-QTALVELLK.499(2+)-502(y4)	26,969
501,79	800,45	20	BSA.LVVSTQTALA.502(2+)-800(b8)	27,079
501,79	913,54	20	BSA.LVVSTQTALA.502(2+)-914(b9)	27,079
507,82	601,39	20	BSA.QTALVELLK.508(2+)-601(y5)	27,344
507,82	502,32	20	BSA.QTALVELLK.508(2+)-502(y4)	27,344
512,25	599,36	20	BSA.LKEC[CM]C[CM]DKPLLEK.512(3+)-599(y5)	22,51
512,25	1003,47	20	BSA.LKEC[CM]C[CM]DKPLLEK.512(3+)-1292(y10)	22,51
519,22	609,31	20	BSA.DDPHAC[CM]YSTVFDK.519(3+)-609(y5)	22,831
519,22	860,25	20	BSA.DDPHAC[CM]YSTVFDK.519(3+)-508(y4)	22,831
526,59	699,25	20	BSA.LKPDPNTLC[CM]DEFK.526(3+)-811(y6)	23,155
526,59	812,32	20	BSA.LKPDPNTLC[CM]DEFK.526(3+)-698(y5)	23,155
537,25	849,39	20	BSA.SHC[CM]IAEVEK.537(2+)-848(y7)	28,617

Table 10.1 (Continued)

Q1	Q3	Dwell time (min)	Protein/sequence/mass/charge/fragmention	CE
537,25	927,33	20	BSA.SHC[CM]IAEVEK.537(2+)-985(y8)	28,617
547,32	589,33	20	BSA.KVPQVSTPTLVEVSR.547(3+)-589(y5)	24,082
547,32	900,51	20	BSA.KVPQVSTPTLVEVSR.547(3+)-901(y8)	24,082
549,82	768,44	20	aCasein_s2.AMKPWIQPK.550(2+)-768(y6)	29,192
549,82	896,54	20	aCasein_s2.AMKPWIQPK.550(2+)-897(y7)	29,192
554,75	908,4	20	BSA.EAC[CM]FAVEGPK.554(2+)-600(y6)	29,387
554,75	747,39	20	BSA.EAC[CM]FAVEGPK.554(2+)-747(y7)	29,387
570,74	818,44	20	BSA.C[CM]C[CM]TESLVNR.570(2+)-818(y7)	30,069
570,74	588,35	20	BSA.C[CM]C[CM]TESLVNR.570(2+)-588(y5)	30,069
571,86	886,56	20	BSA.KQTALVELLK.572(2+)-887(y8)	30,162
571,86	1014,62	20	BSA.KQTALVELLK.572(2+)-1015(y9)	30,162
582,32	951,48	20	BSA.LVNELTEFAK.582(2+)-951(y8)	30,622
582,32	708,39	20	BSA.LVNELTEFAK.582(2+)-708(y6)	30,622
626,3	881,38	20	aCasein_s2.EQLSTSEENSK.626(2+)-881(y8)	32,557
626,3	794,35	20	aCasein_s2.EQLSTSEENSK.626(2+)-794(y7)	32,557
627,97	948,47	20	BSA.RPC[CM]FSALTPDETYVPK.628(3+)-948(y8)	27,616
627,97	934,46	20	BSA.RPC[CM]FSALTPDETYVPK.628(3+)-933(b8)	27,616
634,34	771,47	20	aCasein_s1.YLGYLEQLLR.634(2+)-771(y6)	32,911
634,34	658,39	20	aCasein_s1.YLGYLEQLLR.634(2+)-658(y5)	32,911
644,25	1108,47	20	BSA.C[CM]C[CM]AADDKEAC[CM]FAVEGPK.643(3+)-978(y9)	28,304
644,25	979,43	20	BSA.C[CM]C[CM]AADDKEAC[CM]FAVEGPK.643(3+)-747(y7)	28,304
653,36	712,44	20	BSA.HLVDEPQNLIK.653(2+)-712(y6)	33,748
653,36	1055,57	20	BSA.HLVDEPQNLIK.653(2+)-1056(y9)	33,748
684,35	827,4	20	aCasein_s2.ALNEINQFYQK.684(2+)-827(y6)	35,111
684,35	940,49	20	aCasein_s2.ALNEINQFYQK.684(2+)-940(y7)	35,111
692,84	920,49	20	aCasein_s1.FFVAPFPEVFGK.693(2+)-920(y8)	35,485
692,84	991,52	20	aCasein_s1.FFVAPFPEVFGK.693(2+)-992(y9)	35,485
693,84	1186,53	20	aCasein_s2.TVDMESTEVFTK.694(2+)-1187(y10)	35,529
693,84	940,46	20	aCasein_s2.TVDMESTEVFTK.694(2+)-940(y8)	35,529
706,25	1041,6	20	aCasein_s2.EQLS[Pho]T[Pho]SEENSK.706(2+)-1041(y8)	36,075
706,25	943,32	20	aCasein_s2.EQLS[Pho]T[Pho]SEENSK.706(2+)-1154(y9)	36,075
706,25	648,5	20	aCasein_s2.EQLS[Pho]T[Pho]SEENSK.706(2+)-874(y7)	36,075

Table 10.1 (Continued)

Q1	Q3	Dwell time (min)	Protein/sequence/mass/charge/fragmention	CE
722,82	1168,5	20	BSA.YIC[CM]DNQDTISSK.722(2+)-1167(y10)	36,782
722,82	892,44	20	BSA.YIC[CM]DNQDTISSK.722(2+)-892(y8)	36,782
733,81	922,43	20	aCasein_s2.TVDMES[Pho]TEVFTK.734(2+)-1266(y10)	37,288
733,81	1168,5	20	aCasein_s2.TVDMES[Pho]TEVFTK.734(2+)-1151(y9)	37,288
733,81	395,2	20	aCasein_s2.TVDMES[Pho]TEVFTK.734(2+)-1151(y9)	37,288
740,4	813,49	20	BSA.LGEYGFQNALIVR.740(2+)-813(y7)	37,578
740,4	1017,58	20	BSA.LGEYGFQNALIVR.740(2+)-1018(y9)	37,578
752,79	1012,37	20	BSA.EYEATLEEC[CM]C[CM]AK.752(2+)-796(y6)	38,08
752,79	1083,42	20	BSA.EYEATLEEC[CM]C[CM]AK.752(2+)-909(y7)	38,08
784,37	717,32	20	BSA.DAFLGSFLYEYSR.784(2+)-717(y5)	39,512
790,93	802,37	20	aCasein_s1.VPQLEIVPNSAEER.791(2+)-802(y7)	39,801
790,93	901,44	20	aCasein_s1.VPQLEIVPNSAEER.791(2+)-901(y8)	39,801
830,91	784,39	20	aCasein_s1.VPQLEIVPNS[Pho]AEER.831(2+)-784(fragment)	41,56
830,91	882,34	20	aCasein_s1.VPQLEIVPNS[Pho]AEER.831(2+)-882(y7)	41,56
875,34	1316,46	20	BSA.YNGVFQEC[CM]C[CM]QAEDK.874(2+)-910(y7)	43,472
875,34	1041,33	20	BSA.YNGVFQEC[CM]C[CM]QAEDK.874(2+)-1039(y8)	43,472
976,48	971,56	20	aCasein_s1.YKVPQLEIVPNS[Pho]AEER.976(2+)-972(b8)	47,965
976,48	882,34	20	aCasein_s1.YKVPQLEIVPNS[Pho]AEER.976(2+)-972(b8)	47,965
976,48	1561,73	20	aCasein_s1.YKVPQLEIVPNS[Pho]AEER.976(2+)-1562(y13)	52
687,955	876,4	20	bCasein.FSQ[Pho]EEQQQTEDELQDK	43
687,955	977,4	20	bCasein.FSQ[Pho]EEQQQTEDELQDK	43

3.2. Designing MRMs

Designing an MRM can be divided into three sections (*see* **Fig. 10.1B** for flowchart).

3.2.1. Selection of Precursor Ions

In most cases, an MRM assay will be designed using existing identifications based on LC-MS/MS experiments (**Figs. 10.1B** and **10.2**). Although an assay can be designed from a number of

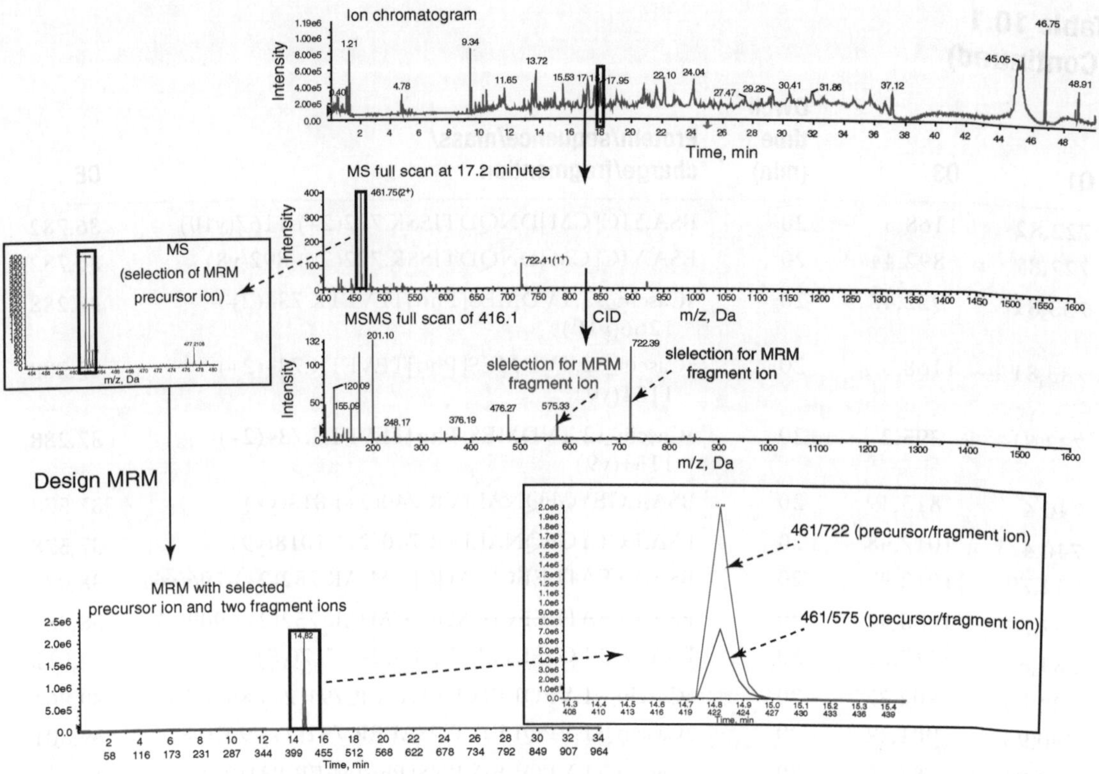

Fig. 10.2 Selection and design of transitions from a LC-MS/MS experiment. Peptides from discovery-based experiments are identified based on their fragmentation pattern. When a specific peptide is eluting from the reverse-phase column the resulting ion current is detected (in the ion chromatogram). Subsequently, the double-charged peptide is selected (*m/z* 461.75) for fragmentation by collision-induced fragmentation (CID). The resulting fragmentation pattern is subsequently used for peptide identification. Transitions are selected from the precursor ion mass to charge (*m/z* 461.75) and the resulting fragment ions (*m/z* 722.39 and 575.33). When the same sample is analyzed by an MRM assay, specific for these particular transitions, all other ions are ignored. As can be seen from the *insert*, the transitions co-elute. In addition, the relative abundance of the signals from the transitions corresponds to the relative abundance of the fragment ions in the MS/MS spectrum.

in silico resources and spectral libraries (8–11), we will focus on establishing an assay using data generated from the biological system of interest. Prior analysis ensures that the sample preparation protocol is suitable for mass spectrometric analysis and that the subset of proteins and peptides selected for monitoring by MRM can be detected by the mass spectrometer.

In addition, by using existing LC-MS/MS data, potential interference from isobaric peptides, splice variants, etc., can better be taken into account. However, the potential for interference from these peptides depends greatly on the sample type that is analyzed. For example, the complexity of a digested cell lysate can mask the presence of isobaric peptides and splice variants, whereas analysis of a purified protein is more likely to reveal if interfering peptides are present.

During the establishment and design of MRMs it is recommended to use several specific peptides from each protein. By selecting several proteotypic peptides (see below) for each protein, it is more likely that a highly selective and sensitive assay can be obtained. Furthermore, in certain cases where some of the selected peptides are unknowingly post-translationally modified, these can more easily be resolved when several peptides have been included. Finally, by including a greater number of peptides, low responding peptides can be discarded during the initial rounds of optimization. Depending on the sample type, we aim toward obtaining minimum three peptides from each protein for the final MRM assay.

There are a number of features that can be used to create a sensitive and specific MRM assay; these include the following:

(a) Selecting peptides that ionize and fragment well, which will provide a more sensitive assay.

(b) Ensuring that the m/z ratio of the selected peptides is compatible with the mass range of the instrument used.

(c) Selecting peptides that are unique to the protein(s), which will be monitored. Such peptides are frequently referred to as proteotypic peptides (9, 12).

(d) If possible, avoiding peptides that contain amino acids that are prone to chemical modification and rearrangements. For instance, peptides containing methionine should be avoided due to its ability to undergo oxidization. Cysteine is typically modified during sample preparation, as such; we usually try to avoid Cys-containing peptides, which avoids potential interference from variations in sample preparation. It is also advisable to deselect peptides that are prone to deamidation (Asn, Glu), as well as peptides with N-terminal Gln or Glu due to their ability to form pyroglutamate. While proline-containing peptides typically fragment very well, care should be taken to ensure that enough fragment ions could be used to enable high-confident identification of the peptides. It is best to avoid peptides that are prone to missed enzymatic cleavages, which thereby avoids interference from sample preparation. Note that trypsin is unable to digest when proline is C-terminal to Arg or Lys, and these peptides can safely be included in the MRM assay.

By excluding peptides with the properties listed above, it is easier to control for external factors such as variability in sample preparation and sample storage. However, in certain cases it is impossible to avoid peptides that are prone to modifications, and in such cases, it will be crucial to monitor both the modified and the unmodified versions of the peptide.

3.2.2. Design of Transitions

For each precursor peptide (precursor ion) three or more peptide fragments are typically selected (fragment ions). Each corresponding pair of precursor and fragment ions is known as a transition. Selection of the fragment ions is based on their *m/z* ratios relative to the parent ion, where selection of fragment ions above the *m/z* ratio of the parent ion is recommended. As can be seen in **Fig. 10.2**, fragment ions above the parent ion typically have better signal-to-noise, and selection of these will ensure greater specificity for the peptide transition. Background signal (chemical noise) is most abundant below the *m/z* value of the precursor ion (**Fig. 10.2**).

Fragment ions from the y-ion series are typically good choices for initial design of transitions. While b-ions can be used, y-ions are more stable fragment ions during QqQ-based CID (13) and are preferential choices. If stable isotopically labeled tryptic peptides are used for AQUA (14) or SILAC (15) experiments, correcting y-type ions for the mass difference will enable the specific detection of the light and heavy labeled peptides. When chemical labels such as mTRAQ and iTRAQ are used both y- and b-type ions can be selected as the chemical label is known to stabilize the b-ions (16).

It is of importance to note that due to differences in the collision cells between mass spectrometers, the relative abundance of selected fragment ions can vary. This is the case when transferring information not only from ion traps to triple quadrupole mass spectrometers but also between quadrupole mass spectrometers from different vendors.

Thus, during the establishment of the MRM assay it is recommended to include several transitions from each peptide and then through rounds of optimization and validation determine which transitions are the most selective and the highest responding.

3.2.3. Conducting the MRM Experiment and Optimizing Transitions

When the mass spectrometric verification of the selected transitions is initiated, ensure that enough starting material is available to perform several rounds of analysis. This will ensure that variations in sample preparation do not affect the optimization of the MRM assay. After the sample has been prepared, aliquots should be made and frozen down. Finally, analyze an aliquot of the sample by LC-MS/MS to ensure the quality of the sample before conducting the MRM.

After the specific transitions have been entered into the data acquisition software, the dwell time should be determined. The dwell time is the amount of time the instrument uses to detect each transition within one duty cycle (see below). The specific settings depend on the instrument used as well as the abundance of the peptides. It is generally advisable to start with a lower number of transitions and have longer dwell time per transition in the initial phase. After the first rounds of analysis, additional transitions can be added and low responding transitions can be removed for subsequent rounds of optimization.

To show that the detected transitions originate from the expected peptides, full MS/MS scans of the precursor ion can be included (17, 18). The setup for such detection depends on the specific instrument and software used. Subsequently, these scans may also be manually inspected to ensure that the fragment ions that were selected are the most abundant following CID. Although it can be difficult to ensure optimal fragmentation for all peptides in the MRM assay, inspecting the full MS/MS scans can be used to optimize settings for fragmentation in the collision cell.

If a specific transition cannot readily be detected, the MRM transition can be optimized with a synthetic peptide of the same sequence. During later stages, an isotopically labeled version of this peptide can be included to control for consistency between the samples and for validation. Since AQUA- or SILAC-labeled peptides have similar properties as the unlabeled peptide, the elution profile and fragmentation pattern will match the unlabeled peptide. Thus, by selecting fragment ions from the y-ion series, the ratio between the non-labeled and the isotopically labeled peptide can be determined under the same experimental conditions and used for quantification (14).

3.3. Controls

After the initial MRM experiment has been set up and transitions have been verified by full scans and/or by isotopically labeled peptides, the MRM assay is ready for the next step. In this phase of the experiment, it is important to incorporate controls for biological, analytical, and chromatographic variability. Addition of an internal control at an early point during sample preparation can be useful in order to track variability in sample preparation. For example, adding a small amount of an extrinsic protein such as an *Escherichia coli* protein to a human lysate prior to digestion can be used to monitor digestion efficiency. Furthermore, adding a small amount of control (for example, a peptide) just before sample analysis can be used to control for reproducibility of the LC systems.

Injections of a blank as well as a control protein digest between sample runs offer two additional controls. Injection of a control sample following the analytical sample can be used to ensure instrument calibration and chromatographic consistency. Since the MRM assay only detects the peptides of interest (**Fig. 10.3**), the effects from contaminants such as polymers will be ignored; however, polymers can greatly affect the chromatographic consistency, and ionization efficiency, which is why control injections are essential. Finally, a blank injection of buffer A while running your specific MRM assay provides an important control for chromatographic carry over.

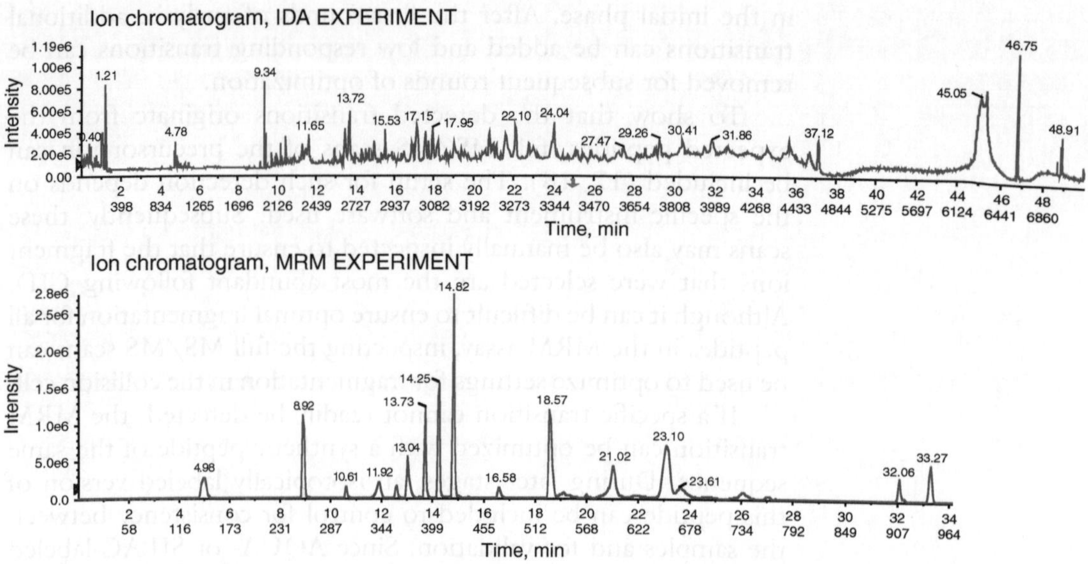

Fig. 10.3 Comparison between a full scan (discovery-based experiment) and an MRM for selected peptides from the same sample. A mixture of BSA (30 fmol) and alpha-casein (60 fmol) was first analyzed in a discovery-type experiment. Second, the same sample was analyzed using an MRM assay (**Table 10.1**). As can be seen, in the MRM assay, only the pre-selected combination of precursor ions and fragment ions are detected in the ion chromatogram.

3.4. Experimental Setup and Optimization

3.4.1. MS Parameter Optimization

There are three groups of parameters that can be adjusted to optimize instrument performance and sensitivity in an MRM experiment. These include instrument-dependent, source-dependent, and scan-based (or experiment-dependent) parameters. For users who are designing their first MRM experiments, the instrument parameter defaults for a nanospray source should be sufficient to get started. Instrument parameters can be classified as the settings that affect the shape and direction of the ion beam as it enters and passes through the mass spectrometer. Source-dependent parameters such as electrospray ionization (ESI) voltage, temperature, and gas settings are typically the same as those used for discovery-based experiments on similar systems. Source parameters are basically settings that control the electrospray ionization as the LC liquid stream enters the nanospray interface and ionizes. The source-dependent parameters tend to be fairly constant from day-to-day, but can vary gradually over time. Source parameters should be checked whenever the ESI emitter is changed or if a decreased intensity is observed for control samples. Source parameters include the physical coordinates of the ESI emitter relative to the orifice of the mass spectrometer; on most instruments, these coordinates can be adjusted manually in the *x*, *y*, and *z* dimensions. To determine the instrument and source parameter settings, a peptide standard, such as Glu-fibrinopeptide, can be infused. Once the mass spectrometer is detecting the peptide, the parameters can be adjusted to optimize signal strength. This infusion standard can also be used to track instrument performance over time.

Scan-based parameters focus on settings that are specific to each of the transitions included in the method. These settings affect the sensitivity, specificity, and duty cycle of the MRM method. Besides the precursor and fragment (Q1/Q3) masses that specify an MRM, the transmission windows for the Q1 and Q3, the collision energy (CE) for each transition, and the dwell time of each measurement can be optimized within the experiment. The transmission window sets the resolution of the quadrupole, that is, the mass range that is transmitted through Q1 and Q3 for a transition. Transmission window settings can be low, high, or unit (**Fig. 10.4**). When a transmission window is set to "unit resolution," the quadrupole transmits 0.7–1 atomic mass unit (amu) of mass range; low resolution is typically set to

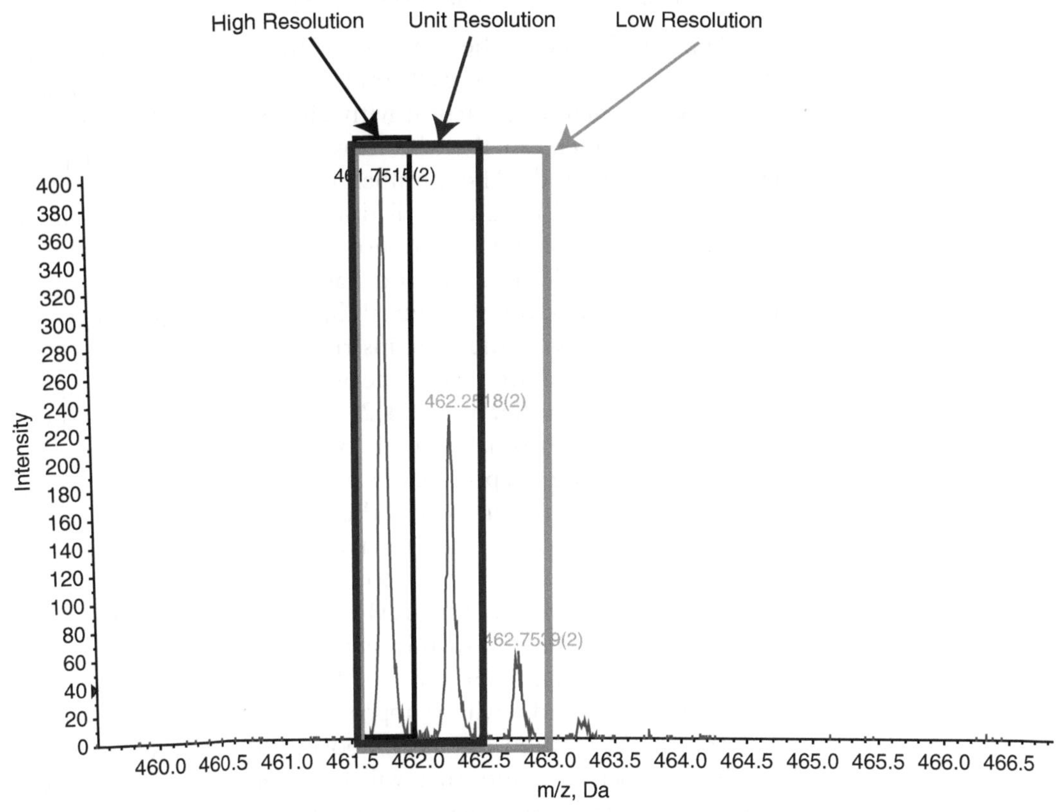

Fig. 10.4 The transmission window sets the resolution of the quadrupole. Q1 settings at high, unit, and low resolution affect the selection of the isotopic envelope from the precursor ion.

transmit 1.5–2.5 amu and high resolution is typically set at 0.3–0.5 amu. As the resolution of the transmission window increases, the transmitted mass window decreases. This results in an overall decreased sensitivity of the scan, while the specificity increases. A high complexity sample such as digested cell lysate, plasma, or serum will require higher selectivity of the MRM assay since the potential for isobaric peptides interference will be high. Settings of 0.7 amu (unit resolution) for Q1 and 0.7 amu (unit) for Q3 are frequently used during initial optimization of the MRM assay.

Peptides with different amino acid sequences, different charge states, and different mass-to-charge ratios (m/z) will fragment optimally at different collision energies (CE). If a software tool is used to select the transitions (such as MRMPilot or PinPoint), the CE for the peptide will automatically be calculated based on the above variables. If the transition of a precursor is selected manually, the following equation can be used to determine a reasonable collision energy (note that these equations are charge-state dependent):

$$[m/z^{2+}\text{: CE} = (m/z)0.044 + 5;\ m/z^{3+}\text{: CE} = (m/z)0.05 + 4;\ \text{and}\ m/z^{4+}\text{: CE} = (m/z)0.05 + 3].$$

The dwell time is the amount of time in milliseconds that the instrument spends on each transmission. When first designing an MRM method, starting with 25 ms/transition is reasonable. This number can then be increased or decreased depending on the method. The number of transitions in an MRM method multiplied by the dwell time will give the approximate duty cycle of the MRM scan (not including instrument overhead). For example, if a method has 100 MRMs with a dwell time of 25 ms each, the theoretical duty cycle for the MRM scan is 2.5 s (2500 ms) (note that instruments from different vendors have different overhead time which is added time due to electronics, etc.). In **Section 3.4.2**, we discuss chromatographic optimization and chromatographic performance. One of the key factors in designing a reproducible and reliable MRM experiment is the number of measurements the mass spectrometer records for a given transition as the target peptide elutes from the LC column. It is generally advisable to measure more than 10 points across an eluting peak in an MRM experiment (19). For an LC-MS experiment where a 45 min gradient is run from 5 to 30% organic buffer using a 75 μm × 12 cm LC column, the peak width of most peptides will be approximately 30–40 s at baseline. For the MRM method discussed above, the duty cycle is 2.5 s, so the number of points that will be measured across a chromatographic peak will equal 12–15. This is an acceptable number of measured points and should allow for good peak definition and integration during sample processing. Peak integration is the process of measuring peak areas for all transitions in an MRM run and is conducted after the experiment has been acquired (*see* **Section 3.5**).

3.4.2. Chromatographic Optimization

MRM assays can be run on any nano-LC setup with the same chromatographic column used in the identification phase of the experiment. The primary factors that can affect the chromatographic reproducibility include the following:

(a) The accuracy of the LC system in delivering the gradient. Using an LC system capable of delivering direct flow is preferential to a system that delivers the desired flow rate using a passive split. Direct flow reduces variation in the elution profile of the peptides between injections. In addition, the LC gradient profile should be consistent for all MRM assays that will be compared, since this will help ensure that the retention times for specific peptides remains consistent.

(b) System re-equilibration: After the gradient is completed and the LC system has been returned to initial mobile-phase conditions, allow 10–15 min of isocratic (constant) run time with the starting mobile-phase condition. This re-equilibration step helps to ensure that the condition of the column (pH and organic composition surrounding the stationary-phase particles in the column) is the same for every new sample that is loaded onto the column, thereby obtaining similar binding conditions for the peptides from run-to-run.

(c) Temperature: Ambient temperature of the LC-MS system should be kept as constant as possible since variation can cause changes in chromatographic elution profile of the peptides.

(d) The type of column used: Home-built packed tips or fritted-packed capillary columns can work very well for both MRM assays and discovery-based experiments. An LC column that is commercially manufactured will typically have a more defined quality control process, which could be preferable in the long run. If a home-built column is used, extra care should be taken during its production. To ensure optimal reproducibility, it is worthwhile considering the following: Pack extra stationary phase (usually C18) in the column and then trim it to a fixed length such as 10 or 12 cm. Should the column get damaged or simply not last for the entire set of MRM experiments, additional columns can be produced that should have very similar chromatographic qualities. Furthermore, ensure that the same chromatographic media is used in all columns for a given project, preferably from the same stock of dry stationary phase.

(e) Sample preparation: Optimize the sample preparation protocol to produce a completely digested sample with a minimal level of contaminants. This will prolong the life span of the LC column and improve reproducibility and robustness of the LC-MS system.

(f) It is also worth noting that reproducible sample delivery by the autosampler is critical. The autosampler should be able to reliably inject the same amount of sample between runs to ensure comparable results. This is very important for quantification.

LC-MS discovery-based experiments are typically set to maximize the number of peptide-triggered MS/MS to provide the highest number of peptide and protein identifications. This is typically accomplished by using a relatively long, shallow LC gradient (90–150 min for a standard nano-LC run with a 75 μm inner diameter × 10 cm column). In effect, peptides take longer to elute, which increases the time window where peptides can be detected by the mass spectrometer. While this permits the acquisition of a higher number of MS/MS spectra, the actual chromatographic efficiency may suffer. MRM-based experiments can be run with steeper (and faster) gradients because the number of peptide ions that are detected in a similar time period will be significantly lower than for a typical discovery-based experiment. In addition, a faster gradient will result in sharper chromatographic peaks for most eluting peptides. The lower the number of transitions included in the assay, the shorter the gradient required. For a typical MRM assay, 30–90 min gradients should be sufficient.

3.4.3. Verification and Validation of Transitions

The first rounds of MRM experiments are typically used to verify the selected transitions in the method and to ensure they are detected in the samples. Transitions are also validated at this stage to verify that the transitions for each peptide are in fact detecting the correct species. There are a number of strategies that can be used to verify that the selected transitions are detecting the correct peptide. If the instrument used is a hybrid Triple Quadrupole/Ion Trap (QqQ/Trap) such as a 4000 QTRAP, MRM-triggered MS/MS can be used to verify the sequence of the peptide (17, 18). Alternatively, when using a dedicated triple quadrupole instrument a higher number of transitions can be selected for each peptide; for high-confidence validations all transitions for a given parent should co-elute.

3.4.4. Scheduled MRM

Scheduled MRM (sMRM) associates a specific chromatographic retention time with every peptide transition in the method and is used to increase the number of transitions that can be detected in an MRM assay. During scheduled MRM, the mass spectrometer is specifically set to detect a given transition only within a specified time window centered on the expected chromatographic retention time. Since only a subset of the total peptides will be eluting within any given time window, the duty cycle of the instrument is maximized. Thus, while the number of transitions can be significantly increased, the dwell time for each transition is maintained to uphold the peak definitions. Using the example discussed in **Section 3.4.1**, an MRM method with 100 transitions, each measured with 25 ms dwell times, will have a duty cycle of 2.5 s. If, for example, only four peptides are detected between 12 and 16 min in the LC gradient, scheduled MRM would only need to measure 12 transitions for this 4 min time window (2500 ms/12 transitions = 208 ms). In its simplest form, a scheduled MRM method can be used to increase the dwell times for all transitions while maintaining a fixed duty cycle. Overall, transition dwell times are increased, but chromatographic peak definition is constant. Sensitivity and signal-to-noise are improved, without compromising the number of points measured across the chromatographic eluting peak. Finally, it should be noted that chromatographic reproducibility is very important in order to obtain a useful sMRM.

3.5. Data Processing and Quantification

Data processing of MRM experiments typically requires that the extracted ion current (XIC) from the detected transitions are integrated to determine peak area. Depending on the software used, processed results are summarized in tables that list each transition with its corresponding peak area. Typical analysis involves comparison of the relative abundance of transitions for specific peptides across different sample conditions. However, this can quickly become a daunting task when proteins, each with at least three peptides, need to be analyzed and compared.

When performing comparative analysis several factors should be taken into account (1, 7, 20):

(a) Due to inherent differences in ionization efficiency the relative abundance can only be compared between peptides of identical sequence and modifications.

(b) Calculate technical and biological variability to ensure that the proper numbers of experiment are conducted.

(c) Use calibration curves to ensure that the response of the samples is linear. These analyses can be conducted by comparing the response/XIC of each transition correlates with injection of an increasing amount of sample.

(d) Data normalization can be extremely valuable. By including an internal or external control for normalization, variability in sample processing can better be accounted for. This is very important when comparing samples prepared from multiple different conditions.

(e) Chemical labeling (SILAC, iTRAQ, TMT) can be of great help by permitting the simultaneous analysis of multiple conditions within the same sample. However, labeling efficiency should be taken into account.

(f) Limit of detection and limit of quantification are important standards to obtain comparable data sets. Limit of detection (LOD): signal-to-noise should be above or equal to 3. Limit of quantification (LOQ): signal-to-noise should be above or equal to 10.

Acknowledgments

The authors would like to thank Tony Pawson, Brett Larsen, Vivian Nguyen, Ginny Chen, Rune Linding, Steve Tate, Sarah Robinson, and Marilyn Hsiung for helpful discussion, support, and advice. Claus Jorgensen would like to thank the Lundbeck Foundation for generous support.

References

1. Lange, V., Picotti, P., Domon, B., and Aebersold, R. (2008) Selected reaction monitoring for quantitative proteomics: a tutorial. *Mol. Syst. Biol.* **4**(222), 1–14.
2. Keshishian, H., Addona, T., Burgess, M., Kuhn, E., and Carr, S. A. (2007) Quantitative, multiplexed assays for low abundance proteins in plasma by targeted mass spectrometry and stable isotope dilusion. *Mol. Cell Proteomics* **6**(12), 2212–2229.
3. Keshishian, H., Addona, T., Burgess, M., Mani, D. R., Shi, X., Kuhn, E., Sabatine, M. S., Gerszten RE., Carr SA. (2009) Quantification of cardiovascular biomarkers in pateint plasma by targeted mass spectrometry and stable isotope dillusion. *Mol. Cell Proteomics* **8**(10), 2339–2349.
4. Addona, T., Abbatiello, S. E., Schilling, B., Skates, S. J., Mani, D. R., Bunk, D. M., Spiegelman, C. H., Zimmerman, L. J., Ham, A. J., Keshishian, H., Hall, S. C., Allen, S., Blackman, R. K., Borchers, C. H., Buck, C., Cardasis, H. L., Cusack, M. P., Dodder, N. G., Gibson, B. W., Held, J. M., Hiltke, T., Jackson, A., Johansen, E. B., Kisinger, C. R., Li, J., Mesri, M., Neubert, T. A., Niles, R. K., Pulsipher, T. C., Ransohoff, D., Rodriguez, H., Rudnick, P. A., Smith, D., Tabb, D. L., Tegeler, T. J., Variyath, A. M., Vega-Montoto, L. J., Wahlander, A., Waldemarson, S., Wang, M., Whiteaker, J. R., Zhao, L., Anderson, N. L., Fisher, S. J., Liebler, D. C., Paulovich, A. G., Regnier, F. E., Tempst, P., Carr, S. A. (2009) Multi-site asessment of the precision and reproducibility of multiple rection monitoring-based measurements of proteins in plasma. *Nat. Biotechnol.* **27**(7), 633–641.
5. Wolf-Yadlin, A., Hautaniemi, S., Lauffenburger, D. A., and White, F. M. (2007) Multiple reaction monitoring for robust quantitative proteomic analysis of cellular signaling networks. *Proc. Natl. Acad. Sci. USA* **104**(14), 5860–5865.
6. Mayya, V., Rezual, K., Wu, L., Fong, M. B., and Han, D. K. (2006) Absolute quantification of multisite phosphorylation by selective reaction monitoring mass spectrometry: determination of inhibitory phosphorylation ststus of cyclin-dependent kinases. *Mol. Cell Proteomics* **5**(6), 1146–1157.
7. Picotti, P., Bodenmiller, B., Mueller, L. N., Domon, B., and Aebersold, R. (2009) Full dynamic range proteome analysis of S. cerevisiae by targeted proteomics. *Cell* **138**(4), 795–806.
8. Fusaro, V. A., Mani, D. R., Mesirov, J. P., and Carr, S. A. (2009) Prediction of high-responding peptides for targeted protein assays by mass spectrometry. *Nat. Biotechnol.* **27**(2), 190–198.

9. Mallick, P., Schirle, M., Chen, S. S., Flory, M. R., Lee, H., Martin, D., Ranish, J., Raught B., Schmitt R., Werner T., Kuster B., Aebersold R. (2007) Computational prediction of proteotypic peptides for quantitative proteomics. *Nat. Biotechnol.* **25**(1), 125–131.
10. Walsh, G. M., Lin, S., Evans, D. M., Khosrovi-Eghbal, A., Beavis, R. C., and Kast, J. J. (2009) Implementation of a data repository-driven approach for targeted proteomics experiments by multiple reaction monitoring. *Proteomics* **72**(5), 838–852.
11. Desiere, F., Deutsch, E. W., Nesvizhskii, A. I., Mallick, P., King, N. L., Eng, J. K., Aderem, A., Boyle R., Brunner, E., Donohoe, S., Fausto, N., Hafen, E., Hood, L., Katze, M. G., Kennedy, K. A., Kregenow, F., Lee, H., Lin, B., Martin, D., Ranish, J. A., Rawlings, D. J., Samelson, L. E., Shiio, Y., Watts, J. D., Wollscheid, B., Wright, M. E., Yan, W., Yang, L., Yi, E. C., Zhang, H., Aebersold, R. (2005) Integration with the human genome of peptide sequences obtained by high-throughput mass spectrometry. *Genome Biol.* **6**(1), R9.
12. Kuster, B., Schirle, M., Mallick, P., and Aebersold, R. (2005) Scoring proteomes with proteotypic peptide probes. *Nat. Rev. Mol. Cell Biol.* **6**(7), 577–583.
13. Sherwood, C. A., Eastham, A., Lee, L. W., Risler, J., Vitek, O., and Martin, D. B. (2009) Correlation between y-type ions observed in ion trap and triple quadrupole mass spectrometers. *J. Proteome Res.* **8**(9), 4243–4251.
14. Gerber, S. A., Rush, J., Stemman, O., Kirschner, M. W., and Gygi, S. P. (2003) Absolute quantification of proteins and phosphoproteins from cell lysates by tandem MS. *Proc. Natl. Acad. Sci. USA* **100**(12), 6940–6945.
15. Ong, S. E., Blagoev, B., Kratchmarova, I., Kristensen, D. B., Steen, H., Pandey, A., and Mann, M. (2002) Stable isotope labeling by amino acids in cell culture, SILAC, as a simple and accurate approach to expression proteomics. *Mol. Cell Proteomics* **1**(5), 376–386.
16. Ross, P. L., Huang, Y. N., Marchese, J. N., Williamson, B., Parker, K., Hattan, S., Khainovski, N., Pillai, S., Dey, S., Daniels, S., Purkayastha, S., Juhasz, P., Martin, S., Bartlet-Jones, M., He, F., Jacobson, A., Pappin, D. J. (2004) Multiplexed protein quantification in Saccharomyces cerevisiae using amine-reactive isobaring tagging reagents. *Mol. Cell Proteomics* **3**(12), 1154–1169.
17. Unwin, R. D., Griffiths, J. R., and Whetton, A. D. (2009) A sensitive mass spectrometric method for hypothesis-driven detection of peptide post-translational modifications: multiple reaction monitoring-initiated detection and sequencing (MIDAS). *Nat. Protoc.* **4**(6), 870–877.
18. Unwin, R. D., Griffiths, J. R., Leverentz, M. K., Grallert, A., Hagan, I. M., and Whetton, A. D. (2005) Multiple reaction monitoring to identify sites of protein phosphorylation with high sensitivity. *Mol. Cell Proteomics* **4**(8), 1134–1144.
19. Heftmann, E. (2004) *Chromatography: fundamentals and applications of chromatography and related differential migration methods – Part A: Fundamentals and techniques*, 6th ed. Elsevier Science, The Netherlands.
20. Bansal, S., and DeStefano, A. (2007) Key elements of bioanalytical method validation for small molecules. *AAPS J.* **9**(1), E109–E114.

Links

AB marketing info: MRMpilot: http://www3.appliedbiosystems.com/cms/groups/psm_marketing/documents/generaldocuments/cms_047953.pdf

Multiquant: http://www3.appliedbiosystems.com/cms/groups/psm_marketing/documents/generaldocuments/cms_047952.pdf

ThermoFisher tool for MRM design: http://www.thermo.com/pinpoint

The GPM: http://www.thegpm.org

The peptide atlas: http://www.peptideatlas.org

Chapter 11

Proteome-Wide Quantitation by SILAC

Kristoffer T.G. Rigbolt and Blagoy Blagoev

Abstract

Ongoing improvements in instrumentation, fractionation techniques, and enrichment procedures have dramatically increased the coverage of the proteome achievable via LC-MS/MS-based methodologies, opening the call for approaches to quantitatively assess differences at a proteome-wide scale. Stable isotope labeling by amino acids in cell culture (SILAC) has emerged as a powerful and versatile approach for proteome-wide quantitation by mass spectrometry. SILAC utilizes the cells' own metabolism to incorporate isotopically labeled amino acids into its proteome which can be mixed with the proteome of unlabeled cells and differences in protein expression can easily be read out by comparing the abundance of the labeled versus unlabeled proteins. SILAC has been applied to numerous different cell lines and the technique has been adapted for a wide range of experimental procedures. In this chapter we provide detailed procedure for performing SILAC-based experiment for proteome-wide quantitation, including a protocol for optimizing SILAC labeling. We also provide an update on the most recent developments of this technique.

Key words: SILAC, quantitative proteomics, mass spectrometry, LC-MS/MS, labeling, isotope.

1. Introduction

In recent years the depth to which the proteome can be examined has increased significantly. Stable isotope labeling by amino acids in cell culture (SILAC) is contributing to this field and has enabled quantitative comparison of entire proteomes (1, 2). SILAC provides a powerful and accurate technique for relative proteome-wide quantitation by mass spectrometry (3). The basic principle of SILAC is that any cell line auxotroph to a given amino acid will take up this amino acid from the growth medium and metabolically incorporate it into its proteome. This concept is

P.R. Cutillas, J.F. Timms (eds.), *LC-MS/MS in Proteomics*, Methods in Molecular Biology 658,
DOI 10.1007/978-1-60761-780-8_11, © Springer Science+Business Media, LLC 2010

exploited by passaging parallel cell cultures in media containing either a normal (light) or a heavy form of a selected amino acid, most frequently lysine and arginine (*see* **Note 1**), where the light and heavy version differ only in mass as a result of different isotopic content. Once fully labelled, cell lines can be subjected to different perturbations and pooled together before performing any preparative procedures, thus eliminating quantitation inaccuracies introduced during sample handling. Finally, the proteome of the two differently labelled cell cultures are analyzed in the same LC-MS/MS run and the origin of the proteins or peptides can readily be discriminated based on the mass shift introduced by the different isotopic content conveyed by the labelled amino acid (*see* **Fig. 11.1A**). Since its development, SILAC has been applied to essentially all frequently used cell lines and also to adult and embryonic stem cells (4–8).

The versatility of SILAC has been demonstrated by a wide range of applications (9, 10). Recently de Godoy et al. (2) characterized the complete yeast proteome and used SILAC to compare the protein expression between the haploid and the diploid state, which enabled the authors to compare the expression of signalling pathways affected as a result of the pheromone response. SILAC has also been widely used to acquire information about the tem-

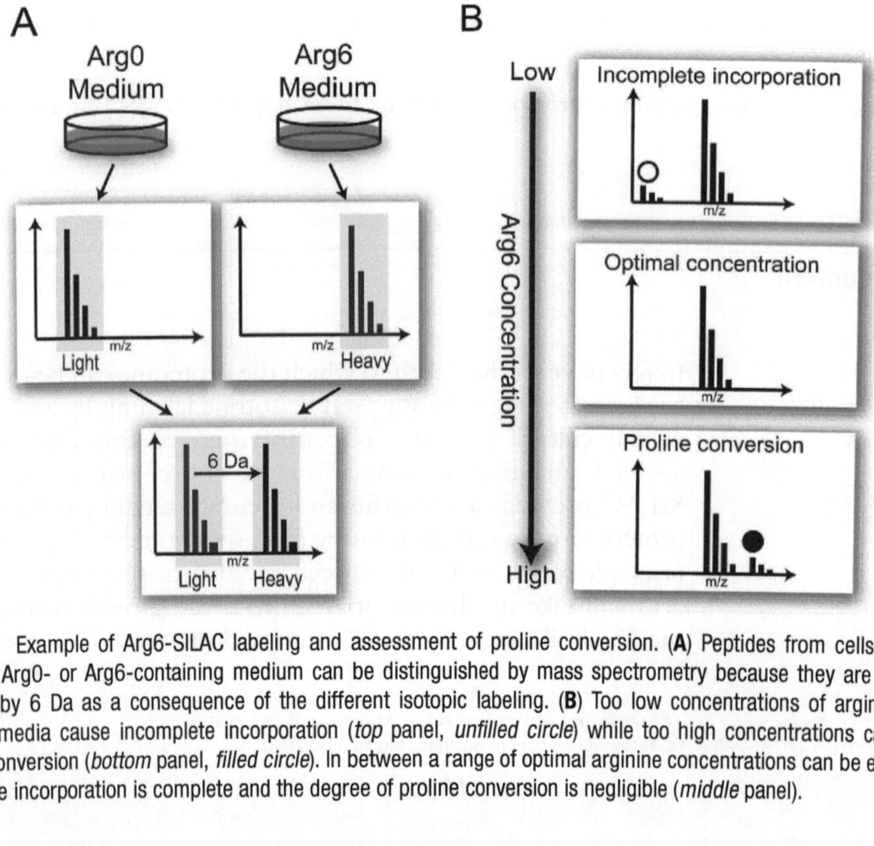

Fig. 11.1 Example of Arg6-SILAC labeling and assessment of proline conversion. (**A**) Peptides from cells passaged in either Arg0- or Arg6-containing medium can be distinguished by mass spectrometry because they are separated in mass by 6 Da as a consequence of the different isotopic labeling. (**B**) Too low concentrations of arginine in the labeling media cause incomplete incorporation (*top* panel, *unfilled circle*) while too high concentrations can lead to proline conversion (*bottom* panel, *filled circle*). In between a range of optimal arginine concentrations can be established where the incorporation is complete and the degree of proline conversion is negligible (*middle* panel).

poral behaviour of signalling pathways by subjecting labelled cells to a time course of activation. This strategy has been used to profile the time course of phosphorylation changes following epidermal growth stimulation, either by specifically targeting phosphotyrosines or by addressing changes at a global scale (11–13). When investigating protein–protein interactions SILAC has also emerged as a useful tool. As discussed further in **Chapter 16** of this volume, for this purpose SILAC is used to discriminate between specific biological interactions and unspecific binding, an application that was initially demonstrated for the activated EGFR complex (14). Another example is the use of SILAC to identify proteins interacting specifically with Histone H3 depending critically on the methylation status of lysine 4 (15). The only stringent limitation of SILAC is that its application to tissue samples is not as straightforward as for its use to cells grown in culture however, significant progress in this field has recently been made (16, 17) (*see* **Note 2**). In this chapter, we describe a protocol for labelling of proteins with SILAC reagents and for their analysis by LC-MS/MS.

2. Materials

2.1. Cell Culture

1. For all buffers and solvents listed in this chapter use Milli-Q water, unless specified otherwise.
2. Dulbecco's Modified Eagle Medium (DMEM) deficient in lysine and arginine (Invitrogen, Carlsbad, CA).
3. Dialysed Fetal Bovine Serum (FBS) (Gibco-Invitrogen, Carlsbad, CA).
4. Normal "light" amino acids: L-lysine (Lys0) and L-arginine (Arg0) hydrochloride (Sigma Chemicals, Copenhagen, Denmark).
5. Stable isotope-labeled "heavy" amino acids: L-arginine-$^{13}C_6$ hydrochloride (Arg6) (Cambridge Isotope Labs, Andover, MA), L-arginine-$^{13}C_6$, $^{15}N_4$ hydrochloride (Arg10) (Sigma-Isotec, St. Louis, MO), L-lysine-4,4,5,5-d_4 hydrochloride (Lys4) (Sigma-Isotec; cat. no. 616192), and L-lysine-$^{13}C_6$, $^{15}N_2$ hydrochloride (Lys8) (Sigma-Isotec).
6. L-Glutamine (200 mM stock solution), penicillin/streptomycin (10,000 U/10,000 μg stock solution), and trypsin–EDTA solution (trypsin, 200 mg/L and Versene–EDTA 500 mg/L) (all from Gibco-Invitrogen).

7. Sterile phosphate-buffered saline (PBS) (Cambrex Bio Science Copenhagen ApS, Denmark).
8. Filter units, MF75TM series (Nalge Nunc International, NY).
9. Adherent HeLa cells, epithelial adenocarcinoma (American Type Culture Collection (ATCC), Manassas, VA).
10. EGF human recombinant (PeproTech, Rocky Hill, NJ). Dissolve at concentration 1 mg/mL and store aliquots at −20°C.

2.2. Cell Lysis, SDS-PAGE, and Digestion

1. Lysis solution: 6 M urea (Invitrogen)/2 M thiourea (Invitrogen,). Store aliquots at −80°C.
2. Cell scrapers (Sarstedt, Newton, NC).
3. Coomassie plus – The Better Bradford™ Assay Reagent (Pierce, Rockford, IL).
4. Dithiothreitol (DTT) and iodoacetamide (Sigma–Aldrich) for reduction and alkylation of proteins, respectively. Store aliquots of 100 mM DTT and 550 mM iodoacetamide at −20°C. Iodoacetamide is light sensitive.
5. NuPAGE LDS 4X LDS Sample Buffer and NuPAGE® Novex 10% Bis-Tris gel system with MOPS running buffer for polyacrylamide gel electrophoresis (Invitrogen).
6. Colloidal Blue Stain (Invitrogen, Carlsbad, CA).
7. Absolute ethanol (EtOH) (Merck, Darmstadt, Germany).
8. Ammonium bicarbonate (ABC) (Sigma Chemicals).
9. Sequence grade-modified trypsin (Promega, Madison, WI).

3. Methods

3.1. Optimization of Labeling Conditions

Before engaging a full-scale SILAC experiment with a particular cell line we recommend to establish the optimal concentrations of the arginine and lysine in the SILAC labeling media for this cell line by performing an initial titration experiment. The concentrations of the amino acids in the media affect four primary issues: efficiency of amino acid incorporation, growth rates of the cell culture, the degree of arginine to proline conversion (*see* **Note 3**), and finally the experiment costs, of which the labeled amino acids constitute a significant part. **Table 11.1** provides optimal amino acid concentrations for a number of commonly used cell lines. Nonetheless, this table should only be used as a guideline as cell lines and culture conditions may vary between laboratories.

Table 11.1
Dilutions of amino acids stocks for some commonly used cell lines

Cell line	Media	Arg stock dilution[a]	Lys stock dilution[a]	References
HeLa	DMEM	1:3300	1:3000	(12)
HeLa S3	RPMI	1:1000	1:3300	(13)
Jurkat	RPMI	1:1000	1:3300	–
HEK293	DMEM	1:2000	1:2000	(30)
MCF7	DMEM	1:1500	1:1500	(31)
NIH 3T3	DMEM	1:3000	1:2000	(23)
3T3-L1	DMEM	1:3000	1:2000	–
C2C12	DMEM	1:3000	1:2000	(3)
hMSC	DMEM	1:3000	1:2000	(7)

[a]Suggested dilutions starting from the 84 mg/mL arginine and 146 mg/mL lysine stock solutions described in Step 1, **Section 3.1.1**.

3.1.1. Media preparation

The key element to a successful SILAC experiment is to use growth media containing exclusively one form of the amino acid(s) used for labeling. To achieve this it is necessary to minimize, or ideally eliminate, sources of undesired variants of the labeling amino acids. Thus, specialized media components are required, namely isotopically labeled amino acids, culture media devoid of these amino acids, and dialysed serum (if the cell line requires addition of serum in the culture media). We recommend using SILAC media with three different concentrations of arginine and lysine to titrate the amino acid concentrations.

1. Start out by preparing stock solutions of lysine and arginine. Stock solutions are made by dissolving the amino acids in sterile PBS to a final concentration of 84 and 146 mg/mL for Arg0 and Lys0, respectively (1000X normal DMEM concentrations). When preparing stock solutions of the heavy amino acids the mass difference should be taken into account to achieve equimolar concentrations of the amino acids; to this end dissolve Arg6 to 86.4 mg/mL and Lys4 to 149 mg/mL. The heavy amino acids are delivered in small quantities and therefore small volumes of stock solutions are usually prepared from these, but the light amino acid (Lys0 and Arg0) stock solutions can be prepared in larger volumes, filter these using a sterile 0.2 μm syringe filter. The amino acid stock solutions can be stored for >3 months at 4°C.

Do not filter the heavy amino acid solutions as the loss of material and cost of the heavy amino acids make this unadvisable. Instead, the growth media including the diluted amino acids should be filtered as described in the following steps.

2. Assemble three 150 mL filter units inside a sterile laminar flow bench. Transfer 80 mL of the SILAC media preparation without anything added to the upper chamber of each of the three filter units.
3. To produce three distinct labeling media, add lysine and arginine from the stock solutions to the media solution in a pairwise fashion as indicated below:
   a. 80 μL from the Arg6 stock (1:1000 dilution) and 27 μL from the Lys4 stock (1:3000 dilution).
   b. 40 μL from the Arg6 stock (1:2000 dilution) and 40 μL from the Lys4 stock (1:2000 dilution).
   c. 27 μL from the Arg6 stock (1:3000 dilution) and 80 μL from the Lys4 stock (1:1000 dilution).
4. Filtrate the three media preparations into a collecting flask using suction in a flow bench and discard the upper section of the filter units.
5. Supplement each of the filtrated media with 8 mL dialysed bovine serum, 800 μL of 100X penicillin/streptomycin and 800 μL Glutamax. Store the complete media at 4°C.

3.1.2. Cell Culture and Lysis

A major requirement for accurate proteome-wide quantitation is to have the entire proteome of the cellular system fully labeled. To achieve this, cells need to be cultured in the SILAC labeling media for at least five passages. It is advisable to carefully plan the cell culture from the start of the labeling procedure to cell harvest in order to avoid preparing excessive amounts of the costly labeling media. Below a procedure for labeling cells resulting in one 10 cm dish per amino acid dilution is given, the amount of protein extracted from these cells will provide sufficient material for a thorough assessment of the labeling properties.

1. From a dish with cells at 80–90% confluency aspirate the media and rinse the cells once with 3 mL PBS, aspirate PBS and detach the cells with 1 mL trypsin/EDTA solution. Transfer 300 μL from the cell suspension into three separate dishes each containing one of the three different SILAC media for amino acid titration.
2. Passage the cells for at least five cell doublings.
3. When cells at the last passage reach 80–90% confluency wash the cell twice with 3 mL ice-cold PBS.
4. Scrape the cell using a sterile cell scraper and transfer cells to a sterile tube suitable for centrifugation, store on ice.

5. Harvest cells by centrifugation at 4°C, discard supernatant.
6. Resuspend cell pellet in four volumes of 6 M urea/2 M thiourea at room temperature (*see* **Note 4**).
7. Estimate protein concentrations on aliquots from the three lysates with Bradford assay. If dilution of the urea is required dilute the aliquots, not the total lysates.

3.1.3. Digestion, Fractionation, and Analysis

1. Reduce and alkylate proteins by addition of DTT to a final concentration of 10 mM and 30 min incubation at room temperature followed by addition of iodoacetamide (*see* **Note 5**) to a final concentration of 55 mM and 20 min incubation at room temperature in the dark.
2. From each lysate take out a volume corresponding to approximately 100 μg of total protein.
3. Dilute four times in LDS sample buffer, incubate for 5 min at 95°C, votex and centrifuge briefly in a bench-top centrifuge.
4. Separate each of the lysates on a NuPAGE® Novex 10% Bis-Tris gel with the MOPS buffer system.
5. Visualize proteins by Colloidal Blue staining.
   To reduce keratin contamination Steps 6–19 should, if possible, be performed within a laminar flow bench, wearing gloves and lab coat at all times. Steps 7–19 are performed in parallel for each of the gel slices.
6. Slice out three equal bands from the three lanes and transfer these to nine separate tubes. Select slices in regions spanning the gel lane and avoid areas with one prominent protein band.
7. Cut each gel band into pieces of approximately 1 mm × 1 mm.
8. Wash the gel cubes with 200 μL 50% EtOH in 50 mM ammonium bicarbonate (ABC), incubate for 20 min at room temperature with agitation, and discard solution.
9. Repeat Step 8. All blue staining should be removed at this point, otherwise repeat the washing step one more time.
10. Dehydrate the gel cubes by adding 200 μL 100% EtOH, incubate for 10 min and discard solution.
11. Swell gel cubes by addition of 200 μL 50 mM ABC, incubate for 20 min, and discard solution.
12. Repeat Step 10.
13. Add trypsin solution (12.5 ng/μL) to the gel cubes and incubate on ice for 15 min. The volume of solution to add depends on the total volume of the cubes, use only just enough to cover the cubes. Any excess of trypsin might

impede the following analysis by clouding the peptides originating from the sample during LC-MS/MS analysis.

14. Add 100–200 μL of 50 mM ABC and incubate over night at 37°C. The volume of ABC also depends on the total volume of the cubes, use as much as to cover approximately 1 mm above the cubes.
15. The following day spin down the solution in a bench-top centrifuge and transfer it to a new tube (this solution already contains some of the peptides from the digested proteins).
16. To extract the rest of the peptides, add 100 μL of 30% EtOH/3% TFA solution to the gel cubes, incubate for 10 min and pool solution with that from the preceding step. (Do not mix solutions from the nine parallel samples.)
17. Repeat Step 16.
18. Repeat Step 16 again but instead of 30% EtOH/3% TFA solution use 100% EtOH.
19. Dry down peptides in a speed-vac, until all EtOH has evaporated.
20. Prepare your sample for MS analysis. This step depends on the mass spectrometry setup and will not be described in this chapter.
21. Analyse the samples by LC-MS/MS.
22. To evaluate the efficiency of labeling estimate the ratio between the labeled and the unlabeled peptides for both arginine and lysine containing peptides (*see* Step 12 in **Section 3.2.2** for description of peptide quantitation). Generally the average incorporation efficiency should be above 95%.
23. Perform similar evaluation of the degree of Arg6 to Pro5 conversion by estimating the ratio between the peak from the peptide containing the normal Pro0 and the peak from the Pro5 peptide which is shifted 5 Da in the mass spectrum. As a general guideline the degree of proline conversion should not exceed 5%.
24. Based on the considerations above select the arginine concentration which provides close to full incorporation while still keeping the degree of Pro conversion at an appropriate low level (*see* **Note 6** and **Fig. 11.1B**), select the lowest lysine concentration giving close to full incorporation.

3.2. Proteome-Wide Quantitative Proteomics

After having established the optimal labeling parameters one can proceed to perform a proteome-wide comparative experiment. The section below depicts a typical workflow of SILAC experiment with two labeling states; however, the SILAC methodology can easily be expanded to three different labeling states to compare simultaneously three different conditions (*see* **Note 7**). As an example we describe an experiment comparing changes in protein expression levels in HeLa cells as a result of 24 h stimulation with epidermal growth factor (EGF) (*see* **Fig. 11.2**).

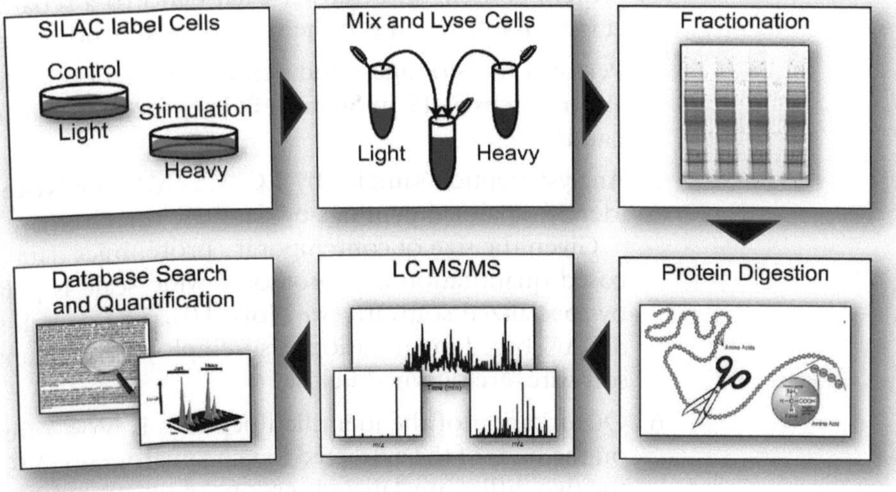

Fig. 11.2 Schematic overview of a typical SILAC-based quantitative proteomics experiment. Cells cultured in parallel are labeled with different versions of the labeling amino acid(s) and mixed together. The combined cell lysate is fractionated by, e.g., 1D gel electrophoresis and proteins are digested to peptides which are analysed by LC-MS/MS and identified by database searching. The identified peptides are then quantified and used to calculate the protein ratio.

3.2.1. Media Preparation and Cell Culture

1. Prepare the SILAC labeling media as in **Section 3.1.1** using the optimal amino acids concentrations established from the titration experiment described above.
2. From a dish with HeLa cells at 80–90% confluency aspirate the media, rinse cells once with 3 mL PBS, and detach the cells with 1 mL trypsin/EDTA solution. Transfer 300 μL from the cell suspension into two dishes containing either Lys0/Arg0 or Lys4/Arg6 media.
3. Passage the two cultures in parallel into the corresponding SILAC media for five passages.
4. When 50% confluent at the fifth passage, supplement the growth media of the Lys4/Arg6-labeled cells with 150 ng/mL EGF. Leave the cells grown in Lys0/Arg0 media untreated.
5. After 24 h of incubation wash the cell twice with 3 mL ice-cold PBS.
6. Harvest cells as described in Steps 5–7 in **Section 3.1.2**.

3.2.2. Digestion, Fractionation, and Analysis

1. Reduce and alkylate proteins as described in Step 1 in **Section 3.1.3** (*see* also **Note 5**).
2. Mix the two lysates 1:1 based on Bradford assay concentration estimates. Take out a fraction corresponding to 100 μg of total protein and separate it by SDS-PAGE as described in Steps 3–5 in **Section 3.1.3** (*see* **Note 8**).
3. Slice the gel lane into 20 bands with approximately equal protein content and transfer each band to a separate tube. It is advisable to cut out the very intense bands individually.
4. Perform trypsin digest and peptide extraction as described above (Steps 7–19 in **Section 3.1.3**) for each of the 20 gel bands.
5. Analyse peptide samples by LC-MS/MS (*see* **Note 9**) and identify analysed peptides by database searches.

   Given the size of contemporary proteomics data SILAC-based quantitation at a proteome-wide scale is performed by specialized software (*see* **Note 10**). The overall steps in SILAC-based quantitation, also employed by quantitation software, are outlined below.
6. Quantitation of the identified peptides is most often done on the basis of the extracted ion chromatogram (XIC) from the two differently labeled versions of the peptide. The XIC is the contribution of a compound at a given m/z to the total ion chromatogram (TIC) and is described by three attributes: the m/z value, the elution time range, and the intensity of the peak across the elution time. It is the area under these three-dimensional peaks which are used for quantitation (*see* **Fig 11.3A**), where it is the ratio of the areas under the light and heavy isotope clusters that reflect the SILAC ratio (18). In most cases it is only the monoisotopic peak that is used to quantify the peptide.
7. From the quantitation of each individual peptide a SILAC ratio for the protein expression is calculated which in this experiment reflects the expression change of the corresponding protein as a result of the EGF stimulation. The SILAC ratio of any given protein can be determined simply by calculating the mean of all peptides from this protein. Alternatively a weighted average can be calculated, where the peptides identified with the highest intensity is given most weight since these will be expected to be quantified most accurately (19, 20). The fact that several peptides are used for quantitation of a protein increases the confidence on the observed ratio; however, this might also introduce artifacts to the quantitation as observed in the presence of peptides shared between distinct proteins (21). These

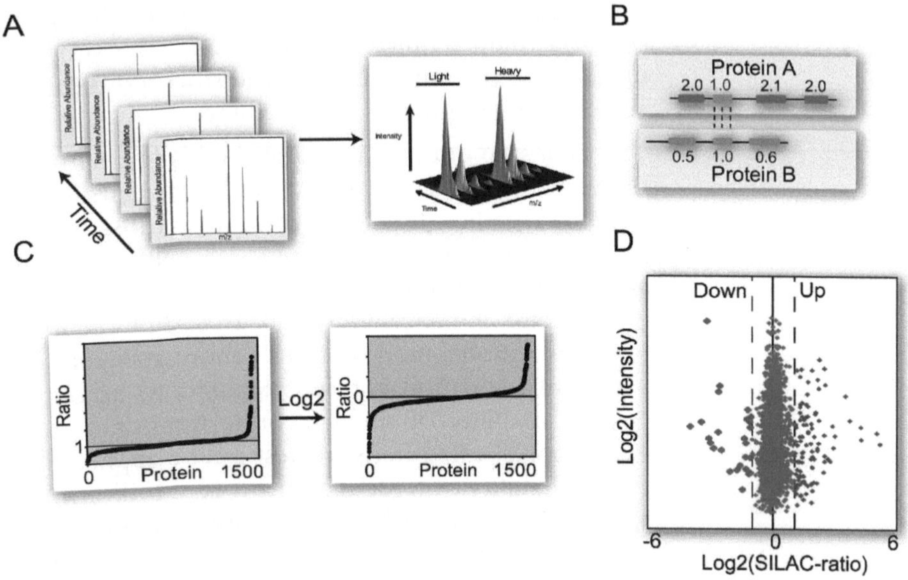

Fig. 11.3 SILAC quantitation and data processing. (**A**) XIC-based quantitation of peptides is performed on a three-dimensional peak defined by the elution time, *m/z* value, and intensity. (**B**) Protein quantitation can be distorted by a peptide shared between two distinct proteins. (**C**) Log2 transformation of SILAC ratios provide a more intuitive overall representation where the data are centered on zero and up- and down-regulated proteins are dispersed between zero and ± infinity. (**D**) Regulated proteins can be observed as significant outliers from the large population of non-regulated proteins (*see* (32) for a recent review).

shared peptides might be observed as outliers, since they do in fact originate from a protein apart from the one being quantified and thus needs to be excluded from the protein quantitation (*see* **Fig. 11.3B**).

8. All software for SILAC quantitation described in this chapter (**Table 11.1**) allow the user to export the data as, e.g., a tab-delimited or comma-separated file which enables the user to process the data further. It is advisable as an initial step to apply a transformation of the SILAC ratios. Log2 transformation of the protein ratios provide a more intuitive representation of the data where a value of zero represents no change and increasing ratios are designated with a positive sign and vice versa (*see* **Fig. 11.3C**). An alternative is the following transformation which also centers the data on zero and has the sign designating if the observed ratio is increasing or decreasing but in contrast to the log2 transformation it does not affect the relative magnitude of the SILAC ratios:

Use (ratio − 1), for the SILAC - ratios > 1 and use (− 1/ratio) +1, for the ratios

9. To account for systematic inaccuracies in the mixing ratio the data often require normalization. The simplest way is to calculate the median of the transformed data and divide all transformed ratios with this value.

10. The final step prior to assessing the biological implications of the data is to define a threshold for regarding a protein as being significantly regulated. Under most circumstances a change in protein expression >1.5-fold can be regarded as significant since the average standard deviation of such experiments is typically around 15%. For quantitation of post translational modifications a more stringent criteria of >twofold regulation is often required to account for the larger variability of single peptide quantitation. Alternatively sophisticated statistical test can be applied to evaluate the probability of a regulation being significant. Different statistical approaches have been suggested, but they all have in common that they test for proteins being significant outliers from the large population of unregulated proteins (22) (*see* **Fig. 11.3D**).

4. Notes

1. In principle any amino acid can be used for labeling although care should be taken when selecting which amino acid to use. By far the most common SILAC amino acids is a combination of lysine and arginine, since this ensures the presence of at least one label in all tryptic peptides, except for the C-terminal peptide. Alternatively, lysine labeling alone in combination with endopeptidase Lys-C digestion or arginine labeling in combination with trypsin digestion can be employed. In the last case the number of peptides amenable to quantitation is dramatically reduced. This is however still a useful approach since most proteins will be identified by a sufficiently large number of peptides to allow accurate quantitation. In experiments targeting phosphorylations or other posttranslational modifications it is crucial to label both arginine and lysine if tryptic digestion is employed to ensure that all peptides can be quantified.

2. Methodologies for applying SILAC to tissue samples have been reported, either by using SILAC labeled cells as a common reference which is mixed in with each of the tissue samples (16) or by using SILAC labeling of an entire organism, exemplified by the 'SILAC mouse' (17).

3. In some cell lines, high concentrations of arginine in the growth media may result in its metabolic conversion into

proline inside the cells (23). Occurrence of proline conversion in cells grown in heavy arginine leads to formation of double-heavy peptides, containing both the proper SILAC amino acid and also a heavy form of proline (*see* **Fig. 11.1B**). For example, unnecessary high amounts of Arg6 in HeLa cells produce a diagnostic satellite peak in the mass spectra of proline-containing peptides at a position 5 Da higher than the normal labeled peptides due to the mass shift of heavy proline (Pro5). This satellite peak can be used to readily asses the presence and level of proline conversion. This amino acid conversion causes an erroneous lower abundance ratio of the proline-containing peptides and should be taken into consideration during quantitation.

4. At elevated temperatures urea reacts with the proteins resulting in carbamylation. Thus, protein solutions in urea should never be heated above room temperature.

5. If an experiment is targeting identification of protein ubiquitination sites, it is advised to exchange iodoacetamide as an alkylation reagent with the less aggressive compound chloroacetamide, since alkylation of lysine residues mimicking the ubiquitinaton GlyGly tag (residual from ubiquitin on the targeted peptide after trypsin digest) has been observed when using iodoacetamide (24).

6. The level of arginine incorporation versus proline conversion is a balance, and for some cell types it might not be possible to establish an optimal arginine concentration which gives a satisfactory incorporation while still keeping proline conversion at a minimum. In these rare cases small amounts of unlabelled proline can be added to the medium as described in (6).

7. The "triple-labeling SILAC" approach is probably the most frequently utilized SILAC methodology (7, 11–13, 15, 25, 26). It is an extremely useful and accurate way to compare three cellular states simultaneously. More than three different labeling states is rarely used because of the increased complexity of the mass spectra introduced when increasing the number of labeling states. In addition, we recommend having minimum 4 Da mass difference between the differently labeled amino acids in order to avoid significant overlap between the isotope clusters from the two differently labeled versions of a peptide. SILAC application with five distinct labellings has however been recently demonstrated (27).

8. As an example of protein fractionation, we describe here SDS-PAGE since this is generally a simple and robust

approach, although there are a number of alternatives. Recently isoelectric focusing either at the protein or at the peptide level has gained significant popularity, a detailed procedure is given in (28). Alternatively, various chromatographic techniques such as, e.g. strong cation exchange can be applied, these have the advantage that they can accommodate much larger sample amounts (in the milligram range) than both SDS-PAGE and isoelectric focusing, a feature of critical importance for example for phosphoproteomics (12). In general, the depth to which the proteome can be sampled depends directly on the extent by which the samples are fractionated, which however in turn requires increased amounts of starting material and an increase in the number of samples for MS analysis.

9. When analyzing a peptide sample for SILAC quantitation by MS, special requirements apply for the instrument method and parameters. The primary issue is that the reliability of a quantitation increases with the number of measurements recorded. Thus the mass spectrometer should be set up to acquire as many precursor spectra as possible during the chromatographic window where the peptide elutes. On the other hand the number of MS/MS events also needs to be optimized to increase the number of identifications. How to set up the proper instrument method depends on the type of instrument. For hybrid instruments, like the LTQ-Orbitrap or LTQ-FT (Thermo), the key issue is to correctly time the number of MS/MS scans to record in the LTQ part during the time elapsed acquiring one full scan in the FT part. The number of MS/MS it is possible to perform varies with the sample type and instrument parameters but in general a fragmentation of the top 5 to top 10 peptides in the full scan allows a sufficient number of precursor full scans without dramatically compromising the number of identifications (2). For stand-alone ion-traps or Q-TOF-type instruments fewer MS/MS spectra per full MS spectra can be acquired and in general the duty cycle should not exceed three or four MS/MS spectra per full MS.

10. A number of software solutions for automated quantitation have been developed during the later years (18). A thorough discussion of all available software solutions is outside the scope of this chapter. Three examples of software with their requirements in terms of MS instrument vendors and search engine are given in **Table 11.2**. All three programs are provided by academia and are free of charge.

    The open-source software MSQuant is the prototypical solution for SILAC-based quantitation which allows the

Table 11.2
Comparison of selected software for SILAC quantitation

Software	Supported instruments	Supported search engine	Supported operating system	Developed by	URL & Reference
MSQuant	LTQ-FT and LTQ-Orbitrap (Thermo), Qstar (Applied Biosystems), and Micromass Q-Tof	Mascot (Matrix Science, London, UK)	Microsoft Windows	Center for Experimental Bioinformatics, University of Southern Denmark, Odense	http://www.cebi.sdu.dk (33)
MaxQuant	LTQ-FT and LTQ-Orbitrap (Thermo),	Mascot (Matrix Science, London, UK)	Microsoft Windows	Proteomics and Signal transduction, Max Planck Institute of Biochemistry, Martinsried, Germany	http://www.maxquant.org (22)
Census	Generic (accept mzXML files as input)	Sequest (Yates Lab)	Generic, runs under Java (Sun)	The Scripps Research Institute, La Jolla, CA, US	http://fields.scripps.edu/census/index.php (29)

user to define a number of critical parameters in an easily operationally display. MSQuant performs automated quantitation of complete proteome-wide SILAC data sets and provide the user an optimal display of the raw data for manual validation. Furthermore, MSQuant contain a number of additional functionalities for assigning and scoring phosphorylation sites or analyzing data from protein correlation profiling-SILAC (PCP-SILAC, *see* **Chapter 15** in this book), as well as several add-on scripts for different types of proteomics data processing. The MaxQuant software (22) performs SILAC quantitation in a highly automated fashion and provides high identification rates. To increase the identification rates, MaxQuant uses elaborate algorithms taking into account the intensity of precursor peaks, chromatographic elution time, and accurate mass measurements from high resolution instruments. In contrast to the two software solutions described above which uses Mascot as a search engine, the Census software (29) relies on Sequest. In addition, Census is flexible in terms of instrument specificity as it uses files in the generalized mzXML format as input.

Acknowledgements

We would like to thank all members of the Center for Experimental BioInformatics (CEBI) for useful discussions, especially Dr. Irina Kratchmarova for the critical reading of the chapter. The research leading to these results has received funding from the European Commission's 7th Framework Programme (grant agreement HEALTH-F4-2008-201648/PROSPECTS), the Danish Natural Science Research Council and the Lundbeck Foundation.

References

1. Cox, J., and Mann, M. (2007) Is proteomics the new genomics? *Cell* **130**, 395–398.
2. de Godoy, L. M., Olsen, J. V., Cox, J., Nielsen, M. L., Hubner, N. C., Frohlich, F., Walther, T. C., and Mann, M. (2008) Comprehensive mass-spectrometry-based proteome quantitation of haploid versus diploid yeast. *Nature* **455**, 1251–1254.
3. Ong, S. E., Blagoev, B., Kratchmarova, I., Kristensen, D. B., Steen, H., Pandey, A., and Mann, M. (2002) Stable isotope labeling by amino acids in cell culture, SILAC, as a simple and accurate approach to expression proteomics. *Mol. Cell. Proteomics* **1**, 376–386.
4. Graumann, J., Hubner, N. C., Kim, J. B., Ko, K., Moser, M., Kumar, C., Cox, J., Scholer, H., and Mann, M. (2008) Stable isotope labeling by amino acids in cell culture (SILAC) and proteome quantitation of mouse embryonic stem cells to a depth of 5,111 proteins. *Mol. Cell. Proteomics* 7, 672–683.
5. Prokhorova, T. A., Rigbolt, K. T., Johansen, P. T., Henningsen, J., Kratchmarova, I.,

Kassem, M., and Blagoev, B. (2009) SILAC-labeling and quantitative comparison of the membrane proteomes of self-renewing and differentiating human embryonic stem cells. *Mol. Cell. Proteomics* **8**, 959–970.
6. Bendall, S. C., Hughes, C., Stewart, M. H., Doble, B., Bhatia, M., and Lajoie, G. A. (2008) Prevention of amino acid conversion in SILAC experiments with embryonic stem cells. *Mol. Cell. Proteomics* 7, 1587–1597.
7. Kratchmarova, I., Blagoev, B., Haack-Sorensen, M., Kassem, M., and Mann, M. (2005) Mechanism of divergent growth factor effects in mesenchymal stem cell differentiation. *Science* **308**, 1472–1477.
8. Van Hoof, D., Pinkse, M. W., Oostwaard, D. W., Mummery, C. L., Heck, A. J., and Krijgsveld, J. (2007) An experimental correction for arginine-to-proline conversion artifacts in SILAC-based quantitative proteomics. *Nat. Methods* **4**, 677–678.
9. Mann, M. (2006) Functional and quantitative proteomics using SILAC. *Nat. Rev. Mol. Cell. Biol.* 7, 952–958.
10. Blagoev, B., and Mann, M. (2006) Quantitative proteomics to study mitogen-activated protein kinases. *Methods* **40**, 243–250.
11. Blagoev, B., Ong, S. E., Kratchmarova, I., and Mann, M. (2004) Temporal analysis of phosphotyrosine-dependent signaling networks by quantitative proteomics. *Nat. Biotechnol.* **22**, 1139–1145.
12. Olsen, J. V., Blagoev, B., Gnad, F., Macek, B., Kumar, C., Mortensen, P., and Mann, M. (2006) Global, in vivo, and site-specific phosphorylation dynamics in signaling networks. *Cell* **127**, 635–648.
13. Dengjel, J., Akimov, V., Olsen, J. V., Bunkenborg, J., Mann, M., Blagoev, B., and Andersen, J. S. (2007) Quantitative proteomic assessment of very early cellular signaling events. *Nat. Biotechnol* **25**, 566–568.
14. Blagoev, B., Kratchmarova, I., Ong, S. E., Nielsen, M., Foster, L. J., and Mann, M. (2003) A proteomics strategy to elucidate functional protein-protein interactions applied to EGF signaling. *Nat. Biotechnol.* **21**, 315–318.
15. Vermeulen, M., Mulder, K. W., Denissov, S., Pijnappel, W. W., van Schaik, F. M., Varier, R. A., Baltissen, M. P., Stunnenberg, H. G., Mann, M., and Timmers, H. T. (2007) Selective anchoring of TFIID to nucleosomes by trimethylation of histone H3 lysine 4. *Cell* **131**, 58–69.
16. Ishihama, Y., Sato, T., Tabata, T., Miyamoto, N., Sagane, K., Nagasu, T., and Oda, Y. (2005) Quantitative mouse brain proteomics using culture-derived isotope tags as internal standards. *Nat. Biotechnol.* **23**, 617–621.
17. Kruger, M., Moser, M., Ussar, S., Thievessen, I., Luber, C. A., Forner, F., Schmidt, S., Zanivan, S., Fassler, R., and Mann, M. (2008) SILAC mouse for quantitative proteomics uncovers kindlin-3 as an essential factor for red blood cell function. *Cell* **134**, 353–364.
18. Mueller, L. N., Brusniak, M. Y., Mani, D. R., and Aebersold, R (2008). An assessment of software solutions for the analysis of mass spectrometry based quantitative proteomics data. *J. Proteome Res.* 7, 51–61.
19. Ong, S. E., Foster, L. J., and Mann, M. (2003) Mass spectrometric-based approaches in quantitative proteomics. *Methods* **29**, 124–130.
20. Bantscheff, M., Schirle, M., Sweetman, G., Rick, J., and Kuster, B. (2007) Quantitative mass spectrometry in proteomics: a critical review. *Anal. Bioanal. Chem.* **389**, 1017–1031.
21. Nesvizhskii, A. I., and Aebersold, R. (2005) Interpretation of shotgun proteomic data: the protein inference problem. *Mol. Cell Proteomics* **4**, 1419–1440.
22. Cox, J., and Mann, M. (2008) MaxQuant enables high peptide identification rates, individualized p.p.b.-range mass accuracies and proteome-wide protein quantitation. *Nat. Biotechnol.* **26**, 1367–1372.
23. Ong, S. E., Kratchmarova, I., and Mann, M. (2003) Properties of 13C-substituted arginine in stable isotope labeling by amino acids in cell culture (SILAC). *J. Proteome Res.* **2**, 173–181.
24. Nielsen, M. L., Vermeulen, M., Bonaldi, T., Cox, J., Moroder, L., and Mann, M. (2008) Iodoacetamide-induced artifact mimics ubiquitination in mass spectrometry. *Nat. Methods* **5**, 459–460.
25. Andersen, J. S., Lam, Y. W., Leung, A. K., Ong, S. E., Lyon, C. E., Lamond, A. I., and Mann, M. (2005) Nucleolar proteome dynamics. *Nature* **433**, 77–83.
26. Kruger, M., Kratchmarova, I., Blagoev, B., Tseng, Y. H., Kahn, C. R., and Mann, M. (2008) Dissection of the insulin signaling pathway via quantitative phosphoproteomics. *Proc. Natl. Acad. Sci. USA* **105**, 2451–2456.
27. Molina, H., Yang, Y., Ruch, T., Kim, J. W., Mortensen, P., Otto, T., Nalli, A., Tang, Q. Q., Lane, M. D., Chaerkady, R., and Pandey, A. (2009) Temporal profiling of the adipocyte proteome during differentiation using a five-plex SILAC based strategy. *J. Proteome Res.* **8**, 48–58.

28. Hubner, N. C., Ren, S., and Mann, M. (2008) Peptide separation with immobilized pI strips is an attractive alternative to in-gel protein digestion for proteome analysis. *Proteomics* **8**, 4862–4872.
29. Park, S. K., Venable, J. D., Xu, T., and Yates, J. R., 3rd. (2008) A quantitative analysis software tool for mass spectrometry-based proteomics. *Nat. Methods* **5**, 319–322.
30. Dobreva, I., Fielding, A., Foster, L. J., and Dedhar, S. (2008) Mapping the integrin-linked kinase interactome using SILAC. *J. Proteome Res.* **7**, 1740–1749.
31. Kristensen, A. R., Schandorff, S., Hoyer-Hansen, M., Nielsen, M. O., Jaattela, M., Dengjel, J., and Andersen, J. S. (2008) Ordered organelle degradation during starvation-induced autophagy. *Mol. Cell. Proteomics* **7**, 2419–2428.
32. Dengjel, J., Kratchmarova, I., and Blagoev, B. (2009) Receptor tyrosine kinase signaling: a view from quantitative proteomics. *Mol. Biosyst.* **5**, 1112–1121.
33. Mortensen, P., Gouw, J. W., Olsen, J. V., Ong, S. E., Rigbolt, K. T., Bunkenborg, J., Cox, J., Foster, L. J., Heck, A. J., Blagoev, B., Andersen, J. S., and Mann, M. (2010) MSQuant, an open source platform for mass spectrometry-based quantitative proteomics. *J. Proteome Res.* **9**, 393–403.

Chapter 12

Quantification of Proteins by iTRAQ

Richard D. Unwin

Abstract

Protein relative quantification is a key facet of many proteomics experiments. Several methods exist for this type of work, some of which are described elsewhere in this volume. In this chapter we will describe the use of isobaric tags for relative and absolute quantification (iTRAQ). These chemical tags attach to all peptides in a protein digest via free amines at the peptide N-terminus and on the side chain of lysine residues. Labelled samples are then pooled and analysed simultaneously. Since the tags are isobaric, labelled peptides do not show a mass shift in MS, instead signal from the same peptide from all samples is summed, providing a moderate increase in sensitivity. Upon peptide fragmentation, sequence ions (b- and y-type) also show this summed intensity which aids sensitivity. However, the distribution of isotopes in the different tags is such that when the tags fragment a tag-specific 'reporter' ion is released. The ratio of signal intensities from these tags acts as an indication of the relative proportions of that peptide between the different labelled samples. This chapter will describe the procedure for labelling and analysing peptide/protein samples using iTRAQ.

Key words: Peptide, protein, iTRAQ, isobaric, relative quantitation, liquid chromatography, mass spectrometry.

1. Introduction

The use of isobaric tags is now a widely utilised strategy for obtaining relative quantification of peptides in up to eight samples simultaneously. The principle of using isobaric tags is an elegant one. Samples are derivatised with one of several tags, all of which have identical overall mass but which vary in terms of the distribution of heavy isotopes around their structure (**Fig. 12.1A**). This means that when samples are pooled and

P.R. Cutillas, J.F. Timms (eds.), *LC-MS/MS in Proteomics*, Methods in Molecular Biology 658,
DOI 10.1007/978-1-60761-780-8_12,

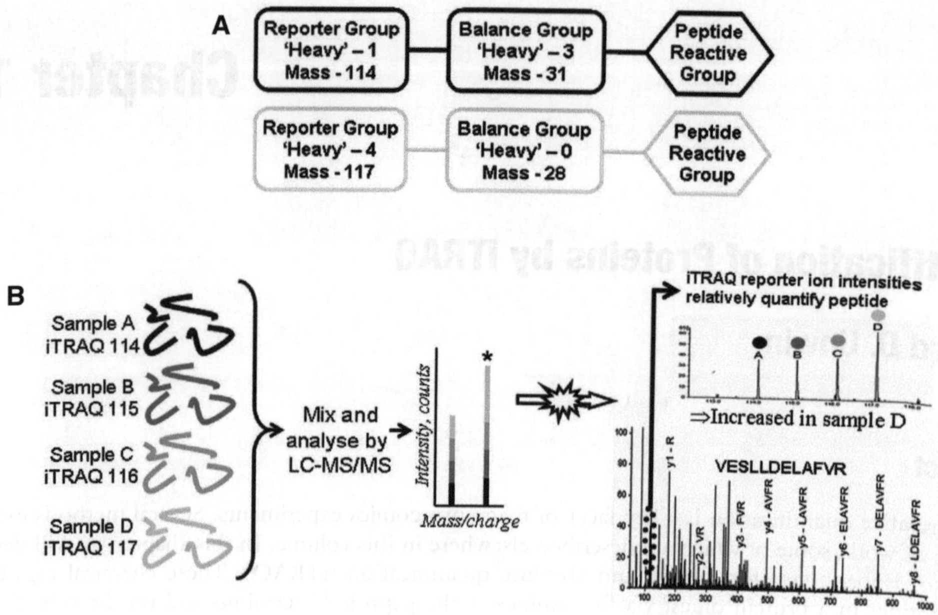

Fig. 12.1 A generic iTRAQ experiment. (**A**) A schematic of two iTRAQ tags, showing three distinct groups, a peptide-reactive group, the reporter group, which allows relative quantification upon MS/MS, and a balance group, which maintains the isobaric nature of the tags. (**B**) An overview of an iTRAQ workflow. Peptides from multiple samples are labelled with iTRAQ tags. These samples are pooled, generating a single ion for each peptide. One putative peptide ion (marked with a * in the figure) is then selected and fragmented. A typical MS/MS spectrum is shown, demonstrating sequence identification, with the reporter ion region expanded to demonstrate how peptide-relative quantification is determined.

analysed simultaneously, the same peptide from the same samples will appear at the same mass in MS. However, upon fragmentation of the peptides by collision-induced dissociation (CID), the peptide fragments provide amino acid sequence information (and therefore peptide identity) and the iTRAQ tag fragments to release a tag-specific reporter ions. The ratios of these reporter ions are representative of the proportions of that peptide in each sample (**Fig. 12.1B**).

This principle was first demonstrated by Thompson et al. (1) who synthesised peptides containing a 'tandem mass tag' and showed that this strategy could indeed be used to obtain relative quantification in tandem MS experiment. A year later, Ross et al. published a similar approach using iTRAQ (isobaric tags for relative and absolute quantification) (2), where they described a tag which contained a reactive moiety-enabling reaction with any peptide (a feature postulated but not shown by Thompson et al.). The remainder of this chapter will deal exclusively with the iTRAQ tagging technology, available through AB Sciex as either a four-channel or eight-channel reagent. The four-channel reagent has reporter ion masses at 114–117, with a small balance group. The eight-channel has reporters at 113–119 and 121 (120 is left blank due to the presence of the phenylalanine immonium ion) and a larger balance group to accommodate the extra isotopes. Practically, we have noticed little difference between the two. Tandem mass tags are available through Thermo Scientific.

iTRAQ has been successful in many experimental settings. The primary advantages over an alternative labelling strategy such as SILAC (*see* Chapter 11 in this volume for a SILAC protocol) are that it is applicable to primary samples, e.g. human biofluids (3–5) and disease tissues (6–8) or primary tissues from animal models (9). The multiplex nature of the reagent is also advantageous over other chemical labelling methodologies such as labelling with ^{18}O during digestion (10) or with ^{13}C-acrylamide (11), making studies on time courses (12) possible, or allowing an increased number of samples to be analysed against the same control simultaneously, saving on sample preparation and analysis time (13). Since iTRAQ also labels all peptides in a sample, it is ideally suited to the quantification of posttranslational modifications (14–16) and can also be used to label whole proteins prior to subsequent protein fractionation steps, if required (17). An alternative to iTRAQ is quantification by label-free approaches (*see* Chapters 9, 10, and 13 for protocols on MS-based label-free quantification of proteins and peptides).

As with many proteomics methods for relative quantification, success in this technique is heavily dependent on sample preparation. The peptide-reactive group on the iTRAQ tags is a *N*-hydroxysuccinimide (NHS) moiety, which will label peptides via free amine groups, namely those at the N-terminus and on lysine side chains. As a result, it is important to ensure that all buffers used to prepare protein prior to labelling should not contain free amines, so common reagents such as Tris and ammonium bicarbonate should be avoided. This chemistry will also react with cysteine, so these should be blocked prior to labelling, and at lower pH can also react with tyrosine residues. Since neither of these reactions goes to completion (i.e. 100% labelling) they should be avoided or at least minimised at all costs with careful design of sample buffer systems. Also, since no method is perfect, it is important to know the technical errors introduced via the sample handling procedure. To test this, it is recommended that each experiment uses all channels with at least two channels taken up by technical or biological replicates. If this is not possible (i.e. a time course experiment), it is recommended that a separate experiment is run containing such replicates to assess inter-experimental variation.

2. Materials

2.1. Sample Preparation and iTRAQ Labelling

1. Triethylammonium bicarbonate (TEAB) supplied as 1 M stock.
2. Sodium dodecyl sulphate (SDS). Dissolved at 0.1% (w/v) in 1 M TEAB and remainder discarded after use.
3. iTRAQ® reagent labelling kit (AB Sciex, Warrington, UK). Store at −20°C. Extremely susceptible to hydrolysis.
4. Tris-(2-carboxyethyl)phosphine (TCEP). 50 mM stock prepared in water and stored at −20°C. *See* **Note 1**.
5. Methylmethanethiosulphate (MMTS; Perbio Scientific, Cramlington, UK). 200 mM stock prepared in isopropanol and stored at −20°C.
6. Sequencing grade-modified trypsin.
7. Lysis buffer: 1 M TEAB with up to 0.1% SDS.

2.2. Sample Analysis by 2D-Liquid Chromatography and Tandem Mass Spectrometry

1. Strong cation exchange chromatography (SCX) buffer A (SCX-A): 10 mM KH_2PO_4 (Sigma), 20% (v/v) acetonitrile (Fisher Scientific, Loughborough, UK), pH 2.7, made with HPLC-grade water (Rathburn Chemicals, Waterburn, UK). Lifetime 2–3 days.
2. Strong cation exchange chromatography (SCX) buffer B (SCX-B): 10 mM KH_2PO_4, 20% (v/v) acetonitrile, 1 M KCl (Sigma), pH 2.7, made with HPLC-grade water. Lifetime 2–3 days.
3. Strong cation exchange chromatography column: (10 × 2.1 cm PolyLC Polysulfoethyl A column, 5 μm beads, 200 Å pore size; Hichrom Ltd., Reading, Berks, UK).
4. Reverse-phase chromatography (RP) buffer A (RP-A): 2% (v/v) acetonitrile, 0.1% (v/v) formic acid, made with HPLC-grade water. Lifetime 2–3 days.
5. Reverse-phase chromatography (RP) buffer B (RP-B): 80% (v/v) acetonitrile, 0.1% (v/v) formic acid, made with HPLC-grade water. Lifetime 2–3 days.
6. Reverse-phase chromatography trap column: Acclaim PepMap C18 (5 μm, 100 Å) in a 300 μm inner diameter × 5 mm column (LC Packings/Dionex, Camberley, UK).
7. Reverse-phase chromatography analytical column: Acclaim C18 PepMap100 (3 μm, 100 Å) in a 75 μm inner diameter × 15 cm column (LC Packings/Dionex, Camberley, UK).

3. Methods

3.1. Sample Preparation and iTRAQ Labelling

1. Prepare protein sample without presence of compounds containing free amine groups (e.g. Tris, ammonium salts). A suggested optimal buffer is 1 M TEAB with up to 0.1% (w/v) SDS. This buffer will efficiently lyse cultured mammalian cells.
2. Assess protein concentration using a suitable protein assay kit. Check the assay kit compatibility with sample buffer.
3. Aliquot equal amounts of protein for labelling – maximum 100 μg – into separate tubes.
4. Equalise the volume with lysis buffer (e.g. 1 M TEAB with up to 0.1% SDS). Minimum volume should be 20 μL. Try to keep the volume as low as possible.
5. Reduce disulphide bonds by adding 0.1 volumes of 'protein reducing buffer' from the iTRAQ kit (50 mM tris-(2-carboxyethyl)phosphine; TCEP). Vortex, pulse spin, and incubate in a heating block at 60°C for 1 h.
6. Alkylate protein by addition of 0.05 volumes of 'cysteine blocking reagent' from the iTRAQ kit (200 mM methylmethanethiosulphate; MMTS). Vortex, pulse spin, and incubate at room temp for 10 min.
7. Digest protein by adding trypsin at a 10:1 substrate:enzyme ratio (i.e. 10 μg trypsin per 100 μg of protein). Lyophilised trypsin should be reconstituted in the supplied buffer and diluted in 1 M TEAB. Ensure that the final concentration of SDS is 0.05% or lower, as concentrations >0.1% will inhibit trypsin activity; add more 1 M TEAB if necessary. Vortex to mix, pulse spin, and incubate at 37°C overnight.
8. Take tryptic digest and, if necessary, reduce the volume in a SpeedVac to around 20 μL. Do not dry.
9. Samples will often dry unevenly. If this is the case, add 1 M TEAB to equalise sample volumes. The maximum volume of sample for labelling is 30 μL.
10. Take iTRAQ vials out of the freezer. Note the batch number and the correction factors on the data sheet supplied in the kit. Allow vials to thaw on the bench for 2–3 min.
11. Spin iTRAQ vials to get all label to the bottom of the tube.
12. Add 60 μL ethanol (4 plex) or isopropanol (8 plex) to the vial of iTRAQ reagent and transfer to sample vial. Wash the iTRAQ reagent vial out with a further 10 μL of solvent and

add this to the sample, resulting in a total of 70 μL organic in the labelling reaction. Repeat for each label/sample.

13. Vortex to mix, pulse spin, and incubate on the bench for at least 1 h (*see* **Note 2**).
14. Samples may slowly form precipitates as addition of iTRAQ reagent may result in decreased solubility. This is not a problem, as this precipitate is solubilised before samples are pooled prior to SCX chromatography. If precipitation forms immediately on addition of the reagent, this could indicate poor digestion and so samples should be assessed by 1D SDS-PAGE for the presence of undigested protein prior to further analysis.
15. Place samples in SpeedVac for 10–15 min to reduce the volume to around 20–30 μL, removing the ethanol.
16. Samples can be stored at −20°C at this stage if required.

OPTIONAL – Remove 10% of each labelling reaction, clean up on SCX cartridge/column as a single fraction, and analyse by mass spectrometry to check labelling efficiency (*see* **Note 3**).

3.2. Sample Analysis by 2D-Liquid Chromatography and Tandem Mass Spectrometry

1. To pool samples, add 100 μL of SCX Buffer A (10 mM KH_2PO_4, 20% ACN, pH 2.7) to each sample. This should help solubilise any precipitate formed during the labelling reaction (if precipitate is still present, add a further 50 μL). Pool labelled peptides into a clean tube.
2. Wash out sample tubes with a further 100 μL SCX Buffer A and add to pooled material, ensuring all material is transferred.
3. Check pH of pooled peptide sample using pH paper. Ensure pH<3.0 by adding small amounts of HCl, if necessary. Spin in a microfuge at full speed for 5 min to remove particulates or and remaining precipitate.
4. Transfer sample to a glass, round-bottomed sample vial and make volume up to 1.6 mL with SCX Buffer A (SCX-A) for loading onto SCX column.
5. Separate peptide sample by strong cation exchange chromatography. Gradient details and flow rate are sample and LC System specific (*see* **Note 4**). We use an ICS-3000 SP pump (Dionex) coupled with a FAMOS autosampler and UV detector. 1.6 mL sample is injected into a 2 mL loop, followed by 200 μL SCX-A. Sample is then loaded from the loop onto the column and washed for at least 30 min with 100% SCX-A at a flow rate of 1 mL/min. UV signal is monitored over this time and washing is not deemed to be complete until the UV trace is flat.

6. The gradient is triggered by a second program, run with the sample loop out of line to prevent peak broadening. A typical gradient is as follows: 0–15% SCX-B in 35 min, 15–30% SCX-B in 5 min, 30–100% B in 5 min, wash at 100% B for 15 min, return to 0% SCX-B, and re-equilibriate column for 15 min, all at flow rate 400 μL/min. Fractions are taken every minute, with peptides usually eluting between 5 min and 45–50 min, yielding 40–45 fractions (**Fig. 12.1B**). These timings/fractionations can be adjusted according to sample and available instrument time.
7. SCX fractions are dried in a SpeedVac to around 10–20 μL and can be stored at −20°C or −80°C until analysis by reverse-phase LC-MS/MS.
8. Resuspend sample in 180 μL RP-A and centrifuge (>10,000 g, 5 min) to remove particulate. Transfer 60 μL to a sample vial for analysis. The protocol described is for use with an ultimate pump, with FAMOS autosampler and SWITCHOS loading pump/valve (all LC Packings). All of the sample is injected into a 100 μL sample loop using a full loop injection and specifying a 20 μL loop in the software. Sample is loaded onto the trapping column and washed with RP-A for 30 min at a flow rate of 30 μL/min. This large amount of washing is critical given the amount of salt in the sample load.
9. Peptides are eluted over an analytical column using an optimised gradient, typically 0% to 40% RP-B over 80 min, followed by 10 min washing with 100% RP-B, followed by re-equilibration with 100% RP-A for 20 min (*see* **Note 4**).
10. Peptides eluting off the LC are infused directly into the mass spectrometer, in this case a QStar XL (Applied Biosystems) via a Microionspray II source (*see* **Note 5**). Source voltages are tuned regularly, but are generally in the region of 100 for the declustering potential, 320 for the focusing potential, source temperature of 120°C, and needle voltage of 2300 V. Analysis is set up to perform a 1 s MS scan, followed by 2 × 1.5 s MS/MS scans on the top two precursors. Each precursor is fragmented up to twice, before exclusion for 2 min. Collision energies are very important in obtaining good iTRAQ data, and a compromise has to be made between obtaining good reporter ions and good peptide sequence ions. We use the linear formula $CE=(m/z \times 0.0575)+7$ for 2+ precursors and $CE=(m/z \times 0.054)+4$ for 3+ precursors. A typical spectrum is shown in **Fig. 12.1**.

3.3. Data Analysis

1. To analyse the data, a number of platforms can be used, but currently the most advanced is the ProteinPilot software from AB Sciex. To set up a search select 'Identify Proteins' from the workflow toolbar. Under the 'Data Sets to Process' window select 'Add' and enter all of the files for a particular run. Peptide data from all data files/SCX fractions is summed and averaged in the final results output.
2. In the 'Process Using' window construct a Paragon Method with all appropriate experimental parameters (alkylation reagent, enzyme, instrument, etc.) and also any enrichment used or whether biological modifications need be considered. These affect the novel PTM discovery tools employed by the Paragon algorithm (18).
3. Run a thorough search for all experiments on complex mixtures (a rapid search is more suited to test experiments where sample is simple and well defined, in our experience). Save the method.
4. To run a false discovery rate calculation, select the PSPEP (Proteomics System Performance Evaluation Pipeline) option.
5. Select where to save the .group file containing the results, and hit Process. The analysis output gives a ratio for each reporter vs a 'Reference' (the reference can be changed under the 'Summary' tab in the results), along with an error factor and p value. The first check should be a QC for the run, determining the efficiency of the iTRAQ labelling reactions. All lysine side chains should be labelled, with around 90% of peptide N-termini also labelled.
6. The point at which a change is deemed 'significant' can be calculated by including replicate samples in the experiment and determining what settings (ratio/p value) a user-defined proportion of the ratios in this comparison (should all be 1:1) exceed (*see* **Note 6**). Typical values in our lab for almost all experiments are ratios greater than 1.25 or less than 0.8 (*see* **Note 7**).

4. Notes

1. All solutions should be made in HPLC-grade water.
2. The iTRAQ reagent tags are extremely unstable in water. It is important to prevent condensation forming in unused tubes, so ensure that a kit is removed from the freezer, reagents taken, and the kit returned as quickly as possible. The reaction is left for 1 h essentially to ensure that all

iTRAQ reagent is hydrolysed. This can be checked, if necessary, by adding a known peptide/protein which would not be otherwise present in the sample after labelling and searching for an iTRAQ-modified version of this protein.

3. The iTRAQ reagent itself is compatible with reverse-phase LC, but will require cleaning up with either a reverse-phase cartridge or StageTip prior to analysis if detergent, e.g. SDS, is used during the sample preparation.
4. LC gradients should be carefully optimised on each LC system and for each sample type to ensure optimal separation and fractionation. Here we describe gradients for a very complex mixture (mammalian whole cell lysate).
5. MS conditions are likely to be instrument specific. Timings, collision energies, and instrument voltages should all be optimised for the system in use, depending on the instrument sensitivity, speed, resolution of the chromatography, etc.
6. At this stage the likelihood is that the investigator is looking to new leads into the biology of a system. Provided that some downstream confirmation experiments are to be performed, it is often better to be inclusive at this stage. It is useful to form two lists of changes – 'definites' and 'possibles'. The possible could include proteins identified with a single (changing) peptide, or those which show a changed ratio but elevated *p* value, possibly due to an outlier peptide. If the possible protein changes fit in with the biology of the system under study or appears as a 'definite' in a replicate experiment, or is confirmed by another means (e.g. western blot), the value of this inclusive strategy becomes obvious. However, for global bioinformatics analyses, e.g. pathway analysis, it is recommended that only the 'definites' be used as the 'possibles' will have a much higher false-positive rate.
7. Remember that when working with ratiometric data it should be transformed into log space. This is because a twofold increase is 2, while a twofold decrease is 0.5, hence the relationship in non-linear (try plotting examples on a graph). In $\log_2$ space, these values become 1 and −1 respectively, with 'no change' having a $\log_2$ ratio of 0.

Acknowledgements

The author would like to thank Prof. Tony Whetton, University of Manchester, for encouragement and advice. This work is partially funded by Leukaemia Research Fund, UK and the NIHR Manchester Biomedical Research Centre.

References

1. Thompson, A., Schafer, J., Kuhn, K., Kienle, S., Schwarz, J., Schmidt, G., Neumann, T., and Hamon, C. (2003) Tandem mass tags: a novel quantification strategy for comparative analysis of complex protein mixtures by MS/MS. *Anal. Chem.* 75, 1895–1904.
2. Ross, P. L., Huang, Y. L. N., Marchese, J. N., Williamson, B., Parker, K., Hattan, S., Khainovski, N., Pillai, S., Dey, S., Daniels, S., Purkayastha, S., Juhasz, P., Martin, S., Bartlet-Jones, M., He, F., Jacobson, A., and Pappin, D. J. (2004) Multiplexed protein quantitation in Saccharomyces cerevisiae using amine-reactive isobaric tagging reagents. *Mol. Cell Proteomics* 3, 1154–1169.
3. Ogata, Y., Charlesworth, M. C., Higgins, L., Keegan, B. M., Vernino, S., and Muddiman, D. C. (2007) Differential protein expression in male and female human lumbar cerebrospinal fluid using iTRAQ reagents after abundant protein depletion. *Proteomics* 7, 3726–3734.
4. Hardt, M., Witkowska, H. E., Webb, S., Thomas, L. R., Dixon, S. E., Hall, S. C. and Fisher, S. J. (2005) Assessing the effects of diurnal variation on the composition of human parotid saliva: quantitative analysis of native peptides using iTRAQ reagents. *Anal. Chem.* 77, 4947–4954.
5. Kristiansson, M. H., Bhat, V. B., Babu, I. R., Wishnok, J. S., and Tannenbaum, S. R. (2007) Comparative time-dependent analysis of potential inflammation biomarkers in lymphoma-bearing SJL mice. *J. Proteome Res.* **6**, 1735–1744.
6. DeSouza, L., Diehl, G., Rodrigues, M. J., Guo, J. Z., Romaschin, A. D., Colgan, T. J. and Siu, K. W. M. (2005) Search for cancer markers from endometrial tissues using differentially labeled tags iTRAQ and cICAT with multidimensional liquid chromatography and tandem mass spectrometry. *J. Proteome Res.* **4**, 377–386.
7. Bouchal, P., Roumeliotis, T., Hrstka, R., Nenutil, R., Vojtesek, B. and Garbis, S. D. (2009) Biomarker discovery in low-grade breast cancer using isobaric stable isotope tags and two-dimensional liquid chromatography-tandem mass spectrometry (iTRAQ-2DLC-MS/MS) based quantitative proteomic analysis. *J. Proteome Res.* **8**, 362–373.
8. Garbis, S. D., Tyritzis, S. I., Roumeliotis, T., Zerefos, P., Giannopoulou, E. G., Vlahou, A., Kossida, S., Diaz, J., Vourekas, S., Tamvakopoulos, C., Pavlakis, K., Sanoudou, D., and Constantinides, C. A. (2008) Search for potential markers for prostate cancer diagnosis, prognosis and treatment in clinical tissue specimens using amine-specific isobaric tagging (iTRAQ) with two-dimensional liquid chromatography and tandem mass spectrometry. *J. Proteome Res.* **7**, 3146–3158.
9. Unwin, R. D., Smith, D. L., Blinco, D., Wilson, C. L., Miller, C. J., Evans, C. A., Jaworska, E., Baldwin, S. A., Barnes, K., Pierce, A., Spooncer, E., and Whetton, A. D. (2006) Quantitative proteomics reveals post-translational control as a regulatory factor in primary hematopoietic stem cells. *Blood* **107**, 4687–4694.
10. Schnölzer, M., Jedrzejewski, P., and Lehmann, W. D. (1996) Protease-catalyzed incorporation of 18O into peptide fragments and its application for protein sequencing by electrospray and matrix-assisted laser desorption/ionization mass spectrometry. *Electrophoresis* **17**, 945–953.
11. Faca, V., Coram, M., Phanstiel, D., Glukhova, V., Zhang, Q., Fitzgibbon, M., McIntosh, M., and Hanash, S. (2006) Quantitative analysis of acrylamide labeled serum proteins by LC-MS/MS. *J. Proteome Res.* **5**, 2009–2018.
12. Williamson, A. J. K., Smith, D. L., Blinco, D., Unwin, R. D., Pearson, S., Wilson, C., Miller, C., Lancashire, L., Lacaud, G., Kouskoff, V., and Whetton, A. D. (2008) Quantitative proteomics analysis demonstrates post-transcriptional regulation of embryonic stem cell differentiation to hematopoiesis. *Mol. Cell Proteomics* 7, 459–472.
13. Pierce, A., Unwin, R. D., Evans, C. A., Griffiths, S., Carney, L., Zhang, L., Jaworska, E., Lee, C. F., Blinco, D., Okoniewski, M. J., Miller, C. J., Bitton, D. A., Spooncer, E., and Whetton, A. D. (2008) Eight-channel iTRAQ enables comparison of the activity of six leukemogenic tyrosine kinases. *Mol. Cell Proteomics* 7, 853–863.
14. Trinidad, J. (2007) Quantitative analysis of synaptic phosphorylation and protein expression. *Mol. Cell Proteomics* 7, 684–696.
15. Wolf-Yadlin, A., Hautaniemi, S., Lauffenburger, D. A., and White, F. M. (2007) Multiple reaction monitoring for robust quantitative proteomic analysis of cellular signaling networks. *Proc. Natl. Acad. Sci. USA* **104**, 5860–5865.
16. Zhang, Y., Wolf-Yadlin, A., Ross, P. L., Pappin, D. J., Rush, J., Lauffenburger, D. A., and White, F. M. (2005) Time-resolved mass spectrometry of tyrosine phos-

phorylation sites in the epidermal growth factor receptor signaling network reveals dynamic modules. *Mol. Cell Proteomics* **4**, 1240–1250.

17. Wiese, S., Reidegeld, K. A., Meyer, H. E., and Warscheid, B. (2007) Protein labeling by iTRAQ: a new tool for quantitative mass spectrometry in proteome research. *Proteomics*, 7, 340–350.

18. Shilov, I. V., Seymour, S. L., Patel, A. A., Loboda, A., Tang, W. H., Keating, S. P., Hunter, C. L., Nuwaysir, L. M., and Schaeffer, D. A. (2007) The paragon algorithm, a next generation search engine that uses sequence temperature values and feature probabilities to identify peptides from tandem mass spectra. *Mol. Cell Proteomics* **6**, 1638–1655.

Chapter 13

Quantification of Proteins by Label-Free LC-MS/MS

Yishai Levin and Sabine Bahn

Abstract

Quantitative proteomic profiling is becoming a widely used approach in systems biology and biomarker discovery. There is a growing realization that quantitative studies require high numbers of non-pooled samples for increased statistical power. We present a descriptive protocol for label-free quantitation of proteins by LC-MS/MS that enables to obtain both quantitative and qualitative information in one study without the need to pool samples or label them.

Key words: Label free, proteomics, LC-MS, MSE, time alignment, nanoLC-MS/MS, quantitation, relative quantitation.

1. Introduction

Proteomics is an evolving field mostly due to new technologies advancing at a rapid rate. These new technologies enable researchers to break new frontiers and reach uncharted territories in the investigation of biological systems. In Clinical Proteomics, for example, these advances are increasingly exploited for the purpose of disease biomarker discovery. Few will doubt that the proteome holds the key to unravelling disease mechanisms, thereby enabling the development of improved diagnostics, the assessment of drug response, drug efficacy, and drug toxicity (1). The discovery of novel biomarkers involves profiling of biological samples in search for disease- or drug-related qualitative and quantitative changes of proteins. Recent advances in analytical instrumentation and bioinformatics now enable the relative quantitation of hundreds of proteins across dozens of samples

P.R. Cutillas, J.F. Timms (eds.), *LC-MS/MS in Proteomics*, Methods in Molecular Biology 658,
DOI 10.1007/978-1-60761-780-8_13,

in a single experiment (2–7), producing information-rich studies that could lead to the discovery of novel disease biomarkers. Obtaining both qualitative and quantitative information also provides deeper and more comprehensive insights into the origin and structure of biological systems by allowing global proteomic profiling. Regardless of the method of choice, quantitative analysis requires an added degree of stringency, especially when analyzing complex samples taken from patients suffering from complex diseases such as neuropsychiatric diseases, which are polygenic, multi-symptomatic, and present with overlapping symptoms.

In order for proteomic profiling experiments to be statistically powered, it is necessary to investigate large sample cohorts.

As discussed in more detail in **Chapters 1** and **2** of this volume, there are several different platforms used for global proteomic quantitative profiling. Among the most popular are gel-based and isotopic-labeling methods, and protocols for these are outlined in **Chapters 11**, **12**, **14**, **15** and **16** of this volume; however, these come with difficulties in meeting some of the requirements for comprehensive and reproducible analysis (6). These requirements are (8, 9) (1) the ability to detect as many proteins as possible; (2) achieving a dynamic range that is wide enough to detect low-abundance proteins; (3) high reproducibility and consistency of the platform performance so that biological differences can be sufficiently distinguished from instrumental ones, along with successful validation of quantitative significance; and (4) the ability to profile and compare a large number of non-pooled samples. The necessity to analyze as many discrete (non-pooled) samples as possible cannot be overestimated. Greater *n*-numbers enable researchers to apply a variety of statistical methods and increase the statistical power of the analysis that are not possible when analyzing only a handful of samples or pooled samples. Furthermore, by analyzing samples in a discrete manner it is possible to investigate subpopulations within a given group as well as verify whether any demographics influence the underlying proteomic profile.

There are several methodologies for label-free relative quantitation of proteins. The basic strategy is similar for all methods where the samples are analyzed sequentially and discretely where neither proteins nor peptides are labeled. Thus the relative quantitation relies on reproducible intensity measurements by the MS. The label-free methods are distinguished by the MS mode of operation, a defining factor for the analysis. There are four types of strategies: the first two rely on acquiring data in data-dependant analysis mode (DDA) and in the second two methods data are acquired in MS survey mode (*see* **Table 13.1**).

A DDA experiment is typically a serial process. The cycle starts by acquiring an MS survey scan followed by the selection of a number of precursor ions for fragmentation (MS/MS) that may

Table 13.1
Four types of label-free quantitation of proteins by LC-MS/MS

Method	MS acquisition mode	Method of quantitation
Spectral counting	Data-dependant acquisition (DDA)	Counting the number of times a peptide was identified
DDA-based ion counting	Data-dependant acquisition (DDA)	Ion intensity of intact peptides detected in the MS scans (survey scans)
MS survey scan-based ion counting	MS survey scan	Ion intensity of intact peptides
MS^E-based ion counting	MS^E	Ion intensity of intact peptides

or may not be at the chromatographic apex. The selected precursor ions are serially isolated for an MS/MS acquisition for an allotted period of time, or until a certain ion current is breached. This cycle of MS and MS/MS acquisitions continues throughout the run time.

The first label-free method, termed 'spectral counting' approach, relies on DDA acquisition and the quantitation is based on how many times a peptide was fragmented. The second method relying on DDA analysis achieves quantitation based on the measured abundance of the intact peptides in the survey scans, which are performed intermittently between MS/MS events. Although this method produces more accurate quantitation than the previous (10) it suffers from two major pitfalls. First, in DDA analysis a survey scan is performed over long time intervals; thus the sampling rate of the eluting peptides is too low to obtain accurate quantitation. The second pitfall is that the identification is not reproducible and most proteins are typically identified with only one or two peptides (11, 12).

The second strategy, which relies on MS survey acquisition, includes two approaches. The first approach requires a blinded analysis in which the molecular weight and intensity of all peptides eluting from the separation column are measured. The MS operates in survey mode (MS scan), at a high sampling rate. The quantitation is based on the extracted ion chromatogram (XIC) of intact peptide intensity measurements. Identification of the peptides and their corresponding proteins is performed by a separate DDA experiment with an 'include list' (switch list) that contains the significantly changing peptides. In this way only the selected peptides are fragmented and thus identified and no other ones.

The second MS survey-based approach, termed data-independent analysis (MS^E) (13), is similar to the previous except that during MS acquisition the collision energy is alternated such that two channels are collected. The first channel includes the abundance measurements of the intact peptides and a second channel for the fragmented peptides. Both are acquired at a high sampling rate. In this type of acquisition the chromatographic profile is reproducibly maintained throughout the run time and thus across the sample set, allowing for the precursor ions to be aligned in time. Once aligned the intensities of precursor ions can be normalized and directly compared across all injections of all samples. This method, therefore, enables relative quantitation and identification of peptides (and thus proteins) in the same experiment without the need to reanalyze samples. Furthermore, data analysis can be performed on unidentified peptides as well as those that were identified.

Whichever method is used, time alignment of data must be performed to summarize the information from all samples (*see* **Fig. 13.1**).

In this chapter we present the methodology for MS^E-based label-free proteomics since in our experience its performance is superior to other methods described in the literature in a number of aspects (14). It enables the reliable identification and

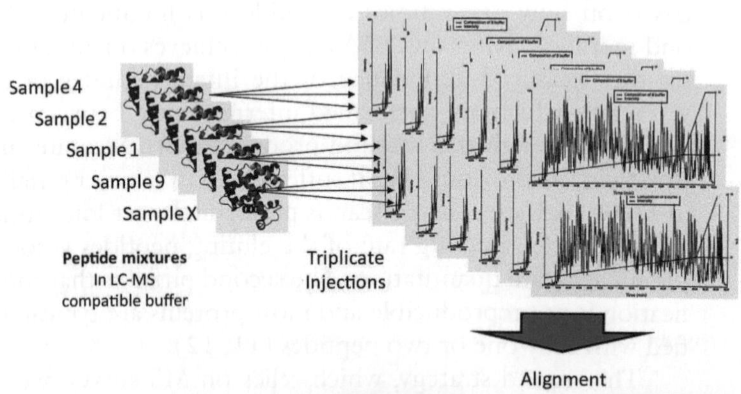

Index	Mass	Retention Time	sample 1_01	sample 1_02	sample 1_03	sample 2_01	sample 2_02	sample 2_03	sample 3_01	sample 3_02	sample 3_03	sample 4_01	sample 4_02	sample 4_03
.	.	.	.	.	.	.	.	.	.	.	.	.	.	.
.	.	.	.	.	.	.	.	.	.	.	.	.	.	.
.	.	.	.	.	.	.	.	.	.	.	.	.	.	.
5879	1915.8466	58.2872	2584	0	3138	5158	4733	4587	3084	3159	0	4635	4202	4936
5880	1835.8864	81.7817	0	5597	0	0	0	0	4303	0	4002	0	0	9349
5882	2027.9971	80.9567	21410	0	19075	17314	14532	11603	58658	53074	54405	18957	19810	25479
5883	1288.7094	56.1242	8213	0	1676	0	1652	7144	5404	4786	0	0	0	28978
5881	807.4214	47.7552	23759	22035	21031	22388	0	35004	0	0	0	0	0	0
5884	3139.469	94.0457	9794	13750	0	16004	17793	14722	19131	19894	16199	13605	9968	11231
.	.	.	.	.	.	.	.	.	.	.	.	.	.	.
.	.	.	.	.	.	.	.	.	.	.	.	.	.	.

Fig. 13.1. A typical workflow for LC-MS^E-based label-free quantitation. Samples are analyzed sequentially and randomly. Each raw data file is processed and searched and then all data files are combined by time alignment of all detected features.

quantification of hundreds of proteins with a high dynamic range across more than three orders of magnitude (3, 15). Furthermore, the reproducibility of the platform is high enough to perform accurate relative quantitation of proteins across dozens of samples at a time, which makes this approach ideal for large-scale biomarker discovery as well as systems biology (4).

The protocol for label-free LC-MS/MS includes three steps: (1) nanoLC- MS^E conditions, (2) data processing/database search, and (3) data analysis. The first part includes information of optimum nano liquid chromatography and mass spectrometry conditions for analysis of complex samples. The second part describes the data processing and database searching parameters. The last part describes the steps needed to analyze a typical data set.

2. Materials

1. Liquid Chromatography: 10kpsi nanoAcquity; trapping column: Symmetry C18 180 μm × 20 mm 5 μm particles; Analytical column: BEH 75 μm × 200 mm 1.7 μm particles (Waters Corp., Milford, MA).
2. Mass Spectrometer: Q-Tof Premier (Waters Corp., Milford, MA).
3. For all preparations and mobile phase HPLC grade water and acetonitrile were used (water – Sigma; acetonitrile – Fisher).
4. Digested yeast enolase (Waters, Milford, MA).
5. Four protein digest mix 1 & 2 (Waters, Milford, MA).
6. Processing, database searching, and time alignment software package.

3. Methods

3.1. Liquid Chromatography-Mass Spectrometry

The protocol for label-free quantitation of proteins by LC-MS is a ‘bottom-up’ approach, which means proteins must be digested into peptides. The analytical process is then performed at the peptide level.

Each biological sample should be injected and analyzed in triplicate followed by a blank injection (to ensure there is no carryover of peptides from one sample to the other in this sequential process). Samples should be run randomly and blinded. For each sample 0.6 μg of total protein digest is loaded using splitless nano Ultra Performance Liquid Chromatography (10kpsi nanoAcquity, Waters, Milford, MA). The autosampler is maintained at 8°C to prevent degradation and evaporation and yet not to cause condensation in the HPLC vial. Buffers used for the mobile phase A: H_2O+0.1% formic acid; B: acetonitrile+0.1% formic acid.

1. Prior to analysis of biological samples a system verification protocol is run using a digested protein standard mix which includes bovine serum albumin, alcohol dehydrogenase, glycogen phosphorylase from rabbit muscle, and yeast enolase. This ensures that the nanoLC-MS system is in peak performance. Initially, five consecutive injections of 25 fmol/μl yeast enolase digest (Waters, Milford) are injected and processed. Minimum system requirements include retention time RSD, intensity RSD, and mass accuracy all based on measurement of the five most intense enolase peptides (see a typical table filled in **Table 13.3** and **Section 4** for further details). Once these requirements are fulfilled two protein mixtures containing four standard proteins are analyzed in triplicate injections to assess the accuracy of relative quantitation (*see* **Table 13.2**).
2. Desalting of the samples is performed online with 100% buffer A for 2 min, using an online reverse-phase C18 trapping column (180 μm i.d., 20 mm length, and 5 μm particle size) (Waters, Milford, MA).
3. The peptides are then separated using a BEH nanoColumn (75 μm i.d., 200 mm length, 1.7 μm particle size) (Waters, Milford, MA), at 300 nl/min with column temperature maintained at 40°C. The gradient used for serum and tissue samples are shown in **Table 13.3**.
4. The nanoUPLC is coupled online through a nanoESI emitter of 7 cm length and 10 μm tip (New Objective, Woburn, MA) to a Quadrupole Time-of-Flight Mass Spectrometer (Qtof Premier, Waters, Milford, MA). Data are acquired in positive V using the MS^E (Expression) mode. In this mode, the quadrupole is set to transfer all ions while the collision cell switches from low to high collision energy intermittently throughout the acquisition time. In the low-energy scans, collision energy is set to 4 eV while in the high-energy scans it is ramped from 17 to 40 eV. This mode enables accurate mass measurement of both intact peptides as well as fragments, and conservation of the chromatographic profile for both intact peptides and fragments. A typical low-energy chromatogram is shown in **Fig. 13.2**.

Table 13.2
A typical 'system verification' table is shown. The verification protocol includes calculation of retention time variation, intensity variation, and mass accuracy

Enolase peptide	1755.9487	1578.801	1416.722	1412.822	1288.7107	1286.7103	Spec
Injection 1 (Intensity)	105069	81385	52072	19656	135892	89877	
Injection 2 (Intensity)	92140	69280	63556	16436	127266	96019	
Injection 3 (Intensity)	85609	81181	55907	17746	115688	95311	
Injection 4 (Intensity)	80341	69139	54935	17040	110997	92092	
Injection 5 (Intensity)	73926	69166	64162	13053	107105	94735	
Average Intensity	11929.43	6621.47	5423.92	2412.51	11931.22	2559.39	
SD Intensity	87417.00	74030.20	58126.40	16786.20	119389.60	93606.80	
%RSD Intensity	**13.65**	**8.94**	**9.33**	**14.37**	**9.99**	**2.73**	**<15%**
Injection 1 (Ret. Time)	27.113	29.4882	21.8523	27.9082	22.4245	25.8871	
Injection 2 (Ret. Time)	27.0873	29.465	21.827	27.8833	22.4033	25.9071	
Injection 3 (Ret. Time)	27.0153	29.3821	21.8025	27.8161	22.3583	25.8134	
Injection 4 (Ret. Time)	27.033	29.4104	21.8319	27.8302	22.3952	25.8255	
Injection 5 (Ret. Time)	27.0033	29.3652	21.7711	27.7893	22.3495	25.7636	
SD Ret. Time	27.05	29.42	21.82	27.85	22.39	25.84	
Average Ret. Time	0.05	0.05	0.03	0.05	0.03	0.06	
	0.18	**0.18**	**0.14**	**0.18**	**0.14**	**0.22**	<2%

Mass accuracy (PPM)	Precursor RMS Mass error	Spec	Mass accuracy (PPM)	Fragment RMS Mass error	Spec
Injection 1	4.9		Injection 1	10.3	
Injection 2	4.1		Injection 2	10.1	
Injection 3	4.7	<10	Injection 3	10.7	<20
Injection 4	4.6		Injection 4	10.8	
Injection 5	4.7		Injection 5	10.7	

Table 13.3
(A) The LC gradient for tissue extracts and cell lines. (B) The LC gradient for depleted serum or plasma

A	Time (min)	Flow (μl/min)	%A	%B	Curve
	Initial	0.3	97	3	Initial
	1	0.3	97	3	Linear
	100	0.3	70	30	Linear
	115	0.3	5	95	Linear
	126	0.3	97	3	Linear
B	**Time (min)**	**Flow (μl/min)**	**%A**	**%B**	**Curve**
	Initial	0.3	95	5	Initial
	1	0.3	95	5	Linear
	80	0.3	70	30	Linear
	90	0.3	5	95	Linear
	104	0.3	5	95	Linear
	105	0.3	95	5	Linear

Mass accuracy is maintained throughout the analysis by the use of a LockSpray apparatus. A reference compound (Glu-Fibrinopeptide B, Sigma, St. Louis, MO) is continuously infused using the LockSpray and scanned intermittently every 30 s. During data processing, the analyte spectra are corrected automatically based on the difference between the detected *m/z* peak and the theoretical *m/z* peak (785.8426 [m+2H]+) of Glu-Fibrinopeptide B.

3.2. Data Processing and Protein Identification

Raw data, acquired in continuum format, are processed using the ProteinLynx Global Server software version 2.3 (also known as Identity^{E}) (Waters, Milford, MA). Both quantitative and qualitative information are produced automatically by the software, using the default parameters.

3.2.1. Quantitative Information

Intensity measurements are obtained by integration of the total ion volume of each extracted, charge state-reduced, deisotoped, and mass-corrected ions across the mass spectrometric and chromatographic volume (**Fig. 13.2**), as opposed to two-dimensional integration of extracted ion chromatograms (XIC). The algorithm calculates the observed mass and intensity measurement deviation for every detected component. The chromatographic area associated with each component is calculated using an integration algorithm similar to the ApexTrack peak integration algorithm provided in the MassLynx software. If a particular component exists in more than one charge state, the

Index	Mass	Retention Time	Intensity											
			sample 1_01	sample 1_02	sample 1_03	sample 2_01	sample 2_02	sample 2_03	sample 3_01	sample 3_02	Sample 3_03	sample 4_01	sample 4_02	sample 4_03
1	1821.9407	81.3128	30026	28035	27818	38122	36191	26999	22124	23621	24762	31701	30763	25922
2	1030.5615	41.6935	9386	8613	8205	11105	11430	8162	6743	7290	7488	10520	10017	7273
3	2256.114	93.3242	5793	5428	5154	7762	7693	5689	4335	4549	5388	7089	6857	4824
4	1755.9407	73.388	58465	61019	81471	19216	18325	14476	15625	18004	17724	30839	29024	23864
5	1578.7996	81.4449	67010	86289	36055	33020	43292	67721	91690	76578	111797	87561	83053	99582
6	1286.7109	67.144	10930	11917	24733	24095	23915	10846	11556	15825	12732	12318	11948	11519

Sum	181609	201302	183437	133318	140845	133893	152073	145867	179891	180028	171662	172983
Normalization factor	1.11	1.00	1.10	1.51	1.43	1.50	1.32	1.38	1.12	1.12	1.17	1.16

normalized data

Index	Mass	Retention Time	sample 1_01	sample 1_02	sample 1_03	sample 2_01	sample 2_02	sample 2_03	sample 3_01	sample 3_02	sample 3_03	sample 4_01	sample 4_02	sample 4_03
1	1821.9407	81.3128	33281	28035	30527	57561	51725	40592	29286	32597	27709	35447	36074	30165
2	1030.5615	41.6935	10403	8613	9004	16767	16336	12272	8925	10060	8379	11764	11747	8464
3	2256.114	93.3242	6421	5428	5656	11721	10995	8553	5739	6277	6030	7926	8041	5613
4	1755.9407	73.388	64805	61019	89406	29014	26191	21764	20683	24847	19833	34483	34036	27770
5	1578.7996	81.4449	74276	86289	39566	49857	61874	101815	121372	105681	125103	97908	97393	115885
6	1286.7109	67.144	12115	11917	27142	36381	34180	16306	15297	21840	14247	13774	14011	13405

Fig. 13.2. The table shows partial and hypothetical data matrices before and after normalization. The *rows* are the detected features (peptides) and the *columns* include the mass, retention time, and intensity in all injections of all samples. In this example, sample 1_02 had the highest sum of intensities; therefore, its normalization factor was set to 1. All the other samples are referenced to it. In actual practice the sum includes all detected features.

corresponding area for any given monoisotopic ion is reported as the summed area from all contributing charge states. The retention time is determined for each reported monoisotopic ion at the moment it reaches its maximum intensity (apex). This process is performed for both the low collision energy and the high-energy scans (saved as separate channels). The ion detection thresholds are set as follows: Low-energy ion detection threshold of 250 counts; high-energy ion detection threshold was set to 100.

3.2.2. Protein Identification

Proteinlynx Global Server version 2.3 (Identity^{E}) is also used for database searches. The database should be chosen based on the biological samples being analyzed. The database should contain species-specific non-redundant sequence entries.

The database search algorithm of the software was described by Li et al. (13). Briefly, the software detects the 250 most abundant peptides and performs an initial pass through the database in order to identify those peptides (with mass tolerance of 10 ppm of precursor ions and 20 ppm for fragment ions). It then calculates the precursor ion mass tolerance, fragment ion mass tolerance, and chromatographic peak widths for these 250 peptides. These peptides are then depleted from the database and the remaining peptides are searched based on these criteria. The cycle continues to the next abundant peptides, which are identified and then temporarily deleted from the database. These tentative peptide identifications are ranked and scored by how well they conform to

14 predetermined models of specific, physicochemical attributes (such as retention time and fragmentation pattern). All tentative peptides are collapsed into their parent proteins utilizing only the highest scoring peptides that contribute to the total protein score. Once a protein has been identified, all top-ranked precursor ions and their corresponding product ions are removed from all other tentatively identified proteins. The remaining unidentified peptides – and tentatively identified proteins – are then re-ranked and re-scored, and the process is repeated until a 4% false-positive rate is reached. The false discovery rate is determined by the number of random or reverse identifications identified (false-positive rate – FPR) divided by the number of correct identifications (true-positive rate – TPR), expressed as a percentage. Therefore the false discovery rate (FDR) is given by FDR=FPR/TPRx100.

All protein identifications were based on at least two peptides. Since this search algorithm is not probability based there is no need for a crude cutoff for selection of individual MS/MS spectra.

3.3. Data Analysis

A typical quantitative analysis includes dozens of samples and thousands of peptides. Thus the data analysis scheme should be automated and unbiased. This can be done using statistical software packages such as the free software R (www.r-project.org). The output of the analysis will highlight the most significantly changing proteins and peptides depending on significance thresholds. There are five major steps to perform: alignment, normalization, filtering, annotation, and combining. The very last step is statistical analysis. Since the choice of statistical methods can vary based on the experimental design and purpose, it is excluded from this protocol. It is, however, recommended that for a comparison of two treatment groups, univariate statistics such as two-tailed, unpaired Student's t test is used after a logarithmic transformation to approximate a normal distribution. For multivariate statistics a partial-least square discriminant analysis (PLS-DA) can be used to find significantly changed peptides or proteins.

3.3.1. Time Alignment

A key step in label-free LC-MS-based quantitation is the time alignment and annotation of data. Since all samples are analyzed sequentially and separately, the data must be combined and summarized. There are several commercially available software packages that can perform this alignment. These include Elucidator© (Rosetta Biosoftware), Progenesis LC-MS© (Nonlinear Dynamics), and Proteinlynx Global Server© (Waters, Milford, MA). The result of the time alignment is a two-dimensional table that includes all detected peptides, their average mass, average retention time, and detected intensity in all replicates of all samples. This table is the basis for all consequent data analysis steps that follow (*see* **Fig. 13.2**). This table can then be annotated using all search results of all analyzed samples. The identity of the peptides is attached based on the accurate mass and retention time.

3.3.2. Normalization

Normalization should be applied to each experiment independently. The normalization should account for experimental variation leaving biological variation unchanged. The choice of normalization method depends on the experimental design. There are two basic options. The first includes normalization based on an exogenous internal standard which is spiked into each sample either before LC-MS analysis, in which case it should be digested first, or in the initial sample preparation stage, in which case it should be spiked as an intact protein. During the normalization step in the data analysis only the signal from the internal standard peptides are used to calculate the normalization factor.

The second method is total ion current normalization. In this method the intensity measurements of all detected peptides in a given sample are used to calculate the normalization factor.

Whichever method is used, the sum of intensities of all peptides in each sample is used to calculate the normalization factor. The sample with the largest sum is set to 1 and all other sums are divided by the intensity of the highest sample (*see* **Fig. 13.2**). Each intensity value is then multiplied by the appropriate normalization factor.

3.3.3. Filtering

The summarized data matrix includes all detected peptides and their corresponding intensity in all replicates of all samples, as mentioned previously. However, for various reasons not all peptides are detected in all replicates or samples. The number of times a given peptide was detected reflects the reliability of the measurement. For example, a peptide that was detected in only one replicate of three can be regarded not reliable and should be excluded. The same is true for peptides that are not detected in several samples. Thus a data set can be filtered to exclude those peptides that are not detected in at least two of three replicates for each sample. The second filtering is on the sample level such that a peptide will be included in the down-stream analysis only if it replicates in the majority of samples in each treatment group (i.e., diseased and control). This method of filtering ensures that a qualitative difference, e.g., detection only in one treatment group, will be noted for further analysis. This is shown for peptide 5881 in **Fig. 13.3**.

3.3.4. Annotation

The database search results for all injections are used to annotate the data set. This is an automated process in which the identity of the peptides is matched to the entries in the data matrix based on the accurate molecular weight and retention time measurements. The peptide with the highest identification score of all search results is used for annotation (*see* **Fig. 13.4**). Since not all peptides are fragmented efficiently, identification will be available

Index	Mass	Retention Time	Intensity											
			Group 1						Group 2					
			sample 1_01	sample 1_02	sample 1_03	sample 2_01	sample 2_02	sample 2_03	sample 3_01	sample 3_02	sample 3_03	sample 4_01	sample 4_02	sample 4_03
5879	1159.5985	46.7278	2584	0	3138	5158	4733	4587	3084	3159	0	4635	4202	4936
5880	1835.8864	81.7817	0	5597	0	0	0	0	4303	0	4002	0	0	9349
5882	1755.9407	73.388	21410	0	19075	17314	14532	11603	58658	53074	54405	18957	19810	25479
5883	1288.7094	56.1242	8213	0	1676	0	1652	7144	5404	4786	0	0	0	28978
5881	807.4214	47.7552	23759	22035	21031	22388	0	35004	0	0	0	0	0	0
5884	1286.7109	67.144	9794	13750	0	16004	17793	14722	19131	19894	16199	13605	9968	11231

Filtered Data

Index	Mass	Retention Time	Group 1						Group 2					
			sample 1_01	sample 1_02	sample 1_03	sample 2_01	sample 2_02	sample 2_03	sample 3_01	sample 3_02	sample 3_03	sample 4_01	sample 4_02	sample 4_03
5879	1159.5985	46.7278	2584	0	3138	5158	4733	4587	3084	3159	0	4635	4202	4936
5881	807.4214	47.7552	23759	22035	21031	22388	0	35004	0	0	0	0	0	0
5882	1755.9407	73.388	21410	0	19075	17314	14532	11603	58658	53074	54405	18957	19810	25479
5883	1288.7094	56.1242	8213	0	1676	0	1652	7144	5404	4786	0	0	0	28978
5884	1286.7109	67.144	9794	13750	0	16004	17793	14722	19131	19894	16199	13605	9968	11231

Fig. 13.3. The table shows partial and hypothetical data matrices before and after filtering. In this example samples 1 and 2 belong to one treatment group and the other two samples belong to a different treatment group. The filtering criteria are detection in at least two of three technical replicates and at least one sample in any one of the treatment groups. Feature no. 5880 is the only one that does not satisfy these criteria. Feature 5881 does fulfil the criteria since it is detected in samples 1 and 2 which are part of the same treatment group.

Index	Mass	Retention Time	sample 1_01	sample 1_02	sample 1_03	sample 2_01	sampl e 2_02	sampl e 2_03	sampl e 3_01	sampl e 3_02	sampl e 3_03	sampl e 4_01	sampl e 4_02	sampl e 4_03
5879	1159.599	46.7278	2584	0	3138	5158	4733	4587	3084	3159	0	4635	4202	4936
5881	807.4214	47.7552	23759	22035	21031	22388	0	35004	0	0	0	0	0	0
5882	1755.941	73.388	21410	0	19075	17314	14532	11603	58658	53074	54405	18957	19810	25479
5883	1288.709	56.1242	8213	0	1676	0	1652	7144	5404	4786	0	0	0	28978
5884	1286.711	67.144	9794	13750	0	16004	17793	14722	19131	19894	16199	13605	9968	11231

Mass	1286.711	1288.709	1755.9407	807.4214	1159.599
RT	67.144	56.1242	73.388	47.7552	46.7278
Score	176.38	192.87	250.53	77.81	33.51
Sequence	NVNDVIAPAFVK	VNQIGTLSESIK	TAGIQIVADDLTVTNPK	TFAEALR	IGSEVYHNLK
Protein	P00924|ENO1_YEAST Enolase 1	P00924|ENO1_YEAST Enolase 1	P00924|ENO1_YEAST Enolase 1	P00924|ENO1_YEAST Enolase 1	P00924|ENO1_YEAST Enolase 1

Index	Mass	Retention Time	Sequence	Score	Protein	sample 1_01	sample 1_02	sample 1_03	sample 2_01	sample 2_02	sample 2_03	sample 3_01	sample 3_02	sample 3_03	sample 4_01	sample 4_02	sample 4_03
5879	1159.599	46.7278	IGSEVYHNLK	33.51	P00924|ENO1_YEAST Enolase 1	2584	0	3138	5158	4733	4587	3084	3159	0	4635	4202	4936
5881	807.4214	47.7552	TFAEALR	77.81	P00924|ENO1_YEAST Enolase 1	23759	22035	21031	22388	0	35004	0	0	0	0	0	0
5882	1755.941	73.388	TAGIQIVADDLTVTNPK	250.53	P00924|ENO1_YEAST Enolase 1	21410	0	19075	17314	14532	11603	58658	53074	54405	18957	19810	25479
5883	1288.709	56.1242	VNQIGTLSESIK	192.87	P00924|ENO1_YEAST Enolase 1	8213	0	1676	0	1652	7144	5404	4786	0	0	0	28978
5884	1286.711	67.144	NVNDVIAPAFVK	176.38	P00924|ENO1_YEAST Enolase 1	9794	13750	0	16004	17793	14722	19131	19894	16199	13605	9968	11231

Fig. 13.4. The figure exemplifies the annotation procedure. The individual peptide identifications are matched to the data matrix based on the accurate mass and retention time measurements.

for only a subset of the detected peptides. The proportion of identified peptides versus detected peptides can vary according to the type of sample and preparation.

3.3.5. Combining

Since the analysis includes technical replicates of each sample these replicates must be combined (**Fig. 13.5**). This is done by

Index	Mass	Retention Time	Sequence	Score	Protein	sample 1_01	sample 1_02	sample 1_03	sample 2_01	sample 2_02	sample 2_03	sample 3_01	sample 3_02	sample 3_03	sample 4_01	sample 4_02	sample 4_03
5879	1159.599	46.7278	IGSEVYHNLK	33.51	P00924|ENO1_YEAST Enolase 1	2584	0	3138	5158	4733	4587	3084	3159	0	4635	4202	4936
5881	807.4214	47.7552	TFAEALR	77.81	P00924|ENO1_YEAST Enolase 1	23759	22035	21031	22388	0	35004	0	0	0	0	0	0
5882	1755.941	73.388	TAGIQIVADDLTVTNPK	250.53	P00924|ENO1_YEAST Enolase 1	21410	0	19075	17314	14532	11603	58658	53074	54405	18957	19810	25479
5883	1288.709	56.1242	VNQIGTLSESIK	192.87	P00924|ENO1_YEAST Enolase 1	8213	0	1676	0	1652	7144	5404	4786	0	0	0	28978
5884	1286.711	67.144	NVNDVIAPAFVK	176.38	P00924|ENO1_YEAST Enolase 1	9794	13750	0	16004	17793	14722	19131	19894	16199	13605	9968	11231

Index	Mass	Retention Time	Sequence	Score	Protein	sample 1	sample 2	sample 3	sample 4
5879	1159.599	46.7278	IGSEVYHNLK	33.51	P00924|ENO1_YEAST Enolase 1	1908	4826	2081	4591
5881	807.4214	47.7552	TFAEALR	77.81	P00924|ENO1_YEAST Enolase 1	22275	19131	0	0
5882	1755.941	73.388	TAGIQIVADDLTVTNPK	250.53	P00924|ENO1_YEAST Enolase 1	13495	14483	55379	21415
5883	1288.709	56.1242	VNQIGTLSESIK	192.87	P00924|ENO1_YEAST Enolase 1	3296	2932	3397	9659
5884	1286.711	67.144	NVNDVIAPAFVK	176.38	P00924|ENO1_YEAST Enolase 1	7848	16173	18408	11601

Fig. 13.5. Intensity measurements of technical replicates are averaged.

calculating the mean intensity value including all three replicates such that for a given peptide there is only one intensity value for each sample. An optional step following combining of technical replicates is combine peptide intensities to the protein level by summing the peptide intensities per protein. This should only include peptides that were not modified by a posttranslational modification. This final step, however, depends on the experimental design and purpose.

4. Notes

4.1. System Verification

This protocol should be run prior to every quantitative study as it ensures the operational performance of the nanoLC-MS/MS system.

1. Inject five consecutive runs of 50 fmol digested yeast enolase using the gradient in **Table 13.3**.
2. Process the five raw data files using the data processing and searching software.
3. Create a table as shown in **Table 13.1** and fill it out based on the results from the processing software.
4. If the specifications in **Table 13.1** are met then proceed to inject mix 1 and mix 2 in triplicates using the gradient in **Table 13.2**.
5. Process the six injections using PLGS.
6. Run the 'Expression Analysis' (time alignment) for these two samples. Set the alcohol dehydrogenase (accession P00330) as the internal standard for normalization. The fold changes of the remaining three proteins should be within the specification listed in **Table 13.4**.

Table 13.4
The table shows the results of processing, databank search and alignment of a four protein standard mix as part of the System Verification protocol

Accession	Description	Score	Mix2:Mix1_Ratio	Expected
(P00924)	Enolase 1 (EC 4.2.1.11)	913.23	2.05	2
(P00489)	Glycogen phosphorylase, muscle form (EC 2.4.1.1)	2054.13	0.48	0.5
(P02769)	Serum albumin precursor	1303.02	8.25	8
(P00330)	Alcohol dehydrogenase l (EC 1.1.1.1)	1179.34	1	1

Acknowledgment

The research was kindly supported by the Stanley Medical Research Institute (SMRI). We also thank Psynova Neurotech® for center support and for Ph.D. funding. The service team from Waters Corp. is acknowledged for its technical support and assistance.

We would also like to present our appreciation to Dr. Hassan Rahmoune and Mr. Emanuel Schwarz of the Cambridge Center for Neuropsychiatric Research for the helpful discussions and suggestions.

References

1. Gao, J., Garulacan, L. A., Storm, S. M., et al. (2005) Biomarker discovery in biological fluids. *Methods* **35**(3), 291–302.
2. Chan, K. C., Lucas, D. A., Hise, D., et al. (2004) Analysis of the human serum proteome. *Clin. Proteomics* **1**(2), 101–226.
3. Levin, Y., Schwarz, E., Wang, L., Leweke, F. M., Bahn, S. (2007) Label-free LC-MS/MS quantitative proteomics for large-scale biomarker discovery in complex samples. *J. Sep. Sci.* **30**(14), 2198–2203.
4. Levin, Y., LW, E., Ingudomnukul, E., Schwarz, S., Baron-Cohen, A., Palotás, S., Bahn. (2009) Real-time evaluation of experimental variation in large-scale LC–MS/MS-based quantitative proteomics of complex samples. *J. Chromatogr.* B, **877**, 1299–1305.
5. van der Greef, J., Martin, S., Juhasz, P., et al. (2007) The art and practice of systems biology in medicine: mapping patterns of relationships. *J. Proteome Res.* **6**(4):1540–1559.
6. Wang G, Wu WW, Zeng W, Chou CL, Shen RF. Label-free protein quantification using LC-coupled ion trap or FT mass spectrometry: Reproducibility, linearity, and application with complex proteomes. J Proteome Res 2006;5(5):1214-23.
7. Wang, W., Zhou, H., Lin, H., et al. (2003) Quantification of proteins and metabolites by mass spectrometry without isotopic labeling or spiked standards. *Anal. Chem.* **75**(18), 4818–4826.
8. Mischak, H. A. R., Banks, R. E, Conaway, M., Coon, J. J., Dominiczak, A., Ehrich, J. H. H., Fliser, D., Girolami, M., Goodsaid, F., Hermjakob, H., Hochstrasser, D., Jankowskii, J., Julian, B. A., Kolch, W., Massy, Z. A., Neusuess, C., Novak, J., Peter, K., Rossing, K., Schanstra, J., Semmes, O.J., Theodorescu, D., Thongboonkerd, V., Weissinger, E. M., Van Eyk, J. E., and Yamamoto, T. (2007) Clinical proteomics: a

need to define the field and to begin to set adequate standards. *Proteomics Clin. Appl.* **1**, 148–156.
9. Qian, W. J., Jacobs, J. M., Liu, T., Camp, D. G., 2nd, Smith, R. D. (2006) Advances and challenges in liquid chromatography-mass spectrometry-based proteomics profiling for clinical applications. *Mol. Cell Proteomics* **5**(10), 1727–1744.
10. Old, W. M., Meyer-Arendt, K., Aveline-Wolf, L., et al. (2005) Comparison of label-free methods for quantifying human proteins by shotgun proteomics* S. *Mol. Cell. Proteomics* **4**(10):1487–1502.
11. Carr, S., Aebersold, R., Baldwin, M., Burlingame, A., Clauser, K., Nesvizhskii, A. (2004) The need for guidelines in publication of peptide and protein identification data. Working Group on Publication Guidelines for Peptide and Protein Identification Data*. Mol. Cell. Proteomics, **3**, 531–533.
12. Wilkins, M. R., Appel, R. D., Van Eyk, J. E., et al. (2006) Guidelines for the next 10 years of proteomics. *Proteomics* **6**(1), 4–8.
13. Li, G. H, V., Silva, J., Golick, D., Gorenstein, M., Geromanos, S. (2008) Database searching and accounting of multiplexed precursor and product ion spectra from the data independent analysis of simple and complex peptide mixtures. *Proteomics*, **9**, 1696–1719.
14. Geromanos, S. J. VH, Silva, J., Dorschel, C., Guo-Zhong, L., Gorenstein, M., Bateman, R., Langridge, J. (2008) The detection, correlation and comparison of peptide precursor and products ions from data independent LC-MS with data dependant LC-MS/MS. *Proteomics*, **9**, 1683–1695.
15. Schwarz, E., Levin, Y., Wang, L., Leweke, F. M., Bahn, S. (2007) Peptide correlation: a means to identify high-quality quantitative information in large-scale proteomic studies. *J. Sep. Sci.* **30**(14), 2190–2197.

Part IV

Analysis of Protein Complexes and Organelles

Chapter 14

Protocol for Quantitative Proteomics of Cellular Membranes and Membrane Rafts

Andrew J. Thompson and Ritchie Williamson

Abstract

Proteomic analysis of membrane and membrane raft proteins is complicated by their inherent insolubility, which exacerbates difficulties with in-solution digestion of the proteins prior to ESI-LC-MS/MS. In-gel digestion yields more comprehensive proteomic and protein coverage of membrane/membrane raft samples, for example by LC-MS/MS of protein samples resolved by 1D SDS-polyacrylamide gel electrophoresis. Although this type of analysis can be performed quantitatively by labelling at the protein level, for instance by SILAC, the separation of proteins on a resolving gel complicates the application of other quantitative methods that employ post-digestion labelling techniques. This chapter describes an alternative protocol to prepare membrane or membrane raft protein samples to be isolated, but not separated, as unresolved bands in a gel. Focusing as a single band enables the confident excision of different samples in their entirety, to be digested, labelled, and fractionated for quantitative mass spectrometric analysis.

Key words: Quantitative proteomics, membrane raft, Isobaric tagging, in-gel digestion, strong cation exchange chromatography, LC-MS/MS.

1. Introduction

Isobaric tagging for quantitative mass spectrometry has several advantages over other quantitative MS-based methods including multiplexed analysis of four to eight samples in a single experiment and generation of quantitative information simultaneously with peptide sequencing. Typical workflows involve differential tagging of samples after in-solution digestion. However, this can be problematic for membrane proteomics due to the insoluble nature of the proteins, and the analysis of membrane rafts, which

P.R. Cutillas, J.F. Timms (eds.), *LC-MS/MS in Proteomics*, Methods in Molecular Biology 658,
DOI 10.1007/978-1-60761-780-8_14, © Springer Science+Business Media, LLC 2010

are comprised of detergent-resistant membrane microdomains, presents an even greater challenge. Although several methods have been explored to improve the in-solution digestion of membrane raft proteins(1–4) arguably the most comprehensive and robust analysis was performed by 1D SDS-polyacrylamide gel electrophoresis (SDS-PAGE) coupled to LC-MS/MS (5). Gel-based methods can facilitate membrane proteomics as the samples can be solubilised in strong denaturing conditions and the proteins resolved and isolated in the gel matrix. Chaotropes and detergents used for sample solubilisation can be removed by washes and in-gel digestion of the gel matrix-suspended proteins can be efficiently performed in a concentrated volume without precipitation of the proteins occurring. However, the separation of proteins on 1D SDS-PAGE resolving gels can complicate quantitation using post-digestion chemical labelling, such as isobaric tagging, due to variations in sample running, multiple gel band excisions and in-gel digestions, and final peptide extractions.

To solve this problem, we developed a method to isolate the protein population from membrane and membrane raft samples

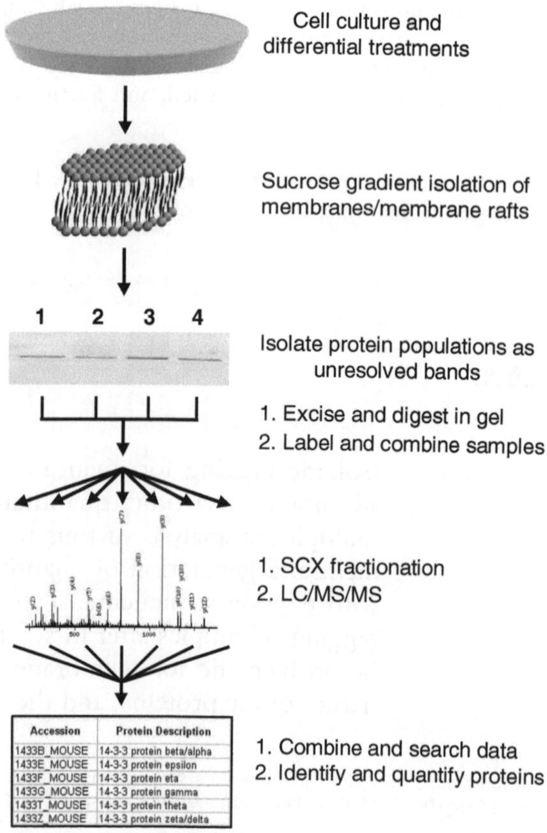

Fig. 14.1 Schematic of the protocol workflow.

as an unresolved band on a 1D SDS-PAGE gel (6). Samples were solubilised effectively in strong detergent and concentrated at the interface of the 4% acrylamide stacking gel and the 20% acrylamide resolving gel. Each membrane protein sample for comparison by quantitative proteomics was then confidently excised as a single band, digested in-gel, and labelled with iTRAQ® reagents for quantitative 2D LC-MS/MS (see **Fig. 14.1** for workflow scheme). Quantitative comparison of samples processed by this gel-based method versus a typical in-solution digestion method revealed a sixfold improvement in the MS ion intensity indicating greatly improved analytical sensitivity. This chapter describes in detail the preparation of membrane and membrane raft samples, confirmation of the integrity of membrane raft isolation by western blot analysis for the raft marker flotillin-1, isolation of the membrane/membrane raft samples as an unresolved band on gel, followed by in-gel digestion, strong cation exchange (SCX) fractionation, and sample desalting in preparation for LC-MS/MS analysis.

2. Materials

2.1. Isolation of Neuronal Plasma Membranes

1. Plasma membrane cell lysis buffer: 10 mM Tris–HCl, pH 7.4, 10 mM NaCl, 3 mM $MgCl_2$,, 10 mM NaF, 2 mM Na_3VO_4, 1 mM EGTA, 1 mM EDTA, 0.2 mM PMSF, protease inhibitor cocktail (Sigma, Gillingham, UK; *see* **Note 1**).
2. Membrane solubilisation buffer: 10 mM Tris–HCl, pH 7.4, 10 mM NaCl, 3 mM $MgCl_2$,, 10 mM NaF, 2 mM Na_3VO_4, 1 mM EGTA, 1 mM EDTA, 0.2 mM PMSF, 0.1% SDS (w/v).
3. Tris-buffered saline (TBS): Prepare a 10X stock of TBS (250 mM Tris–HCl, pH 8.0, 1.4 M NaCl, 50 mM KCl) and store at 4°C. As required, dilute 100 mL of 10X stock with 900 mL of water prior to use.
4. Mini glass homogenisers, 1 mL capacity (Jencons, Leighton Buzzard, UK).

2.2. Isolation of Neuronal Membrane Rafts

1. MES/NaCl buffer: 25 mM MES, 150 mM NaCl, pH 6.5. Store at 4°C.
2. MES buffer: 25 mM MES, pH 6.5. Store at 4°C.
3. Membrane raft cell lysis buffer: 25 mM MES, 150 mM NaCl, 10 mM $MgCl_2$, 10 mM NaF, 2 mM Na_3VO_4, 1 mM EGTA, 5 mM DTT, 0.2 mM PMSF pH 6.5 (*see* **Note 2**),

1% (w/v) CHAPSO (*see* **Note 3**), protease inhibitor cocktail 1 (Sigma).

4. Sucrose gradient solutions: MES buffer containing 10 mM NaF, 2 mM Na_3VO_4, plus 90, 35 or 5% (w/v) sucrose (*see* **Note 4**).
5. Tris/urea buffer: 20 mM Tris, 8 M Urea, pH 7.4, 10 mM NaF, 2 mM Na_3VO_4, 5 mM DTT, 0.2 mM PMSF.
6. Mini glass homogenisers, 1 mL capacity (Jencons, Leighton Buzzard, UK).
7. Ultra-Clear^{TM} centrifuge tubes 14 mm × 89 mm (Beckman Coulter, Fullerton, CA, USA).
8. SW41 swing out rotor and buckets for ultra-centrifugation (Beckman Coulter).

2.3. SDS-PAGE for Western Blot Analysis

1. 4X Protogel Resolving buffer (National Diagnostics, Atlanta, GA, USA): 1.5 M Tris–HCl. 0.4% SDS, pH 8.0.
2. Protogel Stacking buffer (National Diagnostics): 0.5 M Tris–HCl, 0.4% SDS, pH 6.8.
3. Ultrapure Protogel protein sequencing grade acrylamide stock solution (National Diagnostics): 30% (w/v) acrylamide, 0.8% (w/v) bis-acrylamide (37.5:1).
4. *N, N, N', N'*-Tetramethylethylenediamine for electrophoresis (TEMED, Sigma).
5. Ammonium persulphate 98% (Sigma): Freshly prepare a 10% (w/v) solution in water for immediate use.
6. Modified 4X Laemmli buffer (7): 125 mM Tris–HCl, pH 6.8, 40% (w/v) SDS, 20% glycerol, bromophenol blue 0.1% (w/v). Add DTT to 10 mM concentration prior to use.
7. Tris–glycine 10X SDS running buffer (Invitrogen, Paisley, UK).
8. 2-Propanol HPLC grade 99.5% (Sigma): Prepare in a glass vial as a 50% solution in water.
9. SeeBlue®Plus2 prestained marker proteins (Invitrogen).

2.4. Western Blotting for Flotillin-1

1. Wet transfer buffer: 25 mM Tris–HCl (do not adjust pH), 192 mM glycine, 20% v/v methanol.
2. Tris-buffered saline with Tween (TBS-T). Dilute 100 mL 10X stock TBS with 900 mL water and add 2 mL Tween-20.
3. Nitrocellulose membrane 0.45 μm pore (Schleicher & Scheull, Dassel, Germany).
4. Filter paper, extra thick 7.4 × 10 cm (Bio-Rad).
5. Sponge pads (Invitrogen).

6. Membrane blocking buffer (TBS-TM): 3% (w/v) nonfat dry milk in TBS-T.
7. Primary antibody: Flotillin-1 antibody (BD Transduction Laboratories, Lexington, KY, USA).
8. Secondary antibody: Alexa Fluor conjugated goat anti-mouse (Invitrogen).
9. Bovine serum albumin (BSA, Sigma).

2.5. SDS-PAGE for LC-MS/MS Analysis

1. This protocol requires all the reagents listed in **Section 2.3** in addition to those listed here: Fixing solution: Freshly prepare 20 mL of a 7% acetic acid/40% methanol fixing solution per minigel by mixing 1.4 mL glacial acetic acid (Fisher Scientific UK Ltd, Loughborough, UK), 8 mL of HPLC grade methanol (Fisher Scientific UK Ltd), and 10.6 mL water.
2. Colloidal coomassie blue stock solution. This stock solution is a dilution of the 200 mL coomassie blue G colloidal concentrate (Sigma) made up to 1 L with water according to manufacturer instructions and can be stored at 4°C for up to 6 months (*see* **Note 5**). Immediately before use prepare 20 mL of colloidal coomassie staining solution per minigel to stain by diluting 16 mL of the colloidal coomassie blue stock solution with 4 mL HPLC grade methanol (Fisher Scientific UK Ltd). Mix well by vortexing to solubilise dye particulates.
3. Destaining solution 1: Freshly prepare 20 mL of a 7% acetic acid/25% methanol destaining solution per minigel by mixing 1.4 mL glacial acetic acid (Fisher Scientific UK Ltd), 5 mL of HPLC grade methanol (Fisher Scientific UK Ltd), and 13.6 mL water.
4. Destaining solution 2: Freshly prepare 20 mL of a 2% acetic acid/40% methanol destaining solution per minigel by mixing 0.4 mL glacial acetic acid (Fisher Scientific UK Ltd), 5 mL of HPLC grade methanol (Fisher Scientific UK Ltd), and 14.6 mL water.

2.6. In-Gel Digestion

1. Triethylammonium bicarbonate 1 M solution pH 8.0 (TEAB, Sigma): Prepare freshly in a glass vial as a 50 mM and a 100 mM solution in water. The TEAB stock solution is slightly volatile but can be stored at room temperature for up to 1 year if the cap of the stock bottle is sealed with ParaFilm after use. Diluted solutions can be similarly stored for up to 4 weeks.
2. Tris (2-carboxyethyl) phosphine hydrochloride >98% (TCEP, Sigma): Prepare as a 500 mM stock solution by dissolving 143 mg of TCEP hydrochloride in 1.0 mL of

water. This stock solution can be stored as 20 μL single-use aliquots at –80°C for up to 6 months.

3. Methylmethanethiolsulfonate 97% (MMTS, Sigma, *see* **Note 6**).
4. Acetonitrile HPLC grade (Fisher Scientific UK Ltd).
5. Trypsin, protein sequencing grade from bovine pancreas (Roche Diagnostics, Burgess Hill, UK).
6. Trifluoroacetic acid HPLC grade (TFA, Fisher Scientific UK Ltd): Prepare in a glass vial as a 0.1% (v/v) solution in water.

2.7. Labelling of Peptides with iTRAQ® Tags for Quantitation

1. iTRAQ® reagents are provided in a kit (Applied Biosystems, Warrington, UK). This kit includes other reagents which can be used in accordance with the manufacturer's instructions in place of the iTRAQ® labelling protocol described in this chapter.
2. Triethylammonium bicarbonate 1 M solution pH 8.0 (TEAB, Sigma): Prepare fresh as a 0.5 M solution by diluting 1:1 with water.
3. Ethanol HPLC grade (Fisher Scientific UK Ltd).

2.8. SCX Chromatography

1. SCX cartridge and cartridge holder (Applied Biosystems).
2. SCX load buffer: Prepare 500 mL of 5 mM phosphoric acid/25% acetonitrile SCX load buffer pH 2.5 by mixing 288 μL of 85% phosphoric acid solution (Orthophosphoric acid, Ultra grade 85%, Sigma) with 374.7 mL water and 125 mL acetonitrile (HPLC grade, Fisher Scientific UK Ltd). This can be stored at room temperature for up to 6 months.
3. SCX elution buffer: Prepare 250 mL of 5 mM phosphoric acid/350 mM potassium chloride/25% acetonitrile SCX elution buffer pH 2.5 by dissolving 6.52 g of potassium chloride (Biochemika Ultra grade >99.5%, Sigma) in 250 mL SCX load buffer. This can be stored at room temperature for up to 6 months.
4. Formic acid 98% (Fisher Scientific UK Ltd): Freshly prepare in a glass vial as a 10% (v/v) solution in water.

2.9. Sample Desalting

1. Trifluoroacetic acid HPLC grade (TFA, Fisher Scientific UK Ltd): Freshly prepare in a glass vial as a 0.1% (v/v) solution in water.
2. Formic acid 98% (Fisher Scientific UK Ltd): Freshly prepare in a glass vial as a 0.1% (v/v) solution in water.
3. Acetonitrile HPLC grade (Fisher Scientific UK Ltd): Freshly prepare in a glass vial as a neat solution and as a 60% (v/v) solution in water.

3. Methods

The preparation of membrane and membrane rafts is described herein with respect to neuronal cell cultures. However, we have found the methods to be equally applicable to other samples including clinical tissue. Similarly, the application of the subsequent gel-based sample isolation, in-gel digestion, quantitative labelling, and fractionation methods for quantitative proteomics is also not restricted to analysing only membrane and membrane raft proteins, although these methods demonstrate particular benefits for analysing these poorly soluble samples.

3.1. Isolation of Neuronal Plasma Membranes

1. This protocol assumes the use of dissociated primary neuronal cell cultures grown on 10 μg/mL poly-L-Lysine (Sigma) coated 10 cm tissue culture dishes at a seeding density of 9×10^6 cells/dish (cortical) or 3×10^6 cells/dish (hippocampal). Cells were allowed to mature for 7 days (cortical neurons) and 14 days (hippocampal neurons) in culture. Other sample types including other cell cultures or tissue samples are also compatible with the membrane (**Section 3.1**) and membrane raft (**Section 3.2**) isolation procedures. Harvest the cells by placing the dishes on ice and gently aspirate the medium using a pipette. Rinse the cells once in ice-cold TBS, aspirate the TBS, and immediately add 1 mL of ice-cold plasma membrane cell lysis buffer and scrape (*see* **Note 7**). Transfer the lysate to a pre-chilled 1 mL glass homogeniser and incubate on ice for 10 min. Homogenise the lysate with 40 strokes of a loose fitting pestle. Pellet nuclei and cell debris by centrifugation at $8000 \times g$ for 5 min at 4°C.
2. Collect the supernatant and centrifuge at $100{,}000 \times g$ for 30 min at 4°C to pellet the membranes. Discard the supernatant and wash the pellet in ice-cold plasma membrane lysis buffer followed by centrifugation at $100{,}000 \times g$ for 30 min at 4°C. Repeat the wash and centrifugation step and then discard the supernatant.
3. Resuspend the final membrane-containing pellet in 100 μL of membrane solubilisation buffer, immediately snap-freeze in liquid nitrogen and store at –80°C.

3.2. Isolation of Neuronal Membrane Rafts

1. As with **Section 3.1**, this protocol also assumes the use of primary neuronal cell cultures. Other sample types including cell cultures or tissue samples are also compatible with this method. All procedures for the isolation of membrane rafts are performed at 4°C. Immediately prior to membrane raft

isolation prepare the membrane raft cell lysis buffer and 90, 35, and 5% (w/v) sucrose gradient solutions (*see* **Note 4**). Transfer 1 mL of the 90% sucrose gradient solution to the bottom of a 12 mL ultra-centrifuge tube and place on ice. Pre-chill the glass homogenisers on ice and the centrifuge buckets in a 4°C refrigerator.

2. Harvest the cells by placing the cell dishes on ice and gently aspirate the medium using a pipette. Wash the cells once in ice-cold TBS, add 1 mL of membrane raft cell lysis buffer to each dish, and scrape the cells (see **Note 7**). Transfer the lysate suspension to a pre-chilled glass homogeniser and place on ice. Homogenise the lysate on ice with 18 strokes of a tight fitting pestle, then place the homogenate on ice for 30 min. Transfer 1 mL of the cell lysate to an ultra-centrifuge tube and mix thoroughly with 1 mL of the 90% sucrose gradient solution, by repeatedly pipetting or vortexing, to afford 2 mL of a 45% sucrose/lysate mix.
3. Prepare a discontinuous sucrose step gradient by carefully layering 6 mL of the 35% sucrose gradient solution over the 45% sucrose/lysate mix followed by 4 mL of the 5% sucrose gradient solution. Avoid mixing the layers by pipetting slowly and continuously down the side of the tube. Carefully place the sample tube and a balance tube into an ultra-centrifuge and centrifuge at 180,000×*g* for 18 h, not including spin-up and spin-down times (*see* **Note 8**), at 4°C.
4. After centrifugation, a light-scattering band will be visible at the 5%/35% sucrose interface representing the membrane rafts. Carefully collect 1 mL fractions from the top of the gradient using a pipette. This is best achieved by inserting the pipette tip just below the meniscus and carefully lowering the tip as liquid is removed. Transfer the fractions to labelled 1.5 mL vials.
5. Concentrate the membrane rafts by diluting fractions 4 and 5 (counting from the top of the gradient) in 10 mL of MES buffer followed by centrifugation at 100,000×*g* for 1 h at 4°C. Discard the MES buffer and solubilise the membrane raft pellet by triturating in 100 μL Tris/urea buffer and then stand the solution at room temperature for 1 h to allow complete solubilisation. The concentrated membrane raft solution can be stored at −80°C for several weeks before analysis.

3.3. SDS-PAGE for Western Blot Analysis

1. These instructions assume the use of the XCell 'Sure Lock' Mini-Cell system (Invitrogen) using 15-well combs. This can be adapted for use with other systems. Ensure the work area is clean, and all vials, tubes, containers and solutions

are dust free. Thoroughly rinse all apparatus with water before use.

2. Prepare a 1.0 mm thick, 10% resolving gel by mixing 2.60 mL of 4X Protogel resolving buffer, with 3.33 mL Protogel acrylamide stock solution, 3.97 mL water, 100 μL of 10% ammonium persulphate, and 10 μL TEMED. This solution is sufficient to cast two gels, but volumes can be doubled for the simultaneous casting of up to four gels. Pour the gel solution into the gel cassette, leaving ~2 cm from the top for the gel loading wells (~0.5 cm) and stacking gel (1–1.5 cm). Stand the cassette upright and level, and overlay with 50% isopropanol solution to ensure the top of the gel sets evenly. The gel should polymerise in ~30 min (*see* **Note 9**).

3. Pour off the isopropanol and rinse the top of the gel twice with water. Gently shake excess water from the inside of the gel cassette. Blot the top corner of the cassette with lint-free tissue if necessary.

4. Prepare a 1.0 mm thick, 4% stacking gel by mixing 1.5 mL of Protogel stacking buffer, with 0.78 mL acrylamide/bis solution, 3.66 mL water, 90 μL of 10% ammonium persulphate solution, and 5 μL TEMED. Pour the stacking gel and insert a 15-well comb ensuring no air bubbles are present. The stack should be 1–1.5 cm in height with an additional 0.5 cm for the loading wells and should set in 30–40 min.

5. Prepare tris–glycine gel running buffer by diluting 100 mL of 10X stock with 990 mL water with mixing.

6. Carefully remove the well comb from the stacking gel, and rinse the wells gently with water then running buffer.

7. Assemble the gel tank apparatus. Remove the white tape from the base of the gel cassette, insert the cassette into the tank, and a second minigel cassette or the blank spacer cassette into the second cassette position. Fill the tank with running buffer (~800 mL), connect the power pack, and test the system at 150 V. If bubbles appear at the platinum electrodes, the system is assembled correctly.

8. Prepare sucrose gradient fractions for SDS-PAGE by mixing 15 μL of sample from each 1 mL fraction with 5 μL of 4X Laemmli buffer and incubate at 100°C for 10 min (*see* **Note 10**). Allow samples to cool to room temperature and then centrifuge for 10 s.

9. Load Laemmli buffered blanks in the outside lanes. Load pre-stained molecular weight markers in lane 2 and fractions 1 to 12 in lanes 3 to 14. Run the gel at 150 V until the dye front runs off the bottom of the gel. This typically takes 60–80 min.

3.4. Western Blotting for Flotillin-1

1. Samples separated by SDS-PAGE are transferred onto nitrocellulose membranes electrophoretically. These directions assume the use of the XCell 'Sure Lock' Blot Module Kit (Invitrogen) and can transfer two gels at a time. This can be adapted for use with other kits.
2. Two extra thick filter papers per gel are cut to size and soaked in transfer buffer along with sponge pads for blotting immediately prior to electrophoresis. Soak the nitrocellulose in water.
3. The gel unit is disconnected from the power supply unit and disassembled. The stacking gel is discarded and the resolving gel is then laid on top of the nitrocellulose membrane. The nitrocellulose membrane is laid on top of one of the wetted filter papers and the other wetted filter is placed on top of the gel creating a filter paper-gel-nitrocellulose-filter paper sandwich. Place two sponge pads in the cathode core of the blot module. Place the filter paper-gel-nitrocellulose-filter paper sandwich on top of the sponge pads orientated such that the gel is closer to the cathode than the nitrocellulose. Place enough pre-soaked sponge pads on top of the assembly so that the assembly is of the same depth as the cathode core. If transferring two gels, place one sponge pad on top of the assembly then place the second filter paper-gel-nitrocellulose-filter paper sandwich on top followed by more pre-soaked sponge pads as required. Place the anode core on top of the assembly and secure in place in the buffer tank.
4. Fill the inner blot module with transfer buffer and the outer buffer chamber with water.
5. Place the lid on the tank and activate the power supply. Transfer is accomplished by electrophoresis at 35 V constant for 2 h.
6. Once transfer is complete, remove the blot module from the buffer tank and carefully disassemble it. Remove the nitrocellulose membrane and place it in a staining dish. The pre-stained molecular weight markers should be clearly visible on the nitrocellulose.
7. Incubate membrane in 10 mL blocking buffer at room temperature for 1 h on a rotary platform shaker.
8. Discard the blocking buffer and briefly rinse the membrane in TBS-T prior to addition of a 1:2000 dilution of flotillin-1 antibody prepared in TBS-T/2% BSA and left overnight at 4°C on a rotary platform shaker.
9. Remove the primary antibody and wash the membrane three times for 10 min each using 10 mL TBS-T.

10. Prepare the secondary antibody as a 1:10,000 dilution in blocking buffer and add to the membrane for 1 h in the dark on a rotary platform shaker.
11. Discard the secondary antibody and wash the membrane three times for 10 min each with TBS-T.
12. Immunodetection is performed using an Odyssey Infrared Imaging System (Li-Cor Biosciences, Lincoln, NE, USA). Other methods of immunodetection can be applied by altering the secondary antibody and detection system.

3.5. SDS-PAGE for LC-MS/MS Analysis

1. Prepare gels as described in **Section 3.3** except replace the 10% resolving gel with a 20% resolving gel as described here (*see* **Note 11**). Prepare a 1.0-mm thick, 20% resolving gel by mixing 3.25 mL of 4X Protogel resolving buffer, with 8.25 mL Protogel acrylamide stock solution, 0.76 mL water, 125 μL of 10% ammonium persulphate, and 12.5 μL TEMED. This solution is sufficient to cast two gels, but volumes can be doubled for the simultaneous casting of up to four gels. Pour the gel solution into the gel cassette, leaving ~2 cm from the top for the gel loading wells (~0.5 cm) and stacking gel (1–1.5 cm). Stand the cassette upright and level, and overlay with 50% isopropanol solution to ensure the top of the gel sets evenly. The gel should polymerise in 20–30 min.
2. Rinse the resolving gel, prepare the stack and assemble, and test the gel apparatus as described in Steps 3–7 of **Section 3.3**.
3. Prepare equal amounts of the membrane solution (Step 3 of **Section 3.1**) or concentrated membrane raft solution (Step 5 of **Section 3.1**), for example as determined by BCA protein assay (Pierce (Thermo Fisher Scientific UK Ltd), Cramlington, UK), to a final volume of 20 μL. To this add 10 μL 4X Laemmli buffer, and mix thoroughly by repeated aspiration with a pipette (*see* **Note 12**).
4. Load up to four membrane/membrane raft samples for quantitative comparison in alternating lanes, e.g. lanes 2, 4, 6, and 8, of up to 30 μL volume. Load Laemmli buffered blanks between each sample, e.g. lanes 3, 5, 7, and 9 of the same volume as the samples (*see* **Note 13**). Load prestained molecular weight markers in an outside lane, e.g. 1 or 10, also made up with Laemmli buffer and of the same volume as the samples. Run the gel at 150 V until the proteins collect at or near the interface of the 4% stack and 20% gel. This typically takes 20 min and can be visualised using the prestained molecular weight markers. See **Fig. 14.2** for an example.

Fig. 14.2 Example of membrane rafts isolated as unresolved bands on a 1D SDS-PAGE gel for in-gel digestion, quantitative labelling and LC-MS/MS analysis. Membrane raft samples (30 μg in 25 μL) were loaded into lanes 2, 4, 6, and 8, with prestained molecular weight markers in lane 10 and buffered blanks in lanes 1, 3, 5, 7, and 9.

5. Disconnect the power supply and disassemble the electrophoresis apparatus. Remove the gel cassette from the tank and rinse with water. Carefully open the cassette, remove the gel intact, and place into a sterile gel tray (*see* **Note 14**).
6. Add fixing solution and fix the proteins in the gel for 20 min on a rotary platform shaker.
7. Decant the fixing solution, add colloidal coomassie staining solution, and stain the proteins on a rotary platform shaker (*see* **Note 15**).
8. Decant the staining solution from the gel and destain the gel background for 10 min with destaining solution 1.
9. Decant destaining solution 1 from the gel and continue to destain the gel with destaining solution 2 for 1 h or until protein bands are obviously clarified and the gel background is mostly colourless.
10. Decant destaining solution 2 from the gel and replace with water. The gel can be stored at 4°C in the sterile tray overnight.

3.6. In-Gel Digestion

1. Place the gel on a glass pane cleaned with methanol and place on a light box. Carefully excise protein bands from the gel with a sharp scalpel (*see* **Note 16**). Cut each gel band into 1–2 mm^3 pieces and place into a 0.5 mL centrifuge tube.
2. Wash the gel pieces twice with 50 mM TEAB, removing the solutions after washes using a pipette.
3. Dehydrate the gel pieces to small white clumps by washing twice with acetonitrile, again removing the solutions after washes using a pipette, then dry the gel pieces fully in a vacuum centrifuge for 5 min.
4. Prepare a 5 mM TCEP working solution by diluting 10 μL of the 500 mM TCEP stock solution with 990 μL water.

5. Rehydrate the gel pieces with 100 μL of the 5 mM TCEP working solution and incubate on a block heater for 45 min at 55°C to reduce the proteins.
6. Remove the 5 mM TCEP solution using a pipette and dehydrate the gel pieces to small white clumps by washing twice with acetonitrile, removing the solutions after washes using a pipette, then dry the gel pieces fully in a vacuum centrifuge for 5 min.
8. Prepare a 0.1 M MMTS stock solution immediately before use by diluting 10.3 μL of MMTS with 989.7 μL isopropanol in a glass vial (*see* **Note 6**). MMTS is a relatively dense liquid and will collect at the bottom of the vial. Mix the solution thoroughly by careful aspiration with a pipette. Prepare a 10 mM working solution of MMTS by diluting 150 μL of the 0.1 M MMTS stock solution with 1.35 mL 50 mM TEAB.
9. Rehydrate the gel pieces with 100 μL of the 10 mM MMTS working solution and incubate at room temperature for 30 min.
10. Remove the 10 mM MMTS working solution using a pipette.
11. If the gel pieces are not completely destained by this stage, incubate the gel pieces in 50% acetonitrile/50 mM TEAB on a heating block at 37°C with shaking until destained then remove the solution using a pipette.
12. Dehydrate the gel pieces to small white clumps by washing with acetonitrile, remove the wash solutions using a pipette, dry the gel pieces fully in a vacuum centrifuge for 5 min, and then place on ice.
13. Prepare a trypsin stock solution by dissolving 25 μg lyophilised trypsin in a 250 μL solution of 0.1% TFA in water. This solution can be aliquoted and stored at –20°C for up to 6 months. Immediately before use, prepare a trypsin working solution by diluting a 30 μL aliquot of the trypsin stock solution with 200 μL TEAB chilled on ice (*see* **Note 17**).
14. Add just enough trypsin working solution to the gel pieces to cover the estimated volume of the fully hydrated gel pieces. Incubate the suspension on ice for 30 min to rehydrate the gel pieces with trypsin solution. After rehydration, remove any excess trypsin solution using a pipette. Cover the gel pieces with 50 mM TEAB and incubate at 37°C for 4 h (*see* **Note 18**).
15. Transfer the digest solution from the gel piece suspension to a clean centrifuge tube.

16. Dehydrate the gel pieces to small white clumps by washing twice with acetonitrile, and combining the wash fractions with the transferred digest solution. Repeat this step to ensure thorough extraction of peptides from the gel pieces and combine all washes with the digest solution.
17. Evaporate the digest solution to dryness in a vacuum centrifuge.

3.7. Labelling of Peptides with iTRAQ® Tags for Quantitation

1. This protocol describes the labelling of four samples for multiplexed LC-MS/MS and relative quantitation and is essentially performed as per manufacturer instructions (Applied Biosystems). Steps can be adjusted as appropriate for labelling other sample numbers as required. Reconstitute the lyophilised peptide extracts in 30 μL 0.5 M TEAB solution. Thoroughly aspirate the solution and wash down the walls of the sample vial to ensure the entire peptide extract is solubilised.
2. Warm the iTRAQ® reagent vials, e.g. 114, 115, 116, and 117 to room temperature and centrifuge briefly to ensure the reagent solutions are collected at the base of the vial.
3. Add 70 μL of ethanol to each reagent vial, mix by vortexing for 30 s, and then centrifuge for 10 s.
4. Transfer 70 μL of each reagent solution to a peptide extract sample to label samples as desired, e.g. for a four-plex experiment, transfer 70 μL of the 114 reagent solution to sample 1, transfer 70 μL of the 115 reagent solution to sample 2, transfer 70 μL of the 116 reagent solution to sample 3, and transfer 70 μL of the 117 reagent solution to sample 4.
5. Mix the peptide labelling reaction solutions by vortexing for 30 s and then centrifuge for 10 s.
6. Incubate the peptide labelling reaction solutions at room temperature for 1 h.
7. Combine the peptide labelling reaction solutions into one vial, mix by vortexing for 30 s, centrifuge for 10 s, and then evaporate the solvent in a vacuum centrifuge. The sample should concentrate into a dark residue of 5–10 μL volume.

3.8. SCX Chromatography

1. This method is optimised for the fractionation of complex samples containing peptides representing upwards of 50 different proteins. Reconstitute the residue of the combined iTRAQ® labelled samples in 500 μL of SCX loading buffer and mix thoroughly by vortexing.
2. Cut a pH test strip (range pH 1 to pH 6) into ministrips of 2 mm width. Test the pH of the reconstituted iTRAQ®-labelled peptide solution by smearing 5 μL of the solu-

tion across the length of a pH test ministrip. If required, adjust the solution to pH 3.0 or less using 10% formic acid. This can frequently be achieved with a final concentration of 0.1–0.2% formic acid added to the 500 μL reconstituted iTRAQ®-labelled peptide solution.

3. Prepare 15 × 1.5 mL centrifuge tubes each containing 0.5 mL salt fractionation solutions (*see* **Table 14.1**).
4. Assemble the SCX cartridge device and prime the SCX cartridge with 2 mL SCX loading buffer using a 0.5 mL glass syringe equipped with a blunt 25 gauge needle. Flow the solution through the cartridge at ~1 drop per second.
5. Load the reconstituted peptide solution onto the SCX column at a flow rate of 1 drop per second. Return the effluent back into the tube originally containing the reconstituted peptide solution and mix by aspirating briefly with the glass syringe, then load the solution onto the SCX column again to ensure complete sample loading.
7. Elute peptides from the SCX column using a step salt gradient by injecting the 0.5 mL salt fractionation solutions in order of low salt to high salt, i.e. 0–350 mM, collecting the eluent from each fraction in a 2 mL centrifuge tube.

Table 14.1
Preparation of salt fractions for SCX chromatography

Fraction	[KCl] mM	Volume Load buffer (μL)	Volume Elution buffer (μL)
0(wash)	0	500	–
1	30	457	43
2	50	429	71
3	70	400	100
4	80	386	114
5	90	371	129
6	100	357	143
7	110	343	157
8	120	329	171
9	130	314	186
10	140	300	200
11	150	286	214
12	175	250	250
13	225	179	321
14	350	0	500

8. After sample fractionation is complete, freeze the eluent tubes at –80°C for 20 min, and then lyophilise in a vacuum centrifuge.

3.9. Sample Desalting

1. Each fraction from SCX is preferably desalted before LC-MS/MS to ensure optimal analysis. This protocol assumes sample desalting using Millipore C18 ZipTips® and is suitable for recoveries from 10 fmol starting material. Prepare 0.2 mL centrifuge vials for elution by briefly washing with 100 μL of acetonitrile using a pipette. Ensure all acetonitrile has been removed after washing, and then add 4 μL of 60% acetonitrile for later elution of ZipTip® desalted peptides. Close the vial lids to limit solvent evaporation.
2. Reconstitute the lyophilised SCX fraction with 10 μL of 0.1% TFA by rinsing down the sample vial walls and thoroughly aspirating the solution. For higher salt fractions up to 30 μL of 0.1% TFA may be required to completely solubilise the sample.
3. Prime the ZipTip® by gentle aspiration with 3 × 10 μL volumes of acetonitrile.
4. Equilibrate the ZipTip® by gentle aspiration with 3 × 10 μL volumes of 0.1% TFA.
5. Load the sample onto the ZipTip® by 10 cycles of gentle aspiration. For higher volume samples, aspirate half the sample first setting aside the aspirated solution, then aspirate the second half of the solution.
6. Wash the sample-loaded ZipTip® with 6 × 10 μL volumes of 0.1% formic acid.
7. Elute the sample from the ZipTip® in a two stage elution process by loading half (2 μL) of the 60% acetonitrile from the elution vial onto the ZipTip® and gently aspirating twice. Deposit the 2 μL eluent into the elution vial and then load the second half of the 60% acetonitrile from the same elution vial and gently aspirate twice. Combine the eluent halves and evaporate to dryness in a vacuum centrifuge. Evaporation is rapid and should be complete within 15 min.
8. Reconstitute the samples in 0.1% formic acid for LC-MS/MS analysis (see **Note 19**).

4. Notes

1. All water used in this protocol should be deionised and ultra-filtered to have a resistivity of at least 18.2 MΩ-cm and total organic content of less than 5 ppb.

2. PMSF and DTT should only be added to the lysis buffer immediately before use to avoid degradation of the reagents over time.
3. Other detergents can be substituted for the CHAPSO when preparing membrane rafts, e.g. Triton X-100. In our hands, 1% (w/v) CHAPSO isolated a greater percentage of cholesterol in the membrane raft fractions compared to a panel of other detergents.
4. The sucrose in the 90% MES sucrose solution is slow to dissolve. This solution can be prepared quickly as follows: weigh out 18 g of sucrose and place in a 50 mL conical glass/pyrex flask. Add 7 mL of MES buffer, swirl flask, and heat in a microwave on full power for 10 s. Take out flask and swirl, repeat as necessary until all sucrose is dissolved. Be careful not to let the solution boil. While still warm, transfer the solution to a 25 mL glass measuring cylinder and allow to cool to room temperature. Once cooled, add NaF (10 mM) and Na_3VO_4 (2 mM) from 100X stock solutions.
5. The coomassie dye partially precipitates from solution during storage. Before use, always shake the stock solution vigorously to redissolve dye particles. It is strongly recommended not to use colloidal coomassie solutions older than 6 months age as this can result in poor protein staining and poor MS analysis after in-gel digestion.
6. Care should be taken when handling MMTS because of the thiol stench. Handle MMTS in a fume hood and dispose of all solutions and pipette tips used for transfers in standard laboratory bleach immediately after use. Always seal the cap of the MMTS stock vial with ParaFilm after use. This can be stored at room temperature in a fume hood for 2 years.
7. Sufficient material for membrane and membrane raft preparations requires approximately 9×10^6 cells.
8. Spin-up and spin-down times can vary according to the ultracentrifuge used, but some general considerations also apply as follows: Do not allow the centrifuge to accelerate to $180{,}000 \times g$ quickly, as this can disturb the sucrose gradient layers. To prevent this, allow a step-wise acceleration, for example over 45 min. Deceleration should be performed without braking; this can take approximately 3 h depending on the instrument and vacuum.
9. To hold the gel cassettes upright and level, a large office stationary bulldog clip clasped to the bottom of the cassette is usually sufficient. Gels may take longer to set if older TEMED solutions (>6 months) are used. The extent of gel polymerisation can be estimated by inspecting the

remainder of unused gel solution. This will set if left undisturbed and sealed from the atmosphere, e.g. in a capped tube.

10. Immediately before removing a 30 μL aliquot of sample from the respective 1 mL fractions, pass each fraction through a 26 gauge needle syringe twice. From experience proteins in isolated raft fractions tend to aggregate. Passing through a 26 gauge syringe needle gives a more homogenous sample and so the 30 μL aliquot will be more representative of the whole fraction.
11. Preparing a resolving gel at more than 20% acrylamide is not recommended, as the stack can easily tear away from the resolving gel when later removing the gel from the cassette or excising the bands.
12. Do not boil samples solubilised in Tris/urea buffer. Proteins in urea solutions heated above 37°C can become modified by carbamylation.
13. It is essential to load Laemmli buffered blanks of the same volume of the samples on either side of all the samples. This is to prevent band broadening and possible overlap of adjacent samples during electrophoresis.
14. It is important to retain the stacking gel attached to the resolving gel throughout the procedure. This is to ensure the protein samples can be excised consistently and in their entirety as a single band, at or near the interface between the stack and resolving gels.
15. The time to stain is dependent on the sample concentration. For 10–30 μg protein samples, 30 min is usually sufficient, although longer times may be required for less concentrated protein samples. Avoid staining overnight as the dye can become difficult to remove from strongly stained bands during the later in-gel digestion steps.
16. To excise the bands effectively, always use a new sharp scalpel blade for each gel. Cut a line along the breadth of the gel 1–2 mm above the bands in the 4% gel. Low-density gels can be difficult to cut and some gentle sawing through the cut may be required. Cut a similar line 1–2 mm below the bands in the 20% gel. Excise each band by cutting vertically 1–2 mm either side of each band. Use the tip of the scalpel to collect the excised band and dice it into 1–2 mm^3 pieces. The gel band may become sticky if it begins to dry out. To prevent this, periodically wet the scalpel and the excised gel band with a droplet of water during cutting.
17. The trypsin will immediately become activated in neutral or slightly alkaline solution. Use the solution as soon as the aliquot has been diluted with TEAB. Do not stand the solution for later use.

18. Depending on the number of samples to process, the in-gel digestion procedure can take more than half a day to complete and it may be convenient to alternatively digest the samples overnight, at 37°C for at least the initial 2 h.
19. ESI-LC-MS/MS methods are platform dependent, can vary significantly, and should be designed and optimised for the specific instrument type/configuration and laboratory conditions. However, some general considerations for MS analysis of iTRAQ® labelled peptides of complex samples should be considered. Quantitative reporter ions are generated from iTRAQ® labelled peptides by cleavage of an amide bond during MS/MS, analogous to the formation of b and y ions. Higher collision energies than standard can generate more intense reporter ions for quantitation, e.g. for a QToF micromass spectrometer, MS/MS data for peptides are acquired using 30–50 V collision energy depending on peptide mass. For iTRAQ® analysis collision energy is elevated by a further 5 V, similarly to analysing phosphopeptides.

Acknowledgements

The authors would like to thank Dr. Amy Pooler and Dr. Helen Byers for advice and assistance with reviewing the protocols. This work was funded by the Medical Research Council and the Department of Trade and Industry, UK.

References

1. Blonder, J., Hale, M. L., Lucas, D. A., Schaefer, C. F., Yu, L. R., Conrads, T. P., Issaq, H. J., Stiles, B. G., and Veenstra, T. D. (2004) Proteomic analysis of detergent-resistant membrane rafts. *Electrophoresis* **25**, 1307–1318.
2. Li, N., Shaw, A. R., Zhang, N., Mak, A., and Li, L. (2004) Lipid raft proteomics: analysis of in-solution digest of sodium dodecyl sulfate-solubilized lipid raft proteins by liquid chromatography-matrix-assisted laser desorption/ionization tandem mass spectrometry. *Proteomics* **4**, 3156–3166.
3. Blonder, J., Yu, L. R., Radeva, G., Chan, K. C., Lucas, D. A., Waybright, T. J., Issaq, H. J., Sharom, F. J., and Veenstra, T. D. (2006) Combined chemical and enzymatic stable isotope labeling for quantitative profiling of detergent-insoluble membrane proteins isolated using Triton X-100 and Brij-96. *J. Proteome Res.* **5**, 349–360.
4. Bae, T. J., Kim, M. S., Kim, J. W., Kim, B. W., Choo, H. J., Lee, J. W., Kim, K. B., Lee, C. S., Kim, J. H., Chang, S. Y., Kang, C. Y., Lee, S. W., and Ko, Y. G. (2004) Lipid raft proteome reveals ATP synthase complex in the cell surface. *Proteomics* **4**, 3536–3548.
5. Foster, L. J., De Hoog, C. L., and Mann, M. (2003) Unbiased quantitative proteomics of lipid rafts reveals high specificity for signaling factors. *Proc. Natl. Acad. Sci. USA* **100**, 5813–5818.
6. Thompson, A. J., Williamson, R., Schofield, E., Stephenson, J., Hanger, D., and Anderson, B. (2009) Quantitation of glycogen synthase kinase-3 proteins in neuronal membrane rafts. *Proteomics* **9**, 3022–3035.
7. Laemmli, U. K. (1970) Cleavage of structural proteins during the assembly of the head of bacteriophage T4. *Nature* **227**, 680–685.

Chapter 15

Organelle Proteomics by Label-Free and SILAC-Based Protein Correlation Profiling

Joern Dengjel, Lis Jakobsen, and Jens S. Andersen

Abstract

The ability to purify cell organelles and protein complexes on a large scale, combined with advances in protein identification using mass spectrometry, has provided a wealth of information regarding protein localization and function. A major challenge in these studies has been the ability to identify bona fide organelle components from a background of co-purifying contaminants because none of the available biochemical purification protocols afford pure preparations. Since this situation is unlikely to change alternative strategies have been devised to meet this challenge by making use of the information inherent in the fractionation profile of organelles isolated by density gradient centrifugation. In this chapter we describe strategies based on protein correlation profiling and quantitative mass spectrometry to sort out likely candidates. The organelle inventories defined by these methods are suitable to guide future functional experiments.

Key words: Organelle, protein complex, mass spectrometry, proteomics, SILAC, protein correlation profiling.

1. Introduction

Mass spectrometry-based proteomics has become a powerful method to determine the composition of organelles and protein complexes (1, 2). The most widely used approaches involve biochemical or affinity purification of the structure of interest, protein separation by 1D gel electrophoresis, enzymatic digestion of the proteins, followed by separation and analysis of the resulting peptides by liquid chromatography-mass spectrometry (LC-MS). Nevertheless, the analysis of cellular structures remains challenging because such structures are difficult to purify. At best,

P.R. Cutillas, J.F. Timms (eds.), *LC-MS/MS in Proteomics*, Methods in Molecular Biology 658,
DOI 10.1007/978-1-60761-780-8_15, © Springer Science+Business Media, LLC 2010

organelles are enriched but far from pure and the resulting list of proteins obtained by mass spectrometry is crowded with unrelated proteins in the midst of genuine components. As a consequence, the identified proteins provide corroborating rather than unequivocal evidence of organelle association.

To get around this inadequacy it is important to distinguish bona fide organelle components from co-purifying contaminants. A successful strategy to achieve this goal has been the development of quantitative profiling of organelles isolated by density gradient centrifugation (3–5). This strategy is based on algorithms to compare the abundance profile of proteins in subcellular fractions to the profiles of established organelle-associated proteins. Deviation of each protein profile from the marker protein profiles is a measure of its likelihood of being a genuine member of the organelle. Key parameters for successful profiling experiments are the reproducibility and the quality of organelle separation and the accuracy by which the abundance profile of proteins is determined. Various quantitative methods based on mass spectrometry have been applied so far, including label-free quantitation (6), redundant peptide counting (7), stable isotope labeling by ICAT and iTRAQ (8, 9), and stable isotope labeling by amino acid in cell culture (SILAC) (3, 10). These methods assign with a high degree of confidence the core protein inventory of individual organelles but have also been demonstrated for more global analyses of multiple cell organelles under different physiological conditions (8, 11, 12). In this chapter we describe protein correlation profiling (PCP) in the format of label-free and SILAC-based protein quantitation (PCP-SILAC).

2. Materials

2.1. Cell Culture and Lysis

1. Standard culture medium: Dulbecco's Modified Eagle Medium (DMEM) for PCP-SILAC custom culture medium formulated identically to the standard medium but lacking arginine and lysine.
2. Amino acids for PCP-SILAC (13, 14): Normal "light" amino acids, L-lysine (Lys0) and L-arginine (Arg0) hydrochloride; stable isotope-labeled "medium" amino acids, L-lysine-4,4,5,5-d4 hydrochloride (Lys4) and L-arginine-$^{13}C_6$ hydrochloride (Arg6); and "heavy" amino acids, L-arginine-$^{13}C_6$,$^{15}N_4$ hydrochloride (Arg10) and L-lysine-$^{13}C_6$,$^{15}N_2$ hydrochloride (Lys8) (all Sigma-Aldrich, Copenhagen, Denmark).
3. Supplements: Fetal bovine serum (FBS), in the case of PCP-SILAC-dialyzed FBS (Gibco-Invitrogen, Carlsbad, CA);

L-glutamine (200 mM stock solution) (Gibco-Invitrogen); and penicillin/streptomycin (10,000 U/10,000 μg stock solution).

4. Trypsin–EDTA solution (trypsin 200 mg/L and versene–EDTA 500 mg/L) (Gibco-Invitrogen).
5. Sterile phosphate-buffered saline (PBS) (Cambrex Bio Science, Copenhagen ApS, Denmark).
6. Filter units, MF75TM series (Nalge Nunc International, NY).
7. Lysis buffer: 10 mM Tris–HCl (pH=7.5), 150 mM NaCl, 0.5 NP-40, 0.1% β-mercaptoethanol, 1 mM $MgCl_2$, protease inhibitors, EDTA-free (CompleteTM tablets, Roche Diagnostics, Mannheim, Germany). Must be made fresh.
8. Dounce tissue grinders with "tight" pestles. Depending on the sample amount 1 and 7 mL grinders are generally used.

2.2. Cell Fractionation and Gel Electrophoresis

1. Iodixanol OptiPrep® density gradient medium (Sigma-Aldrich) or sucrose solution.
2. Dithiothreitol (DTT) (Sigma-Aldrich).
3. Iodoacetamide (Sigma-Aldrich).
4. SDS loading buffer.
5. In the case of in-solution digestion samples are dissolved in 6 M urea/2 M thiourea.
6. NuPAGE® Novex 4–12% Bis–Tris gel system (Invitrogen).
7. MOPS running buffer (Invitrogen).
8. Colloidal Blue Stain (Invitrogen).
9. Sequence grade-modified trypsin (Promega, Madison, WI) (15).

2.3. Liquid Chromatography (LC)-MS and Data Analysis

1. Hydrophilic buffer A, containing 0.5% acetic acid (Sigma-Aldrich) in water
2. Hydrophobic buffer B, containing 0.5% acetic acid in 80% acetonitrile (Fisher Scientific)
3. STAGE tip for peptide purification (16)
4. Reversed-phase chromatography columns (75 μm ID) filled with Reprosil C18-AQ, 3 mm beads HPLC bulk packing material (Dr Maisch)
5. Protein identification software tools (such as Mascot (MatrixScience)
6. Protein quantitation software (such as MSQuant, http://msquant.sourceforge.net or MaxQuant, http://www.maxquant.org/ (17).

3. Methods

PCP and PCP-SILAC use the information inherent in the fractionation profile of an organelle of interest to distinguish genuine from co-purifying contaminants. The principle of the PCP method is outlined in **Fig. 15.1**. In this method, an organelle is purified from tissue or cell lysates. Protein profiles are obtained by integrating and normalizing peptide ion current signal intensities extracted from LC-MS data. The PCP-SILAC method (**Fig. 15.2**) requires SILAC-labeled proteins and has the advantage that accurate protein profiles are obtained from isotope ratios. The double PCP-SILAC experiment has the additional advantage that protein profiles are obtained from two biological replicates in a single experiment saving measuring time at the mass spectrometer. The protocol described below is generic and does not contain information related to the purification of a particular protein complex or cell organelle. We have successfully applied PCP and PCP-SILAC to the characterization of human centrosomes and autophagosomes which demonstrate the versatility of the two methods. In the protocol presented below, we describe in parallel PCP and PCP-SILAC with indications of steps specific for each method.

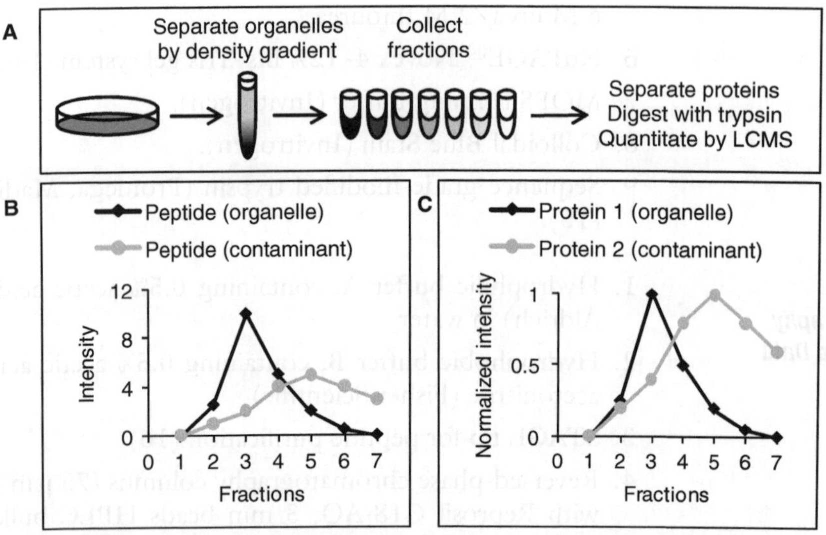

Fig. 15.1 Protein correlation profiling. (**A**) Organelles are purified by density gradient centrifugation from cell lysate. Proteins in each of the collected fractions are separated and digested with trypsin. The resulting peptides are measured by liquid chromatography-mass spectrometry. (**B**) An abundance profile is obtained for each peptide by integrating the peptide ion current in each fraction. (**C**) The profile for each peptide is normalized to the maximum signal and averaged for all peptides identified for a given protein. Marker proteins for an organelle define a consensus profile and deviation of each protein profile from the consensus profile is a measure of its likelihood of being a genuine member of the organelle.

3.1. Organelle Fractionation

1. *PCP*: Prepare lysates from cells or tissues (*see* **Note 1**) and isolate the organelle of interest by density gradient centrifugation (*see* **Note 2**, **Fig. 15.1A**).
2. *PCP-SILAC*: Prepare lysates from three SILAC-labeled cell populations (light, medium, and heavy) and isolate in parallel the organelle of interest by density gradient centrifugation (*see* **Note 3**, **Fig. 15.2A**).
3. Elute the gradients by making a hole in the bottom of the tube with a hot 20 G needle and collect, e.g., 0.5 mL fractions from the bottom of the tube into 2 mL Eppendorf tubes. In the case of gradients of low density, collect fractions from the top or the bottom of the gradient by pipetting or by the use of a gradient collector.

3.2. Determination of Organelle-Containing Fractions

1. Withdraw aliquots (e.g., 50 μL) from each fraction. Determine the distribution of organelle proteins by LC-MS or by the use of organelle-specific antibodies for immunoblot analysis or for immunofluorescent microscopy of organelles sedimented onto a coverslip.
2. *PCP-SILAC*: Combine the 50 μL aliquots of the corresponding fractions for each of the three organelle preparations (light, medium, and heavy) to reduce the number of samples to be tested by LC-MS (**Fig. 15.2B**).
3. Digest the samples in-solution (*see* **Note 4**), purify the resulting peptide on STAGE tips (*see* **Note 5**), and subject the samples to LC-MS analysis and peptide identification as described below.
4. *PCP*: Quantify the abundance of each peptide in each fraction by the integration of the ion current signal intensity extracted from the precursor ion mass spectra.
5. *PCP-SILAC*: Quantify the abundance of each peptide in each fraction by the integration of the ion current signal intensity extracted from the precursor ion mass spectra for each of the three signals corresponding to the light, medium, and heavy isotope clusters.
6. Plot the peptide abundance profiles (**Figs. 15.1B** and **15.2C**) to determine the distribution of organelle proteins.
7. *PCP*: Dilute the sucrose concentration by adding buffer and pellet the organelles by centrifugation. Remove the supernatant after centrifugation. The samples can be stored at this stage.

3.3. Generation of the Internal Standard and Combination of Fractions

1. *PCP-SILAC*: Select organelle-containing fractions, e.g., two to three fractions left and right to the peak fraction (*see* **Note 6**).

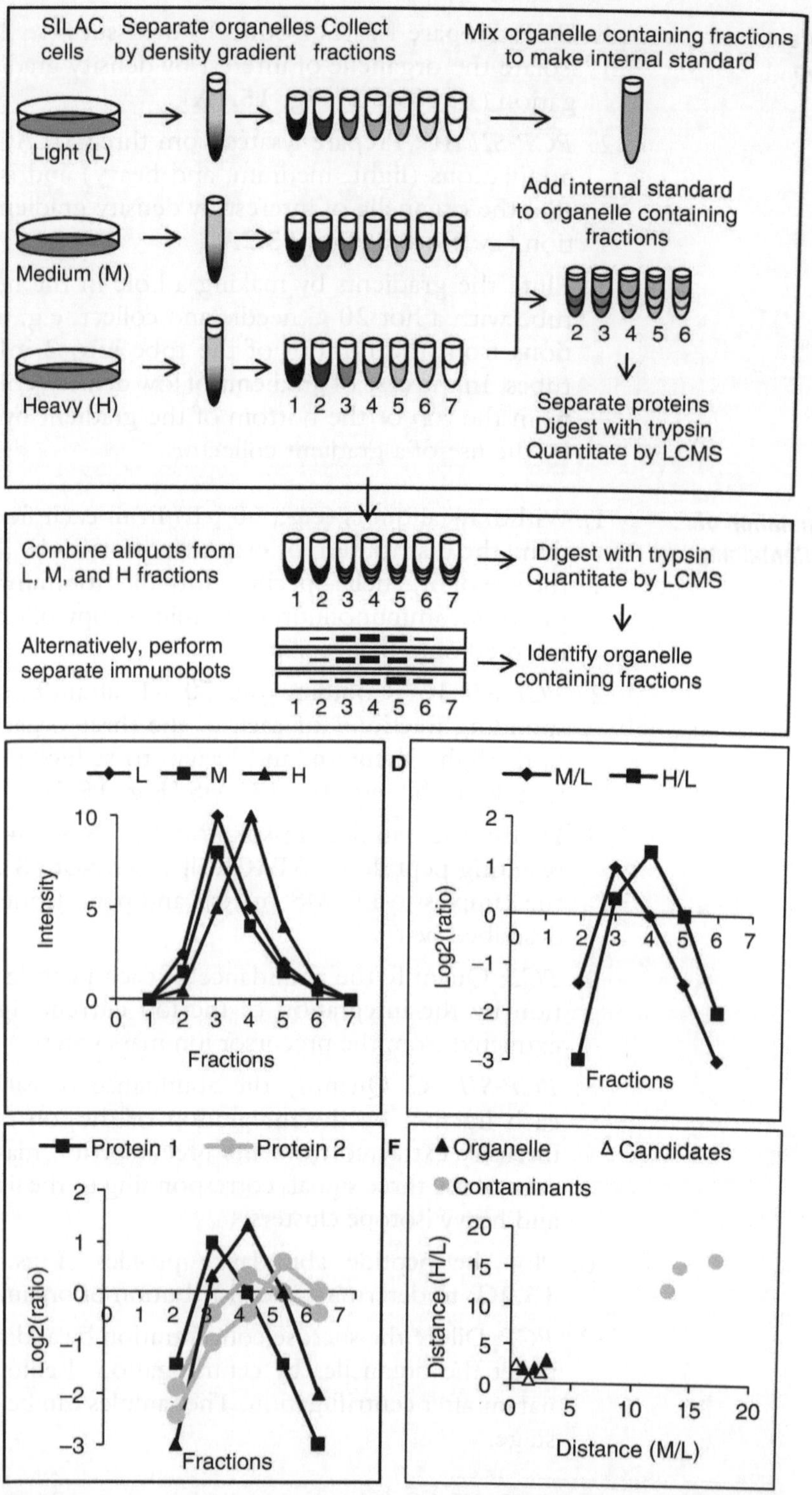

Fig. 15.2 SILAC-based protein correlation profiling. (**A**) Proteins are stable isotope labeled with light, medium, and heavy amino acids in three separate cell cultures. Organelles are purified in parallel from the three cell populations by density gradient centrifugation. Fractions collected from the light labeled preparation are mixed to afford the same internal

2. *PCP-SILAC*: Mix the organelle-containing fractions from the unlabeled cells to generate an internal standard. Dilute the sucrose concentration by adding buffer (*see* **Note 7**), **Fig. 15.2A**.
3. *PCP-SILAC*: Combine the corresponding fractions from the medium and heavy labeled preparations and distribute aliquots of the internal standard into these samples (**Fig. 15.2A**).
4. *PCP-SILAC*: Gently invert to mix the internal standard with the medium and heavy labeled fractions before pelleting the organelles by centrifugation. Remove the supernatant after centrifugation. The samples can be stored at this stage.

3.4. Gel Electrophoresis and LC-MS Analysis

1. Dissolve the pellets in SDS sample buffer, heat the samples for 10 min at 75°C, reduce the disulfide bonds on cysteine in 10 mM DTT for 30 min followed by alkylation in 55 mM iodoacetamide for 20 min. Separate the proteins by electrophoresis on NuPAGE® Bis-Tris 4–12% gradient gels (Invitrogen) and stain the gels with Colloidal Blue to visualize the proteins.
2. Cut the gel lanes into 10–15 slices. Transfer the slices to Eppendorf tubes and cut them into small pieces using a pointed scissor. Wash the gel pieces several times in 50 mM ammonium bicarbonate and 50% acetonitrile and incubated with 12.5 ng/μl trypsin in 50 mM ammonium bicarbonate at 37°C overnight (15). Extract the resulting peptides with 1% TFA, desalt the samples on C18-STAGE tips (*see* **Note 4**), and elute the peptides into a 96-well plate. Remove the organic solvent by vacuum centrifugation and redissolve the samples in 1% TFA.
3. Inject the peptide mixtures onto a 75 ID reversed-phase chromatography column filled with C18-resin. Elute the peptides directly into the mass spectrometer with a 120 min linear gradient of 95% buffer A to 50% buffer B.

Fig. 15.2 (continued) standard for each fraction containing medium and heavy labeled proteins. Proteins in the combined fractions are separated and digested with trypsin. (**B** and **C**) Selection of the organelle-containing fractions for the PCP-SILAC experiment is determined by immunoblot analysis of an organelle marker protein or by peptide ion current measurements as described in **Fig. 15.1** using combined aliquots from the corresponding light, medium, and heavy labeled fractions. (**D**) Relative enrichment profiles for each protein in the medium and heavy labeled organelle preparations are obtained by averaging the medium/light and heavy/light ratios for all peptides identified for the same protein in each fraction. (**E** and **F**) Similar to PCP, the protein enrichment profile can be used to correctly assign organelle proteins directly or by analyzing the distance between profiles for each protein and for a group of known organelle proteins.

4. Operate the mass spectrometer in the data-dependent mode to acquire precursor ion and fragment ion spectra for protein identification and quantitation.

3.5. Determination of Protein Abundance Profiles

PCP and PCP-SILAC experiments necessitate an extensive extraction of data for protein identification and quantitation. A number of software tools fulfill these requirements such as Mascot (MatrixScience) for protein identification and MSQuant or MaxQuant (17) for label-free and SILAC-based protein quantitation.

1. Extract peak lists from the mass spectra and search a sequence database to identify proteins.
2. *PCP*: Quantify the abundance of each peptide ion signal in each fraction by integrating the ion current signal intensity extracted from the precursor ion mass spectra (**Fig. 15.1B**). Normalize the peptide ion abundance profile to the fraction with maximum signal. Calculate the median of the normalized abundance profiles for all peptide ion signals representing the same protein (*see* **Note 8**) and plot the profile (**Fig. 15.1C**).
3. *PCP-SILAC*: Quantify the lysine- and arginine-containing peptides from the precursor ion spectra and calculate the relative abundance of peptides in each fraction as the peak area of the medium/light and heavy/light isotope ratios. Compute the median of $\log_2$-transformed medium/light and heavy/light isotope ratios for all peptides representing the same protein in each fraction and use these values to plot two relative protein abundance profiles for each protein (**Fig. 15.2D**).
4. Compare the protein profiles for each protein with the profiles of organelle marker proteins. Proteins with comparable profiles are likely organelle-associated candidates whereas proteins with dissimilar profiles are likely contaminants (**Fig. 15.2E**).

3.6. Statistical Analysis

The data obtained from PCP and PCP-SILAC experiments can be subjected to statistical analysis to test the likelihood of a given protein to be annotated as an organellar protein. Various methods have been applied for the statistical analysis of protein profile data. The measurement of profile similarity between a given protein and a set of known organelle proteins requires prior knowledge whereas cluster analysis and principle component analysis (5, 12) are unbiased. Here we describe the calculation of the Mahalanobis distance (*see* **Note 9**) between the profile for each protein and the profiles for a group of organelle-associated proteins (**Fig. 15.2F**).

1. Express the peptide ratios in a vector form, $x = (x_1, x_2, \ldots, x_n)$, where each dimension corresponds to a gradient fraction, and n is the total number of fractions. Ratios

of a "tuning set" of a number of known organelle proteins, against which all other proteins is compared, are averaged along each dimension to give a vector of expected values for each fraction, $\mu = (\mu_1, \mu_2, \ldots, \mu_n)$, known as the centroid. The distance of a given observation from this point is a measure of the similarity of that protein's profile to the tuning set.

2. Quantify the similarity as a Mahalanobis distance, d_M, $d_M = \sqrt{(x-\mu)^T \Sigma^{-1}(x-\mu)}$, where Σ is the variance–covariance matrix of the tuning set. The Mahalanobis distance represents the distance of a given point from the center (mean) of the "consensus proteins" in n-dimensional space, normalized by the variance (and covariance) of this group of points. Equal deviations from the centroid in two different dimensions (fractions) may yield different contributions to the Mahalanobis distance, depending upon the breadth of the tuning protein distribution in those dimensions. The metric therefore makes a measure of the similarity of a given point to the consensus set.
3. Plot the Mahalanobis distance. Data from the double PCP-SILAC experiments provide a distance measure for each of the two experiments. Proteins with low values in both experiments are likely organelle-associated proteins (**Fig. 15.2F**).

4. Notes

1. The lysis buffer should contain protease inhibitors to prevent partial digestion of the sample during the purification procedure. However, if, e.g., lysosomes are purified and the organelle distribution should be analyzed by enzymatic activity measurements, protease inhibitors should be omitted. Lysis buffer with detergent should be avoided when isolating membrane enclosed organelles.
2. Discontinuous gradients are typically prepared with solutions of sucrose or iodixanol by underlying solutions of increasing density. Continuous gradients are typically prepared by the use of a gradient mixer, by the centrifugation of iodixanol solutions, or by partial diffusion of a step gradient overnight.
3. Light, medium, and heavy labeled cells refer to cells cultured in SILAC medium with light (Lys0 and Arg0), medium (Lys4 and Arg6), and heavy (Lys8 and Arg10) isotope-labeled amino acids. Cells should be cultured in the SILAC medium for at least six cell divisions to fully incorporate the SILAC amino acids. Incomplete labeling might result

in skewed profiles for proteins associated with the same organelle but having different protein turnover. The same number of cells should be used for each condition to obtain peptide ratios easily quantifiable by MS.

4. In-solution digestion. Dissolve sample in 6 M urea/2 M thiourea, 20 mM ammonium bicarbonate, pH 7.8. Add 1 ug of endoprotease Lys-C per 50 ug of protein. Dilute the concentration of urea to less than 2 M and add 1 ug of endoprotease Lys-C per 50 ug of protein. Incubate at room temperature overnight. Add DTT and incubate for 30 min. Add iodoacetamid and incubate for 20 min. Acidify by adding 1% TFA.
5. STAGE tip purification of peptides (16).
6. PCP-SILAC has the advantage that a variable number of fractions can be selected for the analysis. However, if the coverage of the gradient is extended with the purpose of profiling multiple organelles we anticipate that the relative ratio of enrichment in the peak fraction and the proportion of other proteins introduced by the internal standard will increase to a point where a better choice would be to cover the gradient with several PCP-SILAC experiments.
7. Additions of buffer to the fractions to be combined for the formation of the internal standard ensure a more accurate distribution of internal standard to the gradient fractions and allow pelleting of the organelle after mixing by diluting the sucrose concentration.
8. The PCP method hinges on quantifying the corresponding peptide ion signals in each fraction because peptide ions with different sequences and charge states have incomparable detector responses in mass spectrometry. To quantify peptide ion signals in fractions where the signal is not supported by a fragment ion spectrum identifying the peptide, it is possible to correlate peptide retention between LC-MS experiments to predict the retention of the missing signals.
9. The Mahalanobis distance can be calculated by the use of an Add-in to Microsoft Excel (http://www.tvmcalcs.com/blog/comments/VarianceCovariance_Matrix_Addin_for_Excel_2007/) or by the use of the free software environment R for statistical computing (http://www.r-project.org/).

Acknowledgments

The research leading to these results has received funding from the European Commission's 7th Framework Programme and grant agreements HEALTH-F4-2008-201648/PROSPECTS and HEALTH-F4-2007-200767/APO-SYS. JD was supported by the European Molecular Biology Organization. We thank all CEBI group members for helpful discussions and support.

References

1. Aebersold, R., and Mann, M. (2003) Mass spectrometry-based proteomics. *Nature* **422**, 198–207.
2. Yates, J. R., 3rd, Gilchrist, A., Howell, K. E., and Bergeron, J. J. (2005) Proteomics of organelles and large cellular structures. *Nat Rev Mol Cell Biol* **6**, 702–714.
3. Andersen, J. S., and Mann, M. (2006) Organellar proteomics: turning inventories into insights. *EMBO Rep* 7, 874–879.
4. Yan, W., Aebersold, R., and Raines, E. W. (2008) Evolution of organelle-associated protein profiling. *J Proteomics* **15**, 4–11.
5. Sadowski, P. G., Dunkley, T. P., Shadforth, I. P., Dupree, P., Bessant, C., Griffin, J. L., and Lilley, K. S. (2006) Quantitative proteomic approach to study subcellular localization of membrane proteins. *Nat. Protoc.* **1**, 1778–1789.
6. Andersen, J. S., Wilkinson, C. J., Mayor, T., Mortensen, P., Nigg, E. A., and Mann, M. (2003) Proteomic characterization of the human centrosome by protein correlation profiling. *Nature* **426**, 570–574.
7. Gilchrist, A., Au, C. E., Hiding, J., Bell, A. W., Fernandez-Rodriguez, J., Lesimple, S., Nagaya, H., Roy, L., Gosline, S. J., Hallett, M., Paiement, J., Kearney, R. E., Nilsson, T., and Bergeron, J. J. (2006) Quantitative proteomics analysis of the secretory pathway. *Cell* **127**, 1265–1281.
8. Dunkley, T. P., Hester, S., Shadforth, I. P., Runions, J., Weimar, T., Hanton, S. L., Griffin, J. L., Bessant, C., Brandizzi, F., Hawes, C., Watson, R. B., Dupree, P., and Lilley, K. S. (2006) Mapping the Arabidopsis organelle proteome. *Proc. Natl. Acad. Sci. USA* **103**, 6518–6523.
9. Dunkley, T. P., Watson, R., Griffin, J. L., Dupree, P., and Lilley, K. S. (2004) Localization of organelle proteins by isotope tagging (LOPIT). *Mol. Cell Proteomics* **3**, 1128–1134.
10. Ong, S. E., Blagoev, B., Kratchmarova, I., Kristensen, D. B., Steen, H., Pandey, A., and Mann, M. (2002) Stable isotope labeling by amino acids in cell culture, SILAC, as a simple and accurate approach to expression proteomics. *Mol. Cell Proteomics* **1**, 376–386.
11. Foster, L. J., de Hoog, C. L., Zhang, Y., Zhang, Y., Xie, X., Mootha, V. K., and Mann, M. (2006) A mammalian organelle map by protein correlation profiling. *Cell* **125**, 187–199.
12. Yan, W., Hwang, D., and Aebersold, R. (2008) Quantitative proteomic analysis to profile dynamic changes in the spatial distribution of cellular proteins. *Methods Mol. Biol.* **432**, 389–401.
13. Blagoev, B., and Mann, M. (2006) Quantitative proteomics to study mitogen-activated protein kinases. *Methods* **40**, 243–250.
14. Ong, S. E., and Mann, M. (2007) Stable isotope labeling by amino acids in cell culture for quantitative proteomics. *Methods Mol. Biol.* **359**, 37–52.
15. Shevchenko, A., Tomas, H., Havlis, J., Olsen, J. V., and Mann, M. (2006) In-gel digestion for mass spectrometric characterization of proteins and proteomes. *Nat. Protoc.* **1**, 2856–2860.
16. Rappsilber, J., Mann, M., and Ishihama, Y. (2007) Protocol for micro-purification, enrichment, pre-fractionation and storage of peptides for proteomics using StageTips. *Nat. Protoc.* **2**, 1896–1906.
17. Cox, J., and Mann, M. (2008) MaxQuant enables high peptide identification rates, individualized p.p.b.-range mass accuracies and proteome-wide protein quantification. *Nat. Biotechnol.* **26**, 1367–1372.

6. [illegible] Nature 426, 570–574.

13. [illegible] and Mann, M. (2005) [illegible] 40, [illegible]
14. Ong, S. E., and Mann, M. (2007) [illegible] Methods Mol Biol 359, 37–52.

Chapter 16

Mapping Protein–Protein Interactions by Quantitative Proteomics

Joern Dengjel, Irina Kratchmarova, and Blagoy Blagoev

Abstract

Proteins exert their function inside a cell generally in multiprotein complexes. These complexes are highly dynamic structures changing their composition over time and cell state. The same protein may thereby fulfill different functions depending on its binding partners. Quantitative mass spectrometry (MS)-based proteomics in combination with affinity purification protocols has become the method of choice to map and track the dynamic changes in protein–protein interactions, including the ones occurring during cellular signaling events. Different quantitative MS strategies have been used to characterize protein interaction networks. In this chapter we describe in detail the use of stable isotope labeling by amino acids in cell culture (SILAC) for the quantitative analysis of stimulus-dependent dynamic protein interactions.

Key words: Protein complex, protein interaction, signaling, mass spectrometry, proteomics, SILAC.

1. Introduction

Cellular signaling events are transmitted from the plasma membrane throughout the cytosol and into the nucleus via dynamic interactions of proteins forming multiprotein complexes. Quantitative mass spectrometry (MS)-based proteomics is a powerful tool to determine the composition of such complexes and to follow their dynamics (1, 2). The analysis of protein interaction networks is commonly carried out by a combination of affinity purification and mass spectrometry (AP-MS). Often, affinity-tagged recombinant versions of the proteins of interest are used to decipher the corresponding interaction networks. This has, e.g., been successfully employed for the yeast interactome using flag- and tandem affinity purification (TAP)-tagged proteins (3–5). Due to the experimental setup and the sensitivity of the newest

P.R. Cutillas, J.F. Timms (eds.), *LC-MS/MS in Proteomics*, Methods in Molecular Biology 658,
DOI 10.1007/978-1-60761-780-8_16, © Springer Science+Business Media, LLC 2010

generation of mass spectrometers, contaminant proteins are also often identified and carefully designed control experiments have to be carried out to be able to eliminate these false-positive interaction partners from a data set. Quantitative experiments, either via incorporated stable isotopes or via label-free quantitation, represent another possibility to address the question if proteins are true binders or false-positive contaminants. Different isotope-based labeling strategies exist (6), one of them being stable isotope labeling by amino acids in cell culture (SILAC) (7) where cells are labeled metabolically. This strategy has been employed successfully inter alia for the analysis of epidermal growth factor (EGF), platelet-derived growth factor (PDGF), and insulin-dependent signaling complexes (8–11). Chemical labeling strategies via isotope-coded affinity tags (ICAT) (12) or via isobaric tag for relative and absolute quantitation (iTRAQ) (13) have also been successfully used for the analysis of protein complexes (14, 15). In this chapter we will present a detailed protocol for the quantitative analysis of stimulus-specific signaling complexes using SILAC-based liquid chromatography mass spectrometry (LC-MS/MS), highlighting the possibility to distinguish weak and strong interactions depending on the experimental setup.

2. Materials

2.1. Cell Culture

1. Medium: Dulbecco's Modified Eagle Medium (DMEM) deficient in lysine and arginine (Gibco-Invitrogen, Carlsbad, CA).
2. Stable isotope-labeled "heavy" amino acids: L-arginine-$^{13}C_6$ hydrochloride (Arg6) (Cambridge Isotope Labs, Andover, MA; cat. no. CLM-2265), L-arginine-$^{13}C_6$, $^{15}N_4$ hydrochloride (Arg10) (Sigma-Isotec, St. Louis, MO; cat. no. 608033), L-lysine-4,4,5,5-d_4 hydrochloride (Lys4) (Sigma-Isotec; cat. no. 616192), and L-lysine-$^{13}C_6$, $^{15}N_2$ hydrochloride (Lys8) (Sigma-Isotec; cat. no. 608041).
3. Supplements: Dialyzed fetal bovine serum (FBS); L-glutamine (200 mM stock solution); penicillin/streptomycin (10,000 U/10,000 μg stock solution).
4. Additional: trypsin–EDTA solution (trypsin 200 mg/l and versene-EDTA 500 mg/l); sterile phosphate-buffered saline (PBS); filter units, MF75TM series (Nalge Nunc International, NY); recombinant EGF (PeproTech EC, London, UK).
5. Cell lines:

- Adherent HeLa cells, epithelial adenocarcinoma (American Type Culture Collection (ATCC), Manassas, VA).
- Suspension HeLa S3 cells, epithelial adenocarcinoma (American Type Culture Collection (ATCC), Manassas, VA).

2.2. Cell Lysis, Affinity Purification, and Gel Electrophoresis

1. Lysis buffer:
   - RIPA modified: 50 mM Tris, pH 7.5, 150 mM sodium chloride, 1% NP-40.
   - Protease inhibitors (such as CompleteTM tablets, Roche Diagnostics, Mannheim, Germany).
   - Tyrosine phosphatases inhibitor sodium ortho-vanadate (1 mM), serine/threonine phosphatases inhibitors sodium fluoride (5 mM; Merck), and beta-glycerophosphate (5 mM).
2. Cell scrapers.
3. Immobilized rabbit anti-EGFR antibody sc-03 (Santa Cruz Biotechnology, Heidelberg, Germany) for immunoprecipitation; protein A agarose (Sigma Aldrich, Copenhagen, Denmark) and Poly-Prep columns (Bio-Rad Laboratories, Hercules, CA.) for pre-clearing lysates.
4. Dithiothreitol (DTT) and iodoacetamide for reduction and alkylation of proteins, respectively.
5. NuPAGE® Novex 4–12% Bis–Tris gel system with MOPS running buffer for polyacrylamide gel electrophoresis.
6. Colloidal Blue Stain for visualizing proteins on gel.
7. Sequence grade-modified trypsin (Promega, Madison, WI) for proteolytic digestion of proteins from gel slices.

2.3. LC-MS Analysis

Buffers: buffer A containing 0.5% acetic acid in water and buffer B containing 0.5% acetic acid in 80% acetonitrile, 20% water for STAGE tip purification (16) and LC-MS.

3. Methods

Over the last few years different SILAC-based approaches to study protein–protein interactions have been reported (*see* **Fig. 16.1** and **Table 16.1**). SILAC-based quantitative MS is also ideally suited to determine dynamic, stimulus-specific protein interactions (*see* **Note 1**). Here we focus on the analysis of EGF-dependent protein interactions of EGF receptor (EGFR).

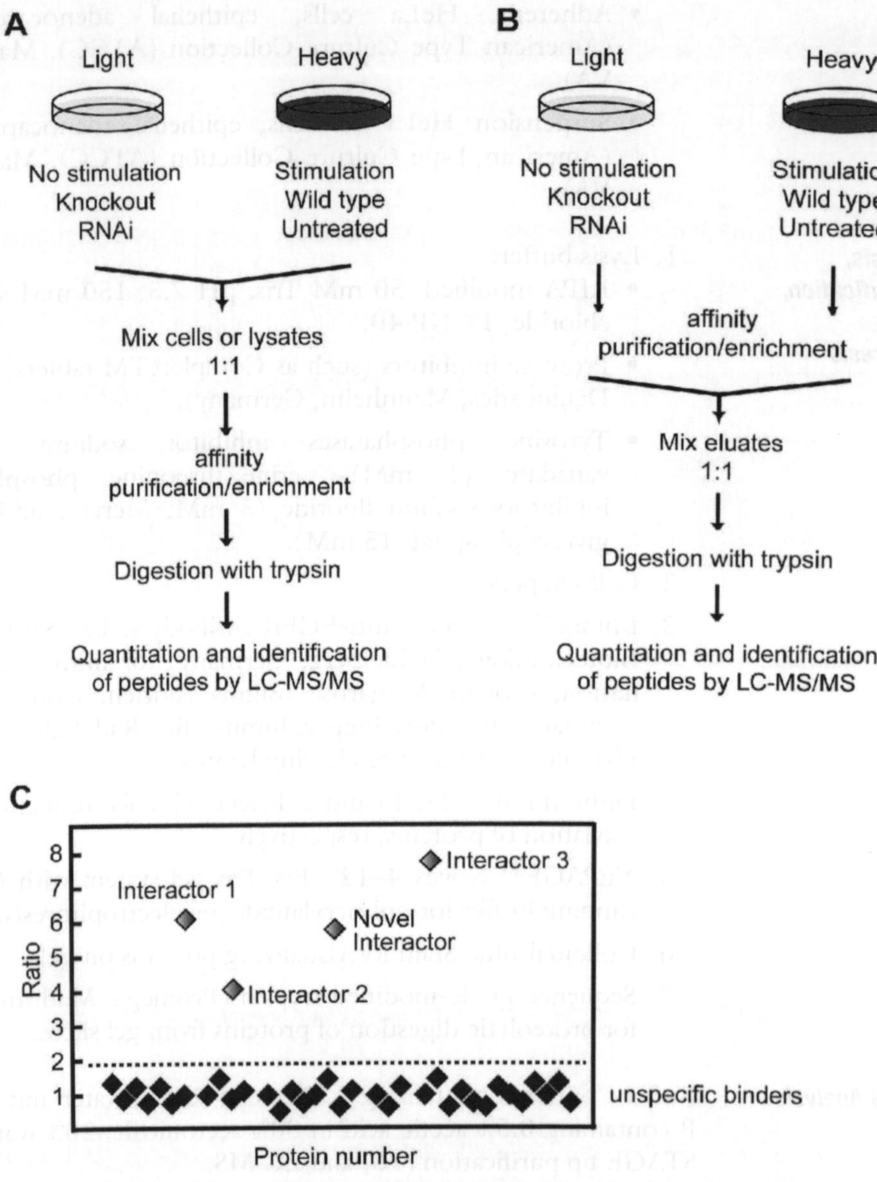

Fig. 16.1 Generic approach for the analysis of protein interactions using quantitative MS. (**A**) Cells are SILAC labeled using Arg0/Lys0 (light) and Arg10/Lys8 (heavy), respectively. Light and heavy labeled cells are differentially manipulated, as required by the experiments set up, mixed at equal proportions, and the protein complexes of interest are purified by a suitable affinity purification procedure. Purified complexes can by digested in-solution and directly analyzed by LC-MS/MS. Alternatively, the eluted protein complexes can first be separated by SDS-PAGE. The whole lane is then cut in slices, proteins are digested using trypsin, and resulting peptide mixtures are analyzed by LC-MS/MS. (**B**) The experimental design follows the one outlined in *panel* (**A**). However, affinity purifications are carried out in parallel using, e.g., either unmodified or PTM-modified version of a specific peptide as bait. This approach is also useful for analysis of weak interactions with fast exchange rates. Performing the purification step in separate tubes allows to preserve the ratios of such dynamic binding partners. However, the standard deviations of the ratios of unspecific binders are generally higher in these experiments. (**C**) Exemplified results. Each of the identified proteins is represented as a diamond and plotted based on the observed SILAC ratio. Specific interaction partners show ratios significantly different from 1 (*gray diamonds*) and can be easily distinguished from the large pool of unspecific binders (*black diamonds*) as those maintain ratios similar to the original mixing ratio.

Table 16.1
Examples of different SILAC-based approaches to study protein–protein interactions[a]

Type of interaction	Type of affinity enrichment	References
Protein–protein	Using recombinant domain as a bait	(8, 29)
Protein–protein	Immunoprecipitation of overexpressed tagged protein	(25, 26, 29–32)
Protein–protein	Immunoprecipitation of endogenous complexes	(9–11, 17, 28, 33, 34)
Sequence-directed Protein–peptide	Using synthetic peptides as a bait: mutated versus control	(35)
PTM-directed Protein–peptide	Using modified synthetic peptides as a bait	(11, 34, 36–39)
Protein–DNA	Synthetic DNA oligonucleotides (mutated versus control)	(40)

[a]For a recent review on the topic *see* (41).

The EGFR signaling network has been extensively studied by our group resulting in identification of several novel components of the signaling cascade and delineation of the site-specific phosphorylation kinetics of many novel phosphorylation sites (8, 9, 17, 18) (*see* **Note 2**). However, the experimental approach provided below is generic and the described protocol can be used as a general outline for the quantitative analysis of stimulus-dependent protein interactions.

3.1. Cell Culture

Cells should be grown for at least five cell doublings in SILAC medium to ensure full incorporation of the labeled amino acids (19) (*see* **Note 3**). We recommend to use four 15 cm dishes per condition (approximately 4–5 $\times$ 10^7 cells).

3.2. Treatment of Cells and Cell Lysis

1. When cells are approximately 70% confluent, replace medium with the respective serum-free SILAC medium. For adherent HeLa cells, 12–14 h of serum deprivation is an optimal time frame.
2. Stimulate Lys4/Arg6- and Lys8/Arg10-labeled cells for a different length of time with 150 ng/ml EGF, leaving

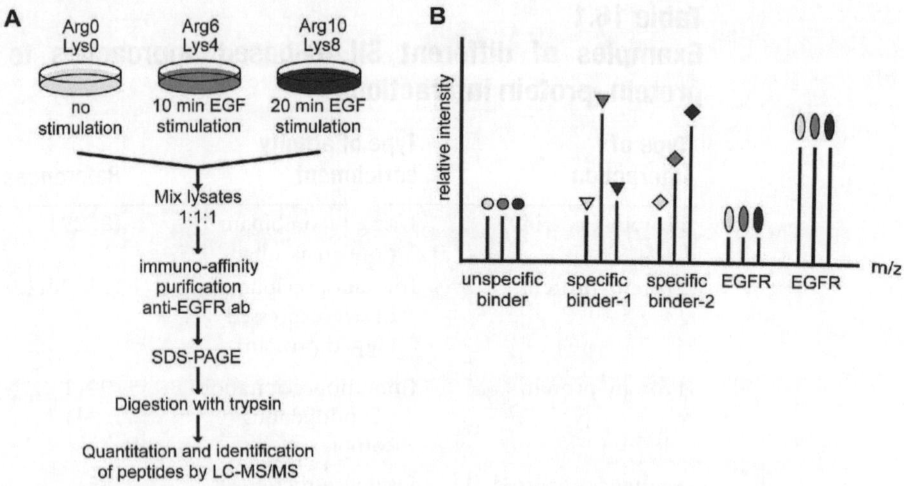

Fig. 16.2: Example for analysis of stimulus-dependent dynamics of protein interactions using quantitative MS. (**A**) Experimental design. Cells are SILAC labeled using Arg0/Lys0, Arg6/Lys4, and Arg10/Lys8. Arg6/Lys4- and Arg10/Lys8-labeled cells are stimulated with EGF for different time periods whereas Arg0/Lys0 cells are left untreated. Afterward cells are mixed and lysed. EGFR and interacting proteins are purified by immuno-affinity enrichment using an anti-EGFR antibody. Precipitates are eluted in SDS loading buffer and separated by SDS-PAGE. The whole lane is cut in slices, proteins are digested using trypsin, and resulting peptide mixtures are analyzed by LC-MS/MS. (**B**) Exemplified results. Peptides derived from EGFR and stimulus-unspecific binders display ratios close to the original mixing 1:1:1 ratio. Stimulus-specific binders show significantly different peptide ratios reflecting the degree of increased association with EGFR at the corresponding time point. Thereby using triple encoding SILAC stimulus-dependent kinetics can be determined.

Lys0/Arg0 cells untreated to serve as control (*see* **Fig. 16.2**) (*see* **Note 4**).

3. Remove medium from the culture dishes and scrape the cells in ice-cold lysis buffer containing both protease and phosphatase inhibitors. Incubate lysates on ice for 10 min. Afterward remove cell debris by centrifugation. Save a small aliquot (corresponding to approximately 30–50 μg of protein) from each lysate and freeze these down (*see* **Note 5**).

3.3. Immunoprecipitation and Gel Electrophoresis

1. Combine lysates and add 200 μl protein A agarose to pre-clear lysates of "sticky" proteins. Rotate for minimally 30 min at 4°C (*see* **Note 6**).
2. Remove protein A agarose by passing lysate through a disposable Poly-Prep column and save a small aliquot corresponding to approximately 30–50 μg of protein (*see* **Note 7**).
3. Add immobilized anti-EGFR antibody into the mixed lysates and rotate for 4–6 h at 4°C (*see* **Note 8**).
4. Wash the beads three times with five volumes of ice-cold lysis buffer. Resuspend the beads in SDS loading buffer and elute precipitated complexes by heating at 75°C for 10 min.

5. Use 10 mM DTT and 55 mM iodoacetamide for the sequential reduction and alkylation of the proteins prior to gel electrophoresis.
6. Load the sample on a NuPAGE® gel. After staining with Colloidal Blue to visualize the proteins, cut each gel lane into slices of equal size and digest proteins with trypsin. We recommend cutting the high-intensity bands separately.
7. Prepare samples for MS analysis (16, 20).

3.4. LC-MS Analysis

1. Comparing the extracted ion currents (XIC) of EGFR peptides allows checking for the mixing ratio. All peptides derived from EGFR should have ratios close to the original 1:1:1 mixing ratio (*see* **Note 9**).
2. Peptides from the stimulus-specific EGFR interaction partners show ratios significantly different from 1, thereby easily distinguishable from unspecific binders (*see* **Fig. 16.2**) (*see* **Note 10**).
3. Triple SILAC allows comparing directly the time-dependent dynamic changes in protein–protein interaction. The observed SILAC ratios reflect the degree of increased association of the corresponding protein with EGFR at the corresponding time point (*see* **Fig. 16.2**).

4. Notes

1. SILAC is an established generic mass spectrometry-based approach for identification and quantitation of genuine interaction partners. It relies on the metabolic incorporation of amino acids that contain distinct stable isotopes into the proteomes of a given cell line. This allows to easily distinguish the peptides originating from the differentially encoded cell populations in the mass spectrometer. After full incorporation the cells are mixed either before or after lysis followed by an optional protein purification/enrichment step. When investigating protein–protein interactions it has been a challenging task to distinguish background binding from functionally relevant interactions. Application of SILAC allows to recognize the specific interaction partners since they will show elevated differential ratios in the mass spectra (*see* **Fig. 16.1**). For example, when using a GST-SH2 domain of Grb2 as a bait to identify both direct and indirect interaction partners following EGF stimulation in total 228 proteins were identified. Only 28 were displaying differential ratios that identify them as selective and specific interaction partners

of EGFR (8). Following this study principle a number of applications using SILAC have been utilized (*see* **Note 11**).

2. Historically, proteomics approaches without the use of stable isotopes have been successfully applied to characterize protein–protein interactions involved in the signaling cascades initiated by growth factors. Typically, protein samples that originate from different stages of cells, e.g. treated or not with a specific growth factor are separated individually on a gel and then the bands that show differential expression are excised, digested with a protease, and identified by MS. For example, in order to identify components of the EGFR signaling pathway, affinity purification using anti-phosphotyrosine antibodies to enrich for tyrosine-phosphorylated proteins followed by one-dimensional gel electrophoresis and analysis by static nanoelectrospray mass spectrometry has been used. This approach has resulted in identification of nine signaling molecules, two of which had previously not been implicated in EGFR signaling (21). Additional experiments have demonstrated that one of these proteins, namely Vav-2, associates directly with both EGFR- and platelet-derived growth factor (PDGF) receptors. Other novel proteins such as STAM2 and Odin have also been found to interact in a similar fashion with growth factor receptors and to be involved in the downstream signaling (21–24).

3. Dialyzed serum has to be used to ensure that only labeled amino acids are present. The content of arginine should be titrated to minimize the conversion to proline. Alternatively, unlabeled proline can be added to the medium. But again its concentration should be titrated to avoid arginine conversion.

4. In order to have peptide ratios easily quantifiable by MS and to keep the quantitation error to a minimum the same amount of cells should be used for each condition, idealy.

5. Samples from the unmixed lysates can be used to check the level of incorporation of the "heavy" amino acids, in case of any troubleshooting.

6. Depending on if the lysates are mixed before the immunoprecipitation or if the immunoprecipitations are carried out separately and the respective eluates are mixed afterward, it is possible to distinguish strong/stable from weak/dynamic interactors (25, 26). If eluates are mixed the mixing ratio might be obscured. According to our experience mixing of the samples at a later stages results in higher variability of the observed ratios. Thus larger thresholds should be used when designating proteins as specific interacting molecules.

7. The use of a column ensures that all protein A agarose particles (together with the "sticky" proteins bound to them) are removed. A sample from the mixed SILAC lysates can be used as a quality control for checking by mass spectrometry the exact mixing ratio.
8. For approximately 1.2×10^8 cells (4×10^7 cells per condition) we recommend to use 75 μg anti-EGFR antibody for immunoprecipitation.
9. If the mixing ratio differs a normalization coefficient has to be calculated. This coefficient should be used for normalizing all other ratios. These results can be double checked by analyzing the saved lysate (*see* **Note 4**).
10. Quantitation errors are usually in the range of 10–20%. Specific binding partner should have protein ratios greater than three times standard deviation.
11. The SILAC approach can be used in virtually every type of interaction study for unambiguous identification of interaction partners. This includes standard protein–bait interaction assays such as GST pull down, affinity purification of overexpressed tagged proteins or of endogenous complexes, and protein–peptide interaction assays (**Table 16.1**). In addition SILAC is applicable in cases when the interaction assay is performed using knockout/RNAi knockdown/mutant versus the wild-type cells or tissues (27, 28).

Acknowledgments

We thank all CEBI group members for helpful discussions and support. The research leading to these results has received funding from the European Commission's 7th Framework Programme (grant agreement HEALTH-F4-2008-201648/PROSPECTS), the Danish Natural Science Research Council, the Danish Medical Research Council, and the Lundbeck Foundation. JD was supported by the European Molecular Biology Organization and by the Excellence Initiative of the German Federal and State Governments.

References

1. Gingras, A. C., Gstaiger, M., Raught, B., and Aebersold, R. (2007) Analysis of protein complexes using mass spectrometry. *Nat. Rev. Mol. Cell. Biol.* **8**, 645–654.

2. Vermeulen, M., Hubner, N. C., and Mann, M. (2008) High confidence determination of specific protein–protein interactions using quantitative mass spectrometry. *Curr. Opin. Biotechnol.* **19**, 331–337.
3. Ho, Y., Gruhler, A., Heilbut, A., Bader, G. D., Moore, L., Adams, S. L., Millar, A., Taylor, P., Bennett, K., Boutilier, K., Yang, L., Wolting, C., Donaldson, I., Schandorff, S., Shewnarane, J., Vo, M., Taggart, J., Goudreault, M., Muskat, B., Alfarano, C., Dewar, D., Lin, Z., Michalickova, K., Willems, A. R., Sassi, H., Nielsen, P. A., Rasmussen, K. J., Andersen, J. R., Johansen, L. E., Hansen, L. H., Jespersen, H., Podtelejnikov, A., Nielsen, E., Crawford, J., Poulsen, V., Sorensen, B. D., Matthiesen, J., Hendrickson, R. C., Gleeson, F., Pawson, T., Moran, M. F., Durocher, D., Mann, M., Hogue, C. W., Figeys, D., Tyers, M. (2002) Systematic identification of protein complexes in Saccharomyces cerevisiae by mass spectrometry. *Nature* **415**, 180–183.
4. Gavin, A. C., Aloy, P., Grandi, P., Krause, R., Boesche, M., Marzioch, M., Rau, C., Jensen, L. J., Bastuck, S., Dumpelfeld, B., Edelmann, A., Heurtier, M. A., Hoffman, V., Hoefert, C., Klein, K., Hudak, M., Michon, A. M., Schelder, M., Schirle, M., Remor, M., Rudi, T., Hooper, S., Bauer, A., Bouwmeester, T., Casari, G., Drewes, G., Neubauer, G., Rick, J. M., Kuster, B., Bork, P., Russell, R. B., Superti-Furga, G. (2006) Proteome survey reveals modularity of the yeast cell machinery. *Nature* **440**, 631–636.
5. Krogan, N. J., Cagney, G., Yu, H., Zhong, G., Guo, X., Ignatchenko, A., Li, J., Pu, S., Datta, N., Tikuisis, A. P., Punna, T., Peregrin-Alvarez, J. M., Shales, M., Zhang, X., Davey, M., Robinson, M. D., Paccanaro, A., Bray, J. E., Sheung, A., Beattie, B., Richards, D. P., Canadien, V., Lalev, A., Mena, F., Wong, P., Starostine, A., Canete, M. M., Vlasblom, J., Wu, S., Orsi, C., Collins, S. R., Chandran, S., Haw, R., Rilstone, J. J., Gandi, K., Thompson, N. J., Musso, G., St, O.nge, P., Ghanny, S., Lam, M. H., Butland, G., Altaf-Ul, A. M., Kanaya, S., Shilatifard, A., O'Shea E, Weissman, J. S., Ingles, C. J., Hughes, T. R., Parkinson, J., Gerstein, M., Wodak, S. J., Emili, A., Greenblatt, J. F. (2006) Global landscape of protein complexes in the yeast Saccharomyces cerevisiae. *Nature* **440**, 637–643.
6. Ong, S. E., and Mann, M. (2005) Mass spectrometry-based proteomics turns quantitative, *Nat. Chem. Biol.* **1**, 252–262.
7. Ong, S. E., Blagoev, B., Kratchmarova, I., Kristensen, D. B., Steen, H., Pandey, A., and Mann, M. (2002) Stable isotope labeling by amino acids in cell culture, SILAC, as a simple and accurate approach to expression proteomics. *Mol. Cell Proteomics* **1**, 376–386.
8. Blagoev, B., Kratchmarova, I., Ong, S. E., Nielsen, M., Foster, L. J., and Mann, M. (2003) A proteomics strategy to elucidate functional protein-protein interactions applied to EGF signaling. *Nat. Biotechnol.* **21**, 315–318.
9. Dengjel, J., Akimov, V., Olsen, J. V., Bunkenborg, J., Mann, M., Blagoev, B., and Andersen, J. S. (2007) Quantitative proteomic assessment of very early cellular signaling events. *Nat. Biotechnol.* **25**, 566–568.
10. Kratchmarova, I., Blagoev, B., Haack-Sorensen, M., Kassem, M., and Mann, M. (2005) Mechanism of divergent growth factor effects in mesenchymal stem cell differentiation. *Science* **308**, 1472–1477.
11. Kruger, M., Kratchmarova, I., Blagoev, B., Tseng, Y. H., Kahn, C. R., and Mann, M. (2008) Dissection of the insulin signaling pathway via quantitative phosphoproteomics. *Proc. Natl. Acad. Sci. USA* **105**, 2451–2456.
12. Gygi, S. P., Rist, B., Gerber, S. A., Turecek, F., Gelb, M. H., and Aebersold, R. (1999) Quantitative analysis of complex protein mixtures using isotope-coded affinity tags. *Nat. Biotechnol.* **17**, 994–999.
13. Ross, P. L., Huang, Y. N., Marchese, J. N., Williamson, B., Parker, K., Hattan, S., Khainovski, N., Pillai, S., Dey, S., Daniels, S., Purkayastha, S., Juhasz, P., Martin, S., Bartlet-Jones, M., He, F., Jacobson, A., Pappin, D. J. (2004) Multiplexed protein quantitation in Saccharomyces cerevisiae using amine-reactive isobaric tagging reagents. *Mol. Cell Proteomics* **3**, 1154–1169.
14. Ranish, J. A., Hahn, S., Lu, Y., Yi, E. C., Li, X. J., Eng, J., and Aebersold, R. (2004) Identification of TFB5, a new component of general transcription and DNA repair factor IIH. *Nat. Genet.* **36**, 707–713.
15. Bai, Y., Markham, K., Chen, F., Weerasekera, R., Watts, J., Horne, P., Wakutani, Y., Bagshaw, R., Mathews, P. M., Fraser, P. E., Westaway, D., St George-Hyslop, P., Schmitt-Ulms, G. (2008) The in vivo brain interactome of the amyloid precursor protein. *Mol. Cell Proteomics* **7**, 15–34.
16. Rappsilber, J., Mann, M., and Ishihama, Y. (2007) Protocol for micro-purification, enrichment, pre-fractionation and storage of peptides for proteomics using StageTips. *Nat. Protoc.* **2**, 1896–1906.
17. Blagoev, B., Ong, S. E., Kratchmarova, I., and Mann, M. (2004) Temporal analysis of phosphotyrosine-dependent signaling

networks by quantitative proteomics. *Nat. Biotechnol.* **22**, 1139–1145.

18. Olsen, J. V., Blagoev, B., Gnad, F., Macek, B., Kumar, C., Mortensen, P., and Mann, M. (2006) Global, in vivo, and site-specific phosphorylation dynamics in signaling networks. *Cell* **127**, 635–648.
19. Blagoev, B., and Mann, M. (2006) Quantitative proteomics to study mitogen-activated protein kinases. *Methods* **40**, 243–250.
20. Shevchenko, A., Tomas, H., Havlis, J., Olsen, J. V., and Mann, M. (2006) In-gel digestion for mass spectrometric characterization of proteins and proteomes. *Nat. Protoc.* **1**, 2856–2860.
21. Pandey, A., Podtelejnikov, A. V., Blagoev, B., Bustelo, X. R., Mann, M., and Lodish, H. F. (2000) Analysis of receptor signaling pathways by mass spectrometry: identification of vav-2 as a substrate of the epidermal and platelet-derived growth factor receptors. *Proc. Natl. Acad. Sci. USA* **97**, 179–184.
22. Pandey, A., Blagoev, B., Kratchmarova, I., Fernandez, M., Nielsen, M., Kristiansen, T. Z., Ohara, O., Podtelejnikov, A. V., Roche, S., Lodish, H. F., Mann, M. (2002) Cloning of a novel phosphotyrosine binding domain containing molecule, Odin, involved in signaling by receptor tyrosine kinases. *Oncogene* **21**, 8029–8036.
23. Pandey, A., Fernandez, M. M., Steen, H., Blagoev, B., Nielsen, M. M., Roche, S., Mann, M., Lodish, H. F. (2000) Identification of a novel immunoreceptor tyrosine-based activation motif-containing molecule, STAM2, by mass spectrometry and its involvement in growth factor and cytokine receptor signaling pathways. *J. Biol. Chem.* **275**, 38633–38639.
24. Kristiansen, T. Z., Nielsen, M. M., Blagoev, B., Pandey, A., and Mann, M. (2004) Mouse embryonic fibroblasts derived from Odin deficient mice display a hyperproliiferative phenotype. *DNA. Res.* **11**, 285–292.
25. Mousson, F., Kolkman, A., Pijnappel, W. W., Timmers, H. T., and Heck, A. J. (2008) Quantitative proteomics reveals regulation of dynamic components within TATA-binding protein (TBP) transcription complexes. *Mol. Cell Proteomics* 7, 845–852.
26. Wang, X., and Huang, L. (2008) Identifying dynamic interactors of protein complexes by quantitative mass spectrometry. *Mol. Cell Proteomics* 7, 46–57.
27. Mertins, P., Eberl, H. C., Renkawitz, J., Olsen, J. V., Tremblay, M. L., Mann, M., Ullrich, A., and Daub, H. (2008) Investigation of protein-tyrosine phosphatase 1B function by quantitative proteomics. *Mol. Cell Proteomics* 7, 1763–1777.
28. Selbach, M., and Mann, M. (2006) Protein interaction screening by quantitative immunoprecipitation combined with knockdown (QUICK). *Nat. Methods* **3**, 981–983.
29. Foster, L. J., Rudich, A., Talior, I., Patel, N., Huang, X., Furtado, L. M., Bilan, P. J., Mann, M., Klip, A. (2006) Insulin-dependent interactions of proteins with GLUT4 revealed through stable isotope labeling by amino acids in cell culture (SILAC). *J. Proteome Res.* **5**, 64–75.
30. Trinkle-Mulcahy, L., Andersen, J., Lam, Y. W., Moorhead, G., Mann, M., and Lamond, A. I. (2006) Repo-Man recruits PP1 gamma to chromatin and is essential for cell viability. *J. Cell Biol.* **172**, 679–692.
31. Dobreva, I., Fielding, A., Foster, L. J., and Dedhar, S. (2008) Mapping the integrin-linked kinase interactome using SILAC. *J. Proteome Res.* 7, 1740–1749.
32. Guerrero, C., Tagwerker, C., Kaiser, P., and Huang, L. (2006) An integrated mass spectrometry-based proteomic approach: quantitative analysis of tandem affinity-purified in vivo cross-linked protein complexes (QTAX) to decipher the 26 S proteasome-interacting network. *Mol. Cell Proteomics* **5**, 366–378.
33. Jin, J., Li, G. J., Davis, J., Zhu, D., Wang, Y., Pan, C., and Zhang, J. (2007) Identification of novel proteins associated with both alpha-synuclein and DJ-1. *Mol. Cell Proteomics* **6**, 845–859.
34. Hinsby, A. M., Olsen, J. V., and Mann, M. (2004) Tyrosine phosphoproteomics of fibroblast growth factor signaling: a role for insulin receptor substrate-4. *J. Biol. Chem.* **279**, 46438–46447.
35. Schulze, W. X., and Mann, M. (2004) A novel proteomic screen for peptide-protein interactions. *J. Biol. Chem.* **279**, 10756–10764.
36. Hanke, S., and Mann, M. (2008) The phosphotyrosine interactome of the insulin receptor family and its substrates IRS-1 and IRS-2. *Mol. Cell Proteomics* **8**, 519–534.
37. Schulze, W. X., Deng, L., and Mann, M. (2005) Phosphotyrosine interactome of the ErbB-receptor kinase family. *Mol. Syst. Biol.* **1**, 2005 0008.
38. Vermeulen, M., Mulder, K. W., Denissov, S., Pijnappel, W. W., van Schaik, F. M., Varier, R. A., Baltissen, M. P., Stunnenberg, H. G., Mann, M., Timmers, H. T. (2007) Selective anchoring of TFIID to nucleosomes by

trimethylation of histone H3 lysine 4. *Cell* 131, 58–69.

39. Christofk, H. R., Vander Heiden, M. G., Wu, N., Asara, J. M., and Cantley, L. C. (2008) Pyruvate kinase M2 is a phosphotyrosine-binding protein. *Nature* 452, 181–186.
40. Mittler, G., Butter, F., and Mann, M. (2009) A SILAC-based DNA protein interaction screen that identifies candidate binding proteins to functional DNA elements. *Genome Res.* 19, 284–293.
41. Dengjel, J., Kratchmarova, I., and Blagoev, B. (2009) Receptor tyrosine kinase signaling: a view from quantitative proteomics. *Mol. Biosyst.* 5, 1112–1121.

Part V

Analysis of Biological Fluids and Clinical Samples

Chapter 17

Analysis of Serum Proteins by LC-MS/MS

Sarah Tonack, John P. Neoptolemos, and Eithne Costello

Abstract

Serum contains a vast array of proteins, some of which are specific to blood whilst others are secreted into blood from tissues and organs. The so-called tissue leakage factors reveal information about the tissue from which they originate and are therefore of great potential importance as disease biomarkers. There are already a number of blood-borne biomarkers in routine clinical use that aid in the diagnosis or management of cancer. However, there is a pressing need for additional markers, and new methods to find them are under development. Here we provide a protocol for serum protein profiling using liquid chromatography tandem mass spectrometry (LC-MS/MS). Included in this procedure, we detail the pre-processing steps of lipid and high-abundance protein removal. These procedures can also be employed up-stream of quantification methods such as isobaric tags for relative and absolute quantification (iTRAQ). **Chapter 12** is devoted to the iTRAQ approach for quantifying proteins, and it is therefore not described in this chapter.

Key words: Serum, serum depletion, strong cation exchange, LC-MS/MS, proteomics.

1. Introduction

Blood is a very attractive source of protein or peptide biomarkers (1). It is minimally invasive and relatively inexpensive to obtain. In addition, tests that can be undertaken routinely by clinical laboratories have been established. This provides the promise that new candidate biomarkers might also be translated to clinical use in the form of assays that are readily amenable to routine widespread use. Serum is the clear liquid obtained after blood has clotted and the clot has been removed. In designing studies aimed at serum biomarker discovery, a number of factors need to be carefully considered. These include decisions as to whether samples from individuals belonging to a distinct disease group

P.R. Cutillas, J.F. Timms (eds.), *LC-MS/MS in Proteomics*, Methods in Molecular Biology 658,
DOI 10.1007/978-1-60761-780-8_17, © Springer Science+Business Media, LLC 2010

will be analysed individually or as a pool of several serum samples from that group. The age and gender of individuals should be matched across groups where possible and appropriate disease control groups should be identified and included in the analysis. Finally, the application of standard protocols for the collection and storage of serum samples is necessary for the production of reliable data (2).

In addition to proteins, serum contains lipids which interfere with mass spectrometry-based analysis. Therefore a method that efficiently and selectively removes lipids (3) is required prior to analysis. The serum proteome itself is complex and contains proteins across a large range of concentrations. A small number of highly abundant proteins account for a large proportion of proteins in serum (4). This makes the analysis of less abundant proteins difficult. A variety of methods are available for the removal of high-abundance proteins, including 'Proteoprep 20 Immunodepletion Kit' by Sigma-Aldrich, Gillingham, UK; 'Multiple Affinity Removal System' by Agilent Technologies UK Ltd., Wokingham, UK; and 'Proteomelab IgY' by Beckman Coulter UK Ltd., High Wycombe, UK). Alternatively, the concentration of low-abundance proteins and reduction of the dynamic range of protein concentrations in serum samples can be attempted using 'ProteoMiner Protein Enrichment', Bio-Rad Laboratories, Hemel Hempstedt, UK (5). In the protocol provided below, we will focus on high-abundance protein depletion using the Sigma-Aldrich Proteoprep 20 column. An overview of the entire procedure is provided in **Fig. 17.1**.

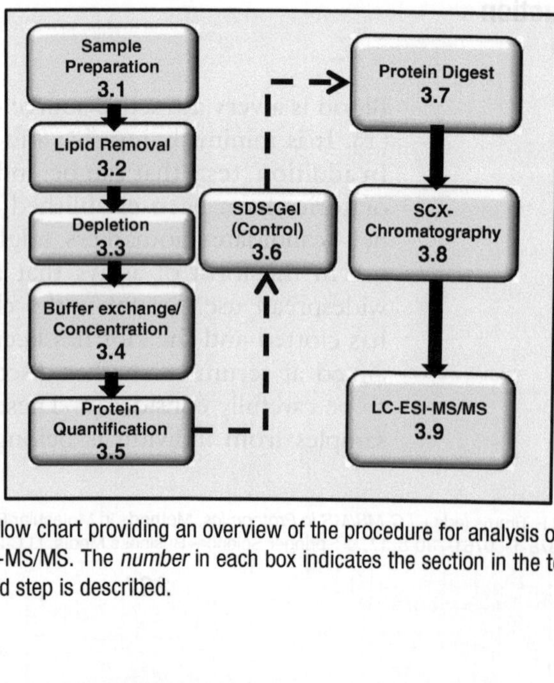

Fig. 17.1 Flow chart providing an overview of the procedure for analysis of serum proteins by LC-MS/MS. The *number* in each box indicates the section in the text, in which the specified step is described.

2. Materials

2.1. For Serum Preparation

1. Serum Z/7.5 mL Sarstedt Monovette
2. Cryogenic 2 mL tubes (Nunc GmbH & Co KG., part of Thermo Fisher Scientific, Langenselbold, Germany)

2.2. For Lipid Removal from Serum Samples

1. Diisopropyl ether (DIPE), butan-2-ol, and ethylenediaminetetraacetic acid (EDTA). Beware: DIPE is a sedative that is highly flammable. Precautions should be taken, including work carried out under a fume hood with appropriate ventilation.
2. Solvent Mix: 60:40 mixture of DIPE and butan-2-ol prepared freshly prior to commencing the delipidation process.

2.3. High-Abundance Protein Depletion

1. Proteoprep 20 Plasma Immunodepletion Kit (Sigma-Aldrich).
2. 1 M HEPES solution, pH 7.5: 119.15 g HEPES (free acid) dissolved in 400 mL MilliQ water. Adjust pH with NaOH pellets until pH reaches 6.8. Make the final adjustment with concentrated NaOH. Finally adjust volume with MilliQ water to 500 mL. Filter sterilize and store at 4°C. Attention: NaOH is highly alkaline. Wear safety glasses.
3. Microsep 1 K Centrifugal Devices (Pall Life Science, Portsmouth, UK).

2.4. Buffer Exchange

1. 0.5 M sodium carbonate solution, pH 8.5: Dissolve 2.64 g sodium carbonate in 25 mL MilliQ water. Adjust pH with concentrated HCl, filter sterilize, and store at 4°C. The solution is stable for up to 1 week. Attention: HCl is a strong fuming acid. Use only under a fume hood with ventilation.
2. Microsep 1 K Centrifugal Devices (Pall Life Science).

2.5. Protein Quantification

1. Bio-Rad Protein Assay Solution (Bio-Rad Laboratories)

2.6. Quality Control – SDS-Polyacrylamide Gel Electrophoresis (SDS-PAGE)

1. 1.5 M Tris–HCl (pH 8.7), 1 M Tris–HCl (pH 6.8), 30% acrylamide/bisacrylamide solution (37.5:1) (beware: non-polymerized acrylamide is neurotoxic), *N*,*N*,*N*,*N*′-tetramethylethylenediamine (TEMED), 10% ammonium persulphate solution (freshly prepared), and 10% sodium dodecyl sulphate (SDS) solution (all chemicals from Sigma-Aldrich).
2. 10X SDS gel running buffer (Laemmli): 30 g Tris-base, 144 g glycine, 10 g SDS (all from Sigma-Aldrich), dissolved in MilliQ water and adjusted to a volume of 1 L.

3. 5X sample buffer (Laemmli): 10% SDS, 50% glycerol, 300 mM Tris–HCl (pH 6.8), 0.05% bromphenol blue. Add dithiothreitol (DTT) to a final concentration of 100 mM prior to use.
4. PageRuler Prestained Protein Ladder (Fermentas UK, York, UK).
5. Silver-Stain Plus Kit (Bio-Rad Laboratories), methanol, acetic acid.

2.7. Trypsin Digest

1. Sequencing grade-modified trypsin (Promega UK Ltd., Southampton, UK).
2. Sodium dodecyl sulphate (SDS), 0.5 M Tris(2-carboxyethyl)phosphine hydrochloride solution (TCEP), and iodoacetamide.

2.8. Strong Cation Exchange Chromatography

1. PolySULFOETHYL A (4.6 × 200 mm inner diameter (i.d.) HPLC column (PolyLC)).
2. Buffer A: 10 mM potassium phosphate (KH_2PO_4), 25% acetonitrile, pH < 3, adjusted with concentrated H_3PO_4.
3. Buffer B: 10 mM potassium phosphate (KH_2PO_4), 1 M potassium chloride, 25% acetonitrile, pH < 3, adjusted with concentrated H_3PO_4.
4. Diluent: 25% acetonitrile, pH < 3, adjusted with 500 mM H_3PO_4.

2.9. LC-MS/MS

1. Buffer A: 5% acetonitrile, 0.05% trifluoroacetic acid.
2. Buffer B: 95% acetonitrile, 0.05% trifluoroacetic acid.

3. Methods

3.1. Serum Preparation

1. Place a freshly acquired blood sample (contained in a Sarstedt Monovette) at 4°C for 15 min, until the blood clots.
2. Spin the Sarstedt Monovette at 800 RCF for 10 min at 4°C.
3. In a sterile environment, pipette the clear serum (upper layer) in aliquots into cryotubes.
4. Store the serum at −80°C.

3.2. Lipid Removal from Serum Samples

The removal of lipids is the first step in the preparation of serum samples for LC-MS/MS analysis (*see* **Note 1**). During the following procedure, serum is maintained at 4°C.

1. Place 500 μL of serum on ice and add 50 μg EDTA.
2. In a 1.5 mL microfuge tube, mix 500 μL of the serum containing EDTA with 1 mL of freshly prepared solvent mix (60:40 mixture of DIPE and butan-2-ol) yielding a one-part serum to two-part solvent mix.
3. Rotate the mixture on a blood cell suspension rotor (end-over-end rotation) at 30 rpm for 30 min at 4°C.
4. Centrifuge the mix at 400 RCF for 2 min at 4°C.
5. Carefully remove the aqueous serum phase (at the bottom of the tube) with a syringe, without disturbing the upper lipid layer, and transfer to a new 1.5 mL microfuge tube.
6. Carefully mix the aqueous phase with two parts DIPE and immediately centrifuge again.
7. Carefully remove and discard the upper solvent layer.
8. Remove the residual solvents from the serum phase by vacuum concentration (1 min, 37°C).
9. Aliquot the serum into 100 μL aliquots in 1.5 mL tubes and store at −80°C until proceeding with the depletion of high-abundance proteins.

3.3. High-Abundance Protein Depletion

The Proteoprep 20 Immunodepletion Kit is designed to remove 20 of the most abundant proteins from blood. This effectively results in the removal of 97–98% of protein from serum samples. The removal of these 20 proteins (albumin, IgG, IgA, IgM, IgD, transferrin, fibrinogen, α_2-macroglobulin, α_1-antitrypsin, haptoglobin, α_1-acid glycoprotein, ceruloplasmin, apolipoprotein A-I, apolipoprotein A-II, apolipoprotein B, complement C1q, complement C3, complement C4, plasminogen, prealbumin) is achieved by the use of immobilized antibodies on agarose beads. One single column allows 100 depletion runs, and each run has a capacity of 10 μL of serum (~500 μg). Depletion is performed according to the manufacturer's protocol. We recommend that depletions should be undertaken in a sterile environment, such as a cell culture hood.

1. Dilute the Proteoprep20 Equilibration buffer and Elution buffer with MilliQ Water (*see* **Note 2**).
2. Defrost one aliquot (100 μL) of delipidated serum on ice and dilute 1:11 by adding 1 mL 1X Equilibration buffer.
3. Add 500 μL of this diluted serum onto a Corning Spin X Centrifuge Filter (0.2 μm) and centrifuge at 2000×*g* for 1 min. Repeat with the second 500 μL and then with the remaining 100 μL. If the filter unit becomes blocked, use a new one and apply the diluted serum again. Combine the flow-through fractions in a 1.5 mL microfuge tube.

4. Now prepare the depletion column by removing the bottom plug (make sure you keep it, as you need it at the end of the procedure) and loosen the upper cap. Place the column in a 2 mL tube. Centrifuge at 2000×g for 30 s.
5. Remove the screw cap and attach the Luer Lock cap onto the column. Using a 20 mL syringe filled with Equilibration buffer slowly push 8 mL of Equilibration buffer through the column.
6. Remove the Luer Lock cap, attach the red cap loosely, and place in a 2 mL tube. Centrifuge at 2000×g for 30 s.
7. Place the column in a fresh 1.5 mL microfuge tube and add 100 μL from the diluted, filtered serum. Ensure that the serum is absorbed into the agarose matrix and not attached to the plastic of the column. Place the red cap loosely on top.
8. Incubate for 15–20 min at room temperature.
9. Centrifuge the column at 2000×g for 30 s. Collect the flow-through in a 1.5 mL tube. (This is the depleted fraction.)
10. Wash the column by applying 100 μL Equilibration buffer and centrifuge immediately at 2000×g for 30 s. Collect this fraction separately (*see* **Note 3**).
11. Repeat Step 10.
12. Attach the Luer Lock and apply 2 mL of 1X Elution buffer by using a 20 mL syringe. Collect the flow-through (this is the eluted fraction, containing the high-abundance proteins). Add 0.5 mL of 1 M HEPES solution, pH 7.4, to adjust the pH every five elution steps.
13. Attach the Equilibration buffer syringe and slowly draw 8 mL of 1X Equilibration buffer through the column.
14. Disassemble the Luer Lock and attach the red cap. Centrifuge at 2000×g for 30 s.
15. Repeat from Step 7 for the desired number of depletions. It is necessary to undertake 20–30 depletions to obtain approximately 300–500 μg of depleted protein. Always store the flow-through fractions (depleted and eluted) at 4°C.
16. Concentrate the depleted and eluted fractions using Microsep Centrifugal devices (at 7500 g in a fixed angel rotor (34°–45°) for approximately 8 h at 4°C. Concentration should be ceased when material from 10 depletion cycles reaches ~100 μL in volume (*see* **Note 4**).
17. Final depletion step: 100 μL of concentrated depleted serum from ten depletions are subjected to a further

depletion step (*see* Steps 8–15). If more than 100 μL are loaded, no washing step is required, samples are loaded onto the beads, eluted by centrifugation, and the column is directly loaded again.

18. Column storage: For short-term storage, after equilibration centrifuge the column at 2000×*g* for 30 s, apply the bottom plug and add 300 μL of 1X Equilibration buffer. For long-term storage prepare the storage solution: 10 μL Proteoprep Preservative Concentrate in 5 mL of 1X Equilibration Buffer. 300 μL is added as for short-term storage. Columns are stored at 4°C.

3.4. Buffer Exchange

Both concentrated fractions (i.e. the depleted-low-abundance and the bound-high-abundance fractions) require buffer exchange prior to digestion with trypsin. Several buffers are suitable for the trypsin digestion step, such as 50 mM ammonium bicarbonate buffer. We use 0.5 M sodium carbonate solution, pH 8.5.

1. Both concentrated fractions are diluted in 2 mL 0.5 M sodium carbonate solution and concentrated using Microsep 1 K Centrifugal Devices as above.

3.5. Protein Quantification

1. Protein concentration is determined using Bio-Rad Protein Assay Reagent (Bio-Rad). For the standard curve known concentrations of albumin (1–10 μg/μL) are used.
2. The low concentration range assay is used in the test tube format. 2 μL of standard or sample is added to 798 μL of MilliQ water. 200 μL of Bio-Rad reagent is added, mixed, and incubated for 10 min at room temperature.
3. The absorbance at the wavelength of 595 nm is measured in a spectrophotometer.

3.6. Quality Control – SDS-Polyacrylamide Gel Electrophoresis (SDS-PAGE)

The protocol is based on the Mini-Protean System (Bio-Rad, Hemel Hempstead, UK).

1. Clean the glass plates to be used and assemble the gel plates in the gel casting unit.
2. Prepare the 10% separating gel by mixing 4 mL MilliQ water, 3.3 mL 30% acrylamide mix, 2.5 mL 1.5 M Tris/HCl (pH 8.8), 100 μL 10% SDS, 100 μL 10% ammonium persulphate, and 4 μL TEMED. Directly after addition of TEMED pour the gel and leave sufficient space for the stacking gel. Overlay the gel carefully with 0.1 % SDS solution. After polymerisation (~30 min) discard the 0.1 % SDS solution, rinse with MilliQ water, and air dry.
3. Prepare the stacking gel by mixing 2.7 mL MilliQ water, 670 μL 30% acrylamide mix, 500 μL 1 M Tris/HCl (pH 6.8), 40 μL 10% SDS, 40 μL 10 % ammonium persulphate,

and 4 μL TEMED. Directly after addition of TEMED pour it on top of the separation gel and insert the comb until polymerized (~30 min).

4. In the meantime prepare the sample by mixing 5–10 μg of depleted fraction with 5X sample buffer (max. loading on a 0.75 mm thick gel is 25 μL). Heat samples at 95°C for 15 min, then cool on ice, and briefly centrifuge (~15 s). The samples are ready for loading.
5. When the stacking gel is polymerized, remove the comb and place the gel into the running chamber. Fill the upper and lower gel chamber with 1X running buffer (10X diluted with MilliQ water) and wash the slots with 1X running buffer using a 1mL syringe with a 25 gauge needle attached.
6. Load the marker (5 μL PageRuler Prestained Protein Ladder) and the samples on the gel. Run the gel at 20 mA through the stacking gel and 25 mA afterwards, until the blue loading dye leaves the separation gel.
7. Stop the run and disassemble the gel unit. Discard the stacking gel and transfer the separation gel carefully into a clean container.
8. Silver staining, using the Silver-Stain Plus Kit, is undertaken according to the Bio-Rad Protocol. The gel is fixed in fixative solution (50% methanol, 10% acetic acid, 10% fixative enhancer, 30% MilliQ water) for 20 min.
9. Rinse the gel several times with MilliQ water, incubating for 10 min each time before replacing with fresh MilliQ water.
10. To stain and develop the gel with developer solution, dissolve 0.5 g development accelerator solution in 10 mL MilliQ Water. Then mix in the following order 7 mL MilliQ water, 1 mL silver complex solution, 1 mL reduction moderator solution, and 1 mL image development reagent. Combine this mix with 10 mL development accelerator solution and immediately apply to the gel after discarding the MilliQ water.
11. When the desired staining is reached, stop the reaction by incubating the gel in 5 % acetic acid solution. The protein pattern from the depleted samples should be similar, but should differ from a normal serum sample containing the highly abundant proteins (*see* **Fig. 17.2**).

3.7. Digestion with Trypsin

1. Add SDS to 200–400 μg of depleted serum in a 1.5 mL microfuge tube to a final concentration of 0.07% using a 2% stock solution (prepared in water and filter sterilized).
2. Add TCEP to a final concentration of 1.78 mM.

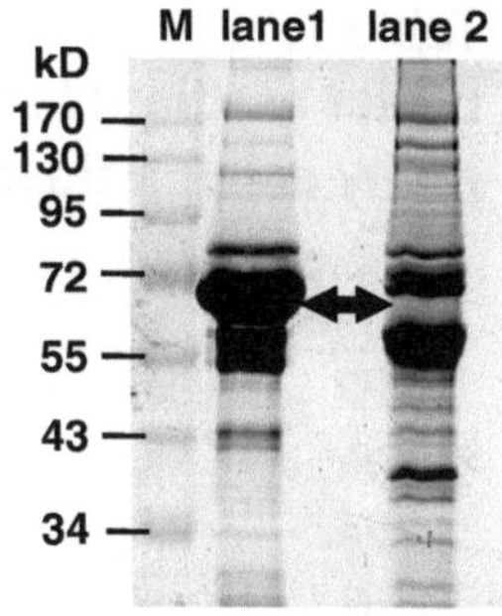

Fig. 17.2 10 μg delipidated serum (*lane 1*) and 10 μg depleted serum (*lane 2*) are separated on a 10 % SDS-Polyacrylamide gel and silver stained using the protocol described in **Section 3.6**. The Fermentas prestained protein ladder is shown (M). The albumin band (~72 kDa) is no longer visible in the depleted fraction (*arrow*). Furthermore, proteins bands not evident in non-depleted serum (*lane 1*) are visible in the depleted fraction. The gel was scanned using a GS800 densitometry scanner (Bio-Rad, Hemel Hempstedt, UK).

3. Mix the sample by vortexing and spin briefly to ensure the sample is at the bottom of the tube.
4. Incubate the sample at 60°C for 1 h and afterwards spin it to the bottom of the tube.
5. Prepare an 84 mM iodoacetamide solution (15.5 mg iodoacetamide in 1 mL MilliQ water; this has to be freshly made) and add it to a final concentration of 2.89 mM.
6. Incubate for 30 min in the dark at room temperature.
7. The trypsin is reconstituted in 20 μL per vial using the manufacturer's supplied buffer. Add 0.2 μL trypsin per μg of serum sample.
8. Mix by vortexing and incubate the serum samples overnight at 37°C.

3.8. Strong Cation Exchange Chromatography

1. 200–400 μg of depleted serum is diluted to 2 mL with diluent and loaded onto the HPLC column.
2. The gradient flow of the HPLC is 1 mL/min. The time gradient is shown in **Table 17.1**.
3. Collect eighty 1 ml fractions using 1.5 mL microfuge tubes.
4. After the run, clean the column with 100 % Buffer B to remove residual proteins and re-equilibrate with 100 % Buffer A.
5. Concentrate the HPLC fractions using vacuum concentration until there are 500 μL. Combine two fractions together (1 min + 2 min, 3 min + 4 min, etc.). Afterwards concentrate the combined fractions until completely dry. Samples may be stored at 4°C until analysed by LC-MS/MS.

Table 17.1
Buffer gradient for strong cation exchange chromatography using a flow rate of 1 mL/min

Time (min)	Buffer A (%)	Buffer B (%)
0.00	100	0
15.00	100	0
60.00	85	15
75.00	50	50
90.00	50	50
90.05	100	0
95.00	100	0

3.9. LC-MS/MS

1. For LC-MS/MS, analysis fractions are re-suspended in 180 μL Buffer A (5% acetonitrile, 0.05% trifluoroacetic acid).
2. The MS/MS system used is dependent on the instrument. This protocol is based on the use of a QSTAR Pulsar i hybrid mass spectrometer (Applied Biosystems, Warrington, UK). Samples were delivered into the instrument by an automated in-line LC (integrated LCPackings System, 5 mm C18 nano-precolumn and 75 mm × 15 cm C18 PepMap column (Dionex, CA, USA)) via a nanoelectrospray source head and 10 μm inner diameter PicoTip (New Objective, Massachusetts, USA).
3. One quarter of the sample (the ideal amount must be determined) is delivered into the QSTAR.
4. A gradient from 5% Buffer A to 48% Buffer B is applied at a flow rate of 300 nL/min for 60 min.
5. MS and MS/MS spectra are obtained using information-dependent acquisition consisting of a 1 s scan for *m/z* 400–2000 and the three most intense ions are selected for 1 s MS/MS scans (IDA, Analyst Software). Software such as ProteinPilot (Applied Biosystems) or other appropriate software is used to analyse the MS data.

4. Notes

1. For delipidation, a quantity of 500 μL serum is ideal, but the quantity can be scaled up or down as appropriate. As this is the first step of serum processing for LC-MS/MS analysis,

single samples or pooled serum samples can be used. Ideally, Eppendorf Lo-bind protein tubes (Eppendorf UK Limited, Histon, UK) should be used throughout the whole protocol.

2. It is best to aliquot the 10X buffers (5 mL for Equilibration and 2.5 mL for elution buffer) into 50 mL Falcon tubes under sterile conditions. The elution buffer may be cloudy, so warming it up for a couple of minutes as 37°C resolves the precipitate. After diluting the 10X to 1X we recommend a filter-sterilization step, using a 0.2 μm syringe filter (Minisart Plus, Satorius) and 50 mL syringe. Further we recommend performing all of the depletion steps under sterile conditions to avoid contamination.

3. In our studies we collect the wash fraction separately and concentrate it. We do not add this fraction to the depleted fraction, as it dilutes it and does not contribute significantly to the protein amount. If desired, the wash fraction may be added to the depleted fraction.

4. In our laboratory, several methods were tested to concentrate the depleted protein fraction. We found that the use of centrifugal filter concentration yielded minimal protein loss and optimum protein quality. However, the alternative approaches of precipitation, different types of centrifugal filter devices, and dialysis can be used.

References

1. Hanash, S. M., Pitteri, S. J., and Faca, V. M. (2008). Mining the plasma proteome for cancer biomarkers. *Nature* **452**, 571–579.
2. Schrohl, A. S., Wurtz, S., Kohn, E., Banks, R. E., Nielsen, H. J., Sweep, F. C., and Brunner, N. (2008). Banking of biological fluids for studies of disease-associated protein biomarkers. *Mol. Cell Proteomics* 7, 2061–2066.
3. Cham, B. E., and Knowles, B. R. (1976). A solvent system for delipidation of plasma or serum without protein precipitation. *J. Lipid Res.* **17**, 176–181.
4. Anderson, N. L., and Anderson, N. G. (2002). The human plasma proteome: history, character, and diagnostic prospects. *Mol. Cell. Proteomics* **1**, 845–867.
5. Sennels, L., Salek, M., Lomas, L., Boschetti, E., Righetti, P. G., and Rappsilber, J. (2007). Proteomic analysis of human blood serum using peptide library beads. *J. Proteome Res.* **6**, 4055–4062.

single samples or pooled serum samples can be used. Ideally, Eppendorf LoBind microtubes (Eppendorf UK Limited, Histon, UK) should be used throughout the whole protocol.

It is best to aliquot the 10X buffers (4 mL for Equilibration and 2.5 mL for Elution buffer) into 50 mL Falcon tubes under sterile conditions. The elution buffer may be [illegible] to warm up to room temp [illegible] as 52 °C [illegible] the precipitate. After diluting the 10X to 1X we recommend a filter-sterilization step using a 0.2 µm syringe filter (Sartorius, Sartorius) and 30 mL syringe. In this we recommend performing all of the depletion steps under sterile conditions to avoid contamination.

In our studies we collect the wash fraction separately and concentrate it. We do not add this fraction to the depleted fraction, as [illegible] and does not contribute significantly to the protein amount. If desired, the wash fraction may be added to the depleted fraction.

References

[illegible]

Chapter 18

Urinary Proteome Profiling Using 2D-DIGE and LC-MS/MS

Mark E. Weeks

Abstract

Proteomic methodologies have been at the forefront of cancer research for several years. The use of proteomic strategies to study all expressed genes aims to discover biomarkers indicative of the physiological state of cancer cells at specific time points, enabling early diagnosis, following cancer development/progression, screening and monitoring the efficacy of new therapeutic agents. Onco-proteomics has the potential to impact on oncology practice by delivering individualised highly selective clinical care.

2D-DIGE (2D difference in gel electrophoresis) enables simultaneous examination and comparison of multiple samples using cyanine dyes to label amino acid residues that are then separated based on charge and mass. These advantages combined with universal availability have until recently made 2D-DIGE a first method of choice in cancer proteome analysis of diverse specimens, including tissues, cell lines, blood and other body fluids.

Key words: Proteomics, 2D-DIGE, urine, cancer, HPLC-Chip/MS system, LC-MS/MS.

1. Introduction

Proteomics encompasses the identification of changes in protein expression, sub-cellular distribution, post-translational modifications and the deciphering of protein–protein interactions. Onco-proteomics aims to isolate, identify and recognise patterns of protein expression that differentiate different physiological states (cancer vs benign, early cancer vs advanced cancer states, etc.) (1). Pancreatic ductal adenocarcinoma accounts for over 213,000 deaths worldwide each year, largely due to late presentation and diagnosis (2). Preliminary studies demonstrate that urine can be a valid source for non-invasive biomarker discovery in patients with pancreatic diseases (3).

P.R. Cutillas, J.F. Timms (eds.), *LC-MS/MS in Proteomics*, Methods in Molecular Biology 658,
DOI 10.1007/978-1-60761-780-8_18, © Springer Science+Business Media, LLC 2010

Urine is an easily and non-invasively obtained bio-fluid that can potentially act as a source of biomarkers. This possibility has been explored by several studies (4, 5); indeed more than 1500 proteins have recently been identified in the urine of healthy donors (6) and this number is likely to increase. Almost half of all urinary proteins are soluble and are the product of the glomerular filtration of plasma (7), a substantial proportion of which arise from extrarenal sources (8). Large anionic proteins such as bile salt-dependent lipase, a 110 kDa pancreatic protein, have also been detected in urine (9), suggesting the 'leakage' of diverse proteins into urine. In addition, the urine protein profile is less complex when compared to plasma, its proteins are thermostable (10) and albumin and uromodulin comprise a lesser proportion of the urinary proteome. Therefore sample processing requires less pre-cleaning/fractionation.

The recent rapid evolution of new technologies has led to a variety of proteomic techniques being available to interrogate the proteome on a large-scale permitting simultaneous study of numerous proteins from multiple biological samples. Broadly speaking these are divided into gel-based and non-gel-based techniques. Gel-based proteomics (2D polyacrylamide gel electrophoresis, 2D-PAGE) has become extremely popular since first described in 1975 by O'Farrell (11). The development of 2D difference in gel electrophoresis using positively charged, amine reactive and molecular weight-matched fluorescent cyanine dyes (Cy2, Cy3 and Cy5) (2D-DIGE) significantly improved accuracy and led to more precise quantitation over a wider dynamic range (12). The increased dynamic detection range increased the sensitivity of the technique with the added advantages of reduced inter-gel variability, number of gels required, accurate spot matching and compatibility with the identification of protein spots using mass spectrometry (MS) (13, 14).

Gel-free strategies such as ICAT (isotope-coded affinity tagging), iTRAQ (isobaric tags for relative and absolute quantitation) and ESI MS/MS (electron spray ionisation tandem mass spectrometry) rely on liquid chromatography (LC) for protein separation interfaced with high-end mass spectrometers for protein identification (15). Such techniques can be automated and have a reduced sample requirement and often identify different subsets of regulated proteins making them complementary to gel-based techniques (15). Developments at the LC interface by a number of leading manufacturers have also greatly increased the discovery range and sensitivity available to research groups in the proteomic field. In fact LC and mass spectrometry (LC-MS/MS) have become central to protein identification and quantification, and many LC-MS/MS workflows have been developed and applied to proteomics research.

The Agilent 1200 Series HPLC-Chip/MS system is a new microfluidic chip-based technology for nanospray LC-MS/MS. The HPLC-Chip integrates the sample enrichment columns of a nanoflow LC system with the intricate connections and nanoelectrospray tip on a reusable biocompatible polymer chip, eliminating traditional fittings, valves and connections typically required in a nanoelectrospray LC-MS system (16).

We detail here the materials, instruments and protocols required for 2D-DIGE analysis and LC-MS/MS mass spectrometric identification of differentially expressed proteins in urine. The DIGE protocols detailed here have been developed and adapted at University College London (UCL, Proteomics Unit) over a number of years (3, 17–19).

2. Materials

(1) Coomassie Protein Assay Reagent (Pierce, Perbio Science Ltd, UK)

(2) CyDye^{TM} DIGE fluors *N*-hydroxy-succinimidyl (NHS) esters of Cy2, Cy3 and Cy5, immobiline dry strip, pH 3–10 NL, 24 cm, dry strip cover fluid, ampholines, pH 3.5–10, pharmalyte, pH 3–10 (Amersham Biosciences, GE Healthcare, NJ, USA)

(3) Anhydrous 99.8% *N,N*-dimethylformamide (Aldrich, Sigma-Aldrich, UK)

(4) Other reagents used: Urea, thiourea, dithiothreitol (DTT), L-lysine, Tris base, phosphate-buffered saline (PBS), IGEPAL-630, bromophenol blue, methanol, glacial acetic acid, 3-[(3-cholamidopropyl)-dimethylammonio]-1-propanesulfonate (CHAPS), sodium dodecyl sulphate (SDS), N,N,N',N'-tetramethylethylenediamine (TEMED), iodoacetamide (IAM), ammonium persulphate (APS), electrophoresis grade agarose, methanol (Fisher Scientific part of Thermo Fisher Scientific, p/a Perbio Science, Northumberland), ethanol (Fisher Scientific) and Milli-Q 18 Ω water (lab resource)

(5) GelCode Blue Safe Protein Stain (Thermo Fisher Scientific, p/a Perbio Science, Northumberland)

(6) Modified porcine trypsin (Promega, Southampton, UK)

(7) LC-MS grade acetonitrile (ACN) and LC-MS grade water, dithiothreitol (DTT), iodoacetamide (IAM) and ammonium bicarbonate

(8) Formic acid ampoules, trifluoro-acetic acid ampoules (Thermo Fisher Scientific, p/a Perbio Science, Northumberland)

(9) Plus One Repel Silane, Plus One Bind Silane, Dry Strip Cover Fluid (Amersham Biosciences, GE Healthcare, NJ, USA)

(10) 37.5:1 30% acrylamide-bis solution (Bio-Rad Laboratories, CA)

2.1. Equipment

The 2D-DIGE equipment used in these experiments was purchased from Amersham Biosciences, GE Healthcare, NJ, USA (similar equipment may be purchased from other manufacturers):

(1) Immobiline Dry Strip Reswelling Tray

(2) Immobiline Dry Strip IEF gels

(3) Multiphor II Electrophoresis Unit

(4) Ettan DALT low-fluorescence glass plates with reference markers

(5) Ettan DALT twelve gel caster, separation unit and power supply

(6) Typhoon 9400 Imager

(7) Ettan Spot Picker

(8) DeCyder differential analysis software

(9) SpeedVac Thermo savant SPD 1010 (Thermo Fisher Scientific, p/a Perbio Science, Northumberland)

(10) Agilent 1200 Series HPLC-Chip/MS system interfaced to an Agilent 6210 TOF mass spectrometer (Agilent Technologies UK Limited, Life Sciences & Chemical Analysis Group, Cheshire)

2.2. Solutions

(1) Lysis buffer: 8 M urea, 2 M thiourea, 4% (w/v) CHAPS, 0.5% IGEPAL-630 (v/v), 10 mM Tris–HCl, pH to 8.3, store at −20°C.

(2) 10% stock detergent solution: 1 ml IGEPAL-630 (v/v) Milli-Q 18 Ω water volume adjusted to 10 ml, store at room temperature.

(3) 40% stock (w/v) CHAPS solution: 40 mg CHAPS in Milli-Q 18 Ω water volume adjusted to 10 ml, store at room temperature.

(4) CyDyes stock solutions: centrifuge at 12,000×*g* for 30 s prior to addition of 10 μl DMF to 1000 pmol/μl, store at −20°C. Before opening the tubes, stock solutions are equilibrated on ice.

(5) L-lysine dye stop solution: 10 mM L-lysine in Milli-Q18 Ω water, store at –20°C.

(6) 1.3 M stock dithiothreitol (DTT) solution: 2 g DTT is dissolved in Milli-Q 18 Ω water and volume adjusted to 10 ml, store at −20°C. This solution should not be heated.

(7) Stock ampholines/pharmalyte solution: Equal volumes of pH 3.5–10 ampholines and pH 3–10 pharmalyte are mixed prior to use, store at 4°C for short periods. These broad pH range IPG buffers can be replaced with narrow range buffers depending on the first dimension pH range.

(8) Stock 0.2% (w/v) bromophenol blue solution: 20 mg of bromophenol blue is diluted in 10 ml with distilled water. The solution is then filtered and stored at room temperature.

(9) 14% PAGE solution: 46.5 ml 30% acrylamide-bis solution, 26.9 ml Milli-Q 18 Ω water (or equivalent), 25 ml 1.5 M Tris–HCL, pH 8.8, 1 ml 10% SDS, 35 μl TEMED (*N,N,N′,N′*-tetramethylethylenediamine) and 0.5 ml 10% APS and gently agitated under mild vacuum for 1 h. The TEMED and APS are added immediately prior to pouring the gel solution into the casting unit. These volumes are sufficient for one 24 cm gel. 10–15% solutions may be used for second dimension separation depending on the mass range separation required.

(10) Bind silane solution: 16 μl Plus One Bind Silane, 400 μl glacial acetic acid, 16 ml ethanol, 3.6 ml Milli-Q 18 Ω water. Sufficient for 12 24 cm × 20 cm plates, cannot be stored.

(11) Equilibration buffer: 6 M urea, 30% (v/v) glycerol, 50 mM Tris–HCl, pH 6.8, 2% (w/v) SDS, store at −20°C for 4–6 weeks.

(12) Agarose overlay 0.5% (w/v): melt 1 g of agarose in 200 ml of 1X SDS electrophoresis buffer in a microwave on low heat. Great care must be taken using a microwave and solution must be allowed to cool to luke warm before moving. Add bromophenol blue solution for a pale blue colour.

(13) Tris glycine-SDS electrophoresis running buffer: 0.025 M Tris Base, 0.192 M glycine and 0.1% sodium dodecyl sulphate.

(14) Gel fixing solution: 35% (v/v) methanol and 7.5% (v/v) acetic acid made to required volume with Milli-Q 18 Ω water, store at 4°C.

(15) Acetonitrile wash solution: 50%/50% acetonitrile and Milli-Q 18 Ω water (v/v), store at room temperature.

(16) Stock ammonium bicarbonate solution: 10 mM ammonium bicarbonate in 100 ml of Milli-Q 18 Ω water, store at 4°C.

(17) DTT solution: 10 mM DTT in 10 mM ammonium bicarbonate stock solution, store at 4°C.

(18) IAM solution: 10 mM IAM in 10 mM ammonium bicarbonate, store at 4°C.

(19) Stock solution of modified porcine trypsin: 40 μl trypsin re-suspension buffer to 20 μg of trypsin provided in vial. This is of low pH and allows suspension to be stored without autolysis occurring. Further dilution and activation can be achieved with the addition of 10 mM ammonium bicarbonate solution. The diluted trypsin can be stored at −20°C.

(20) HPLC Buffer A: 0.1% formic acid in 100% LC-MS grade H_20.

(21) HPLC Buffer B: 5% LC-MS grade H_20, 0.1 % formic acid, made up to 100% with LC-MS grade acetonitrile.

3. Methods

3.1. Sample Collection and Preparation

The standardisation of collection and processing procedures of samples is a vital first step in expression profiling (20). During collection, EDTA-free protease inhibitors should be used to prevent sample degradation, as they do not interfere with Cy dye labelling. Collection and processing methods are sample driven and a variety of cell types (whole tissue, cell lines) and body fluids (saliva, cerebrospinal fluid, plasma and urine) have already been subject to proteomic analysis (21–26); these references can be used to obtain optimal processing conditions of particular sample types.

3.2. Experimental Design

2D-DIGE experiments require careful design to ensure that statistically meaningful data can be obtained. The use of an internal standard permits robust analysis and more precise quantitation. Labelling a pooled internal standard (sample composed of equal aliquots of each sample used in the experiment) with Cy2 dye, while Cy3 and Cy5 are used to label test and reference samples is the most popular experimental design. Statistical analysis can be further improved by increasing the number of gels per experiment, performing both technical and biological replicates (minimum of three per sample group). Differential expression is then calculated as the average fold change (the average

Table 18.1
Example of the experimental design of a simple 2D-DIGE experiment

	Cy2	Cy3	Cy5
Gel 1	100 μg pool	100 μg disease sample, replicate 1	100 μg control sample, replicate 1
Gel 2	100 μg pool	100 μg control sample, replicate 2	100 μg disease sample, replicate 2
Gel 3	100 μg pool	100 μg disease sample, replicate 3	100 μg control sample, replicate 3

spot intensity ratio between differentially labelled spots matching across all three gels) with statistical confidence provided by a *t* test. **Table 18.1** demonstrates an experimental design where 100 μg of proteins from a disease sample (cancer) and control sample (healthy/benign disease) is compared in triplicates (50 μg of each sample are pooled for the internal standard). Dye bias is reduced by interchangeably labelling samples as detailed (**Table 18.1**). Where high biological variability exists (population variance) and in cases where limited quantity of samples is available pooling can be considered in order to generate statistically valid data.

More complex comparisons can be performed by running up to 12 gels simultaneously comparing 24 samples (18). Protein expression profiling experiments using 2D-DIGE are best performed in a dedicated laboratory space (*see* **Note 1**).

3.3. 2D-DIGE Method

(1) The protein concentration of all samples must be accurately determined; three to four replicate assays must be performed for each sample to ensure accurate protein determination. Proprietary solutions such as Pierce Coomassie protein assay reagent (performed according to manufacturer's instructions) are usually sufficient (*see* **Notes 2** and **3**).

(2) A working solution of 200 pmol/μl of Cy dye is made by adding 4 μl of DMF to 1 μl of stock solution (*see* **Note 4**).

(3) The required sample volume (concentration) is aliquoted into low protein bind Eppendorf tubes (100 μg of protein from disease and control samples as shown in the experiment in **Table 18.1**) and a pool of all samples (internal standard) made of a mixture of 150 μg of each disease and control sample is made in one tube (*see* **Note 5**).

(4) Labelling is achieved by the addition of 400 pmol of the appropriate CyDye (Cy3/Cy5) per 100 μg of protein used interchangeably as shown in **Table 18.1** and the internal standard is labelled with Cy2 (1200 pmol/300 μg of protein). Labelled samples are then vortexed prior to incubation on ice in the dark for 30 min (*see* **Note 6**). The labelling reaction is quenched with a 20-fold molar excess of L-lysine with incubation on ice in dark for a further 10 min.

(5) The Cy3- and Cy5-labelled samples are mixed appropriately and 100 μg aliquot of the Cy2-labelled pool is added (to give 300 μg total protein).

(6) Samples are reduced by the addition of 1.3 M DTT to a final concentration of 65 mM.

(7) 9 μl of fresh carrier ampholines/pharmalyte mix are added to a final concentration of 2% (v/v) and 1 μl of 0.2% bromophenol blue is added. Total volume can now be adjusted to 450 μl with lysis buffer. Samples are centrifuged at 13,200×*g* for 5 min at 4°C.

3.3.1. Sample Loading First Dimension IEF (Immobiline^{TM} Drystrips)

(1) The Cy dye-labelled samples are pipetted into the individual wells of an Immobiline re-swelling tray avoiding bubble formation.

(2) The plastic protective cover from Immobiline Drystrip (24 cm, pH 3–11) is removed and the strips are placed gel side down into the sample solution (*see* **Note 7**).

(3) The gel strips and samples are incubated at room temperature for 10 min prior to covering each strip in Immobiline Drystrip cover fluid. This prevents drying of the sample over the re-swelling period. Gels are rehydrated in the dark at room temperature for a period up to 12 h.

3.3.2. First Dimension Separation Using Isoelectric Focussing (IEF)

(1) Once rehydration is completed a pair of forceps is used to remove the strips and drain excess cover fluid, taking extreme care not to touch or damage the gel.

(2) Wicks soaked in a 65 mM solution of DTT are used to improve contact between gel surface and electrodes leading to better separation.

(3) Samples are separated by isoelectric point on a MultiPhore II flatbed system (Amersham, UK) maintained throughout at 17°C for a total of 95 kVh, in accordance with manufacturer's instructions.

(4) A programmed sloped gradient of 0–500 V for 1 h, 500 V for 2 h, 500 V–3500 V for 2 h, 3500 V for 24 h and finally 500 V for 2 h is used. This can be protein load and sample

specific and should be increased for higher protein loads (*see* **Note 8**). Apparatus is covered to exclude light.

3.3.3. Second Dimension Gel Preparation and Separation

(1) Sets of clean low-fluorescence glass plates containing large back plate (with spacer attached) and smaller 'front plates' (Ettan DALT 24 cm gel plates) are selected depending on the number of gels that needs to be run. Spacer size can be varied for 0.5, 1.0 and 1.5 mm gel thickness.

(2) Reference markers are applied to the surface of the smaller plates. These are placed halfway down the plates and 15–20 mm from each edge and are critical for determining coordinates for automated spot picking. 1.5 ml of fresh Bind Silane solution is applied per small plate and surface wiped with lint-free tissue. Plates are left to dry for a minimum of 1 h. The inner surface of the larger spacer plate is treated with Repel Silane, wiped using lint-free tissues and left to dry for 10 min.

(3) Glass plates with the repel and bind surfaces facing each other are assembled in Ettan gel casting unit assuring a tight seal and no leak. Feeding tube and funnel are attached to the caster and the required polyacrylamide gel solution is poured ensuring no air bubbles are introduced. Water-saturated butanol (2 ml) is overlaid on each gel and gels are allowed to polymerise for at least an hour (*see* **Note 9**).

(4) On completion of the first dimension separation, gel strips are equilibrated for 15 min in equilibration buffer containing 65 mM DTT and then 15 min in the same buffer containing 240 mM IAM (*see* **Note 10**).

(5) The equilibrated Immobiline gel strips are then rinsed with 1X SDS electrophoresis buffer prior to being placed onto the top of handmade PAGE gels described in **Section 3.5**. Strips are placed in a molten 0.5% agarose overlay, with the basic end of the strip towards the left-hand side with the bonded plate facing forward (*see* **Note 11**).

(6) The agarose is allowed to cool and set at room temperature.

(7) The DALT twelve electrophoresis tank is filled with SDS running buffer and gels complete with Immobiline Drystrips are inserted into the designated slots in the gel tank. Unoccupied slots are filled with plastic blanks and top tank filled with running buffer.

(8) The Ettan DALT twelve system is run for 16 h at 2.2 W per gel or until the dye front had reached the bottom of the gel (*see* **Note 12**).

3.3.4. Image Capture

(1) Gels (still between glass plates) are rinsed in Milli-Q 18 Ω water prior to image capture (*see* **Note 13**). Gels plates

are aligned on Typhoon scanner and photomultiplier tube (PMT) voltage set to low (500 V) on each channel (Cy2, Cy3 and Cy5).

(2) A preliminary low-resolution scan (1000 μm) for each channel is performed and grey scale pixel values generated. Using ImageQuant software (GE Healthcare) for Typhoon 9400, maximum pixel values in various user-defined, spot-rich regions of each image are obtained by adjusting PMT voltages. Repeated scans may be required to ensure voltage settings achieve maximum pixel values within 10% for each of the three channels.

(3) High-resolution scans (100 μm) are then performed across all gels at these optimal settings.

3.3.5. Image Analysis

(1) Scanned images are analysed using differential analysis software (DeCyder version 5.1, later versions available). Spot boundaries are defined using DeCyder differential in gel analysis (DIA) module, allowing for intra-gel analysis, and standardised abundance for each spot is obtained by comparing Cy3- and Cy5-labelled spots to the internal standard (ratios – Cy3:Cy2 and Cy5:Cy2).

(2) Test spot volumes across all gels (inter-gel analysis) are matched using the DeCyder biological variation analysis (BVA) module. Lists of statistically significant deregulated protein spots between disease and control samples are generated using Student's *t* test or one- or two-way ANOVA.

(3) BVA software is used to define the position of the reference markers and produce *x*–*y* coordinates of each spot of interest; these can be exported as a 'coordinate pick list.'

3.3.6. Post-staining (and Spot Picking)

Post-staining of 2D gels using CCB is compatible with Cy dye labelling and mass spectrometry and allows for improved spot picking (18). **Figure 18.1** shows a post-stained CCB gel image with differentially expressed and identified protein spots in a 2D-DIGE comparison of pooled pancreatic ductal adenocarcinoma, chronic pancreatitis and healthy urine specimens.

3.3.7. CCB Post-staining

(1) After image capture gel plates are separated, individual gels bonded to low-fluorescent glass back plates are washed, fixed and stained with GelCode Blue Safe Protein Stain.

(2) GelCode Blue Safe Protein Stain (400 ml/gel) is added to the washed gels in a new plastic tray and left shaking for several hours to overnight at room temperature (*see* **Note 14**).

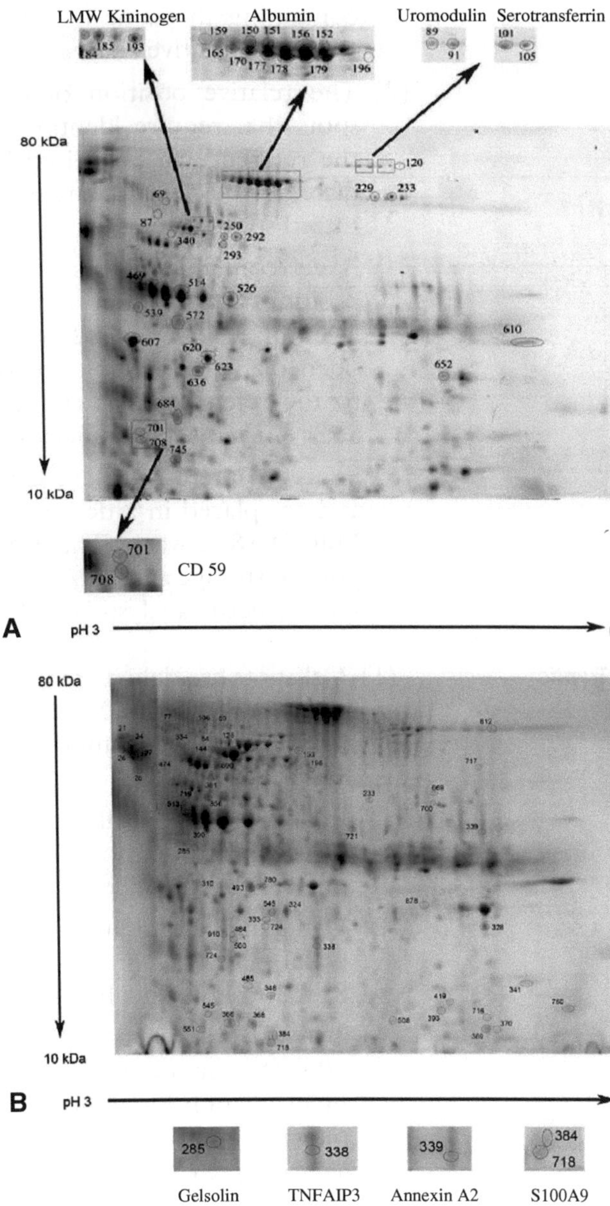

Fig. 18.1 2D-Dige separation of urine samples labelled with spots showing statistically significant differences between sample groups.

3.4. Spot Picking

(1) After completion of CCB staining gels are washed briefly in Milli-Q 18 Ω water (*see* **Note** 15) and scanned using Typhoon 9400 scanner using red laser and no filters (*see* **Note 16**). Scanned CCB-stained gel images are saved as .tiff format files and imported into DeCyder software.

(2) Spots are matched accurately between CCB- and Cy dye-stained image leading to the generation of a pick list

containing the same master spot numbers as provided by BVA quantitative analysis.

(3) The relative position of differentially expressed protein spots that require identification are defined according to the reference markers initially applied on the smaller gel plate. Pick list coordinate file is exported to the Ettan automated spot picker.

(4) A gel containing spots of interest is clamped down into position in the picker and submerged into 1–2 mm deep layer of Milli-Q 18 Ω water. The imported pick list is opened and the spot picking head of the instrument aligned to the reference markers previously defined by DeCyder according to the manufacturer's instructions.

(5) Spots of interest are excised using a 2 mm picking head and are placed in a designated 96-well plate in 200 μl of Milli-Q 18 Ω water. The water from each well is removed prior to storage at −20°C or subsequent digestion and mass spectrometric analysis is performed.

3.5. In-Gel Tryptic Digest

(1) Gel spots are dehydrated and destained by washing three times in 30 μl of 50% acetonitrile (ACN) and then dried in SpeedVac for 10 min.

(2) Destained spots are removed from the 96-well plate and transferred to labelled lo-bind Eppendorf tubes (most 96-well plates are not suited to tryptic digests as they bind proteins and peptides).

(3) Di-sulphide bonds in proteins are reduced by the addition of 10 mM DTT in 10 mM ammonium bicarbonate (pH 8) followed by incubation for 45 min at 50°C.

(4) The DTT is removed and 50 mM solution of IAM in 10 mM ammonium bicarbonate (pH 8) is added followed by incubation for an hour in the dark to alkylate cysteine moieties in proteins.

(5) The IAM solution is removed and the gel pieces washed twice in 50% ACN and dried in SpeedVac for 10 min.

(6) Sufficient trypsin solution to a total of 50 ng (conc. of 5 ng/μl, refer to **Section 2.3**) is added to cover the gel spot. Tubes are then incubated at room temperature for 10 min prior to overlaying each spot with 10–20 μl of 10 mM ammonium bicarbonate.

(7) The samples are incubated at 37°C overnight (not exceeding 16 h).

(8) The samples are centrifuged at 10,000×*g* for 2 min prior to the addition of 5 μl of 50% of ACN/5% trifluoro-acetic acid. Samples are gently agitated for 5 min.

(9) The supernatant is transferred to a fresh tube and sufficient 50% ACN/5% trifluoro-acetic acid to cover gel piece is added again. This process is repeated twice.

(10) The pools of extracted peptides from each gel piece (Steps 9 and 10) are taken to dryness in a SpeedVac prior to resuspension in 5 μl 0.1% TFA solution.

(11) Samples can be stored at −20°C or subject to immediate mass spectrometry analysis.

3.6. Protein Identification by LC-MS/MS

Protein identification by mass spectrometry can be performed using either (or a combination of) peptide mass fingerprinting (PMF) or sequence-specific peptide fragmentation (27). Matrix-assisted laser desorbtion ionisation mass spectrometry (MALDI MS) is fast, relatively accurate, easy to perform and is the most commonly used technique for peptide mass fingerprinting. If PMF fails to unambiguously identify proteins, peptide sequencing needs to be performed (28). The choice of mass spectrometer is generally dependant on resources and the local availability of hardware and expertise.

(1) Prior to use the Agilent 6210 TOF mass spectrometer was calibrated using proprietary Agilent calibration mix and semiautomatic instrument settings in tune mode. The calibration mix was directly infused into the instrument in tune mode (via syringe pump). The updated calibration coefficients were within range and were applied.

(2) Fresh buffers for analytical nanoflow pump were prepared.

(3) An HPLC-Chip was inserted into the instrument. The HPLC-Chip configuration included a 40 nl enrichment column and 150 mm × 75 μm analytical column packed with Zorbax 300SB-C18, 5 μm materials. Prior to use the loading and analytical pumps and lines on Agilent 1200 NanoLC system were purged for 5 min at 2.5 ml/min. The LC stream was switched to waste and column was conditioned. Base line was stable. LC stream was switched to MS. The LC system consisted of a nanoflow pump, a capillary pump for sample loading and a microwell-plate auto sampler with cooler. Complete system control was accomplished using Agilent TOF LC/MS software.

(4) As samples from 2D gels are not complex a relatively short LC gradient method can be used to introduce peptides into source (**Table 18.2**).

(5) 0.5 μl of sample was analyzed using Agilent's microfluidics-based HPLC-Chip for nanoelectrospray LC-MS connected to Agilent's Q-TOF MS.

(6) Instrument settings, Ion Source HPLC-chip, positive mode with centroid data storage selected. Reference mass

Acquisition rate/time	MS	MS/MS
Rate	3 spectra/s	5 spectra/s
Time	333.7 ms/spectrum	200 ms/spectrum
Transient spectrum	1547	2384
Isolation width	–	4 *m/z*
Mass range	300–2500 *m/z*	50–3000 *m/z*

Table 18.2
Agilent NanoLC gradient

Time (min)	% B	Flow rate[a] (nl/min)
0.00	5	0.4
1.00	5	0.4
25.00	60	0.4
30.00	100	0.4
40.00	100	0.4
40.01	5	0.4
Post run 10.00	5	0.4

Stop time 50.00 min.
[a]Note that recommended flow rate is 0.3 nl/min; however, spray stability at low organic is better with this increased flow rate.

correction was enabled on mass 391.2843. No exclusion masses were selected.

Gas temp 325°C, drying gas flow rate 2.5 l/min, V_{cap} 1850 V, Capillary 0.063 mA, Chamber 1.72 mA.

Fragmentor 170 V, Skimmer 65 V, OCT1RF V_{pp} 750 V

Precursor selection 5 maximum precursors per cycle at abs threshold 1000, relative threshold (%) 0.01 with active exclusion enabled; exclude after two spectra, released after 0.1 min. Precursors were sorted by abundance only.

In line with continued method development Agilent Technologies have recently suggested a new recommendation for Q-tof collision energy when performing protein identifications (trypsin digestion), these being slope 3.6 and offset minus 4.8 (−4.8). The new recommendations should improve amino acids coverage %.

(7) Peptides were identified using Spectrum Mill MS Proteomics Workbench software with the SwissProt protein database.

4. Notes

(1) If available the use of a dedicated clean room is highly recommended. To prevent contamination by hair/skin keratins gloves, aprons, facemask and hair caps need to be worn while in the clean room.

(2) Sample preparation methods should remove salts, lipids, detergents and nucleic acids as these can affect isoelectric focusing.

(3) The buffers used during sample preparation need to be compatible with the protein quantitation technique used.

(4) Cyanine dyes are water sensitive and need to be stored in a water-free environment (silica gel may be used). DMF used to make stock solution of cyanine dyes should be 99.5% pure and anhydrous. Resolubilisation of stock dyes should be conducted under a nitrogen atmosphere.

(5) Adding 10% to these figures allows for variations in pipetting and ensures sufficient sample for all replicates.

(6) The method detailed here is for minimal labelling using cyanine dyes.

(7) The length of IPG gel strip, pH range and IEF gradient can be optimised to achieve the best possible resolution for the samples of interest.

(8) Prior to the second dimension separation and BEFORE equilibration IEF gel strips can be stored at −80°C for several weeks in a rigid container.

(9) The gels need to polymerise completely and should preferably be left overnight.

(10) Exceeding the stipulated times during reduction and alkylation of IPG gel strips may contribute to protein loss.

(11) Care should be taken not to introduce air bubbles at the gel interface while inserting the IPG gel strip between the glass plates. The overlaying agarose must solidify prior to commencing second dimension separation.

(12) To ensure continuity use IPG gel strips from the same batch and cast second dimension gels at the same time.

(13) Prior to scanning, both outer gel plate surfaces are wiped clean and dry to ensure optimum scanned image quality.

(14) After applying GelCode Blue Safe Protein Stain gels do not need to be destained to visualise proteins.

(15) GelCode Blue Safe Protein Stain gives very low background and gel does not require destain; however,

washing for several hours in Milli-Q 18 Ω water can improve visualisation of low-abundance proteins.

(16) CCB staining is required as minimal labelling does not define the spot centre where concentration is highest. Defining this centre improves protein identification success rate from in gel tryptic digests of 2D-DIGE gel spots.

References

1. Pastwa, E., Somiari, S. B., Czyz, M., and Somiari, R. I. (2007) Proteomics in human cancer research. *Proteom. Clin. Appl.* **1**(1), 4–17.
2. Jemal, A., et al. (2007) Cancer statistics, 2007. *CA Cancer J. Clin.* **57**(1), 43–66.
3. Weeks, M. E., Hariharan, D., Petronijevic, L., Radon, T. P., Whiteman, H. J., Kocher, H. M., Timms, J. F., Lemoine, N. R.,Crnogorac-Jurcevic, T. (2008) Analysis of the urine proteome in patients with pancreatic ductal adenocarcinoma. *Proteom. Clin. Appl.* **2**(7–8), 1047–1057.
4. Pieper, R., et al. (2004) Characterization of the human urinary proteome: a method for high-resolution display of urinary proteins on two-dimensional electrophoresis gels with a yield of nearly 1400 distinct protein spots. *Proteomics* **4**(4), 1159–1174.
5. Spahr, C. S., et al. (2001) Towards defining the urinary proteome using liquid chromatography-tandem mass spectrometry. I. Profiling an unfractionated tryptic digest. *Proteomics* **1**(1), 93–107.
6. Adachi, J., Kumar, C., Zhang, Y., Olsen, J. V., and Mann, M. (2006) The human urinary proteome contains more than 1500 proteins, including a large proportion of membrane proteins. *Genome Biol.* 7(9), R80.
7. Barratt, J. and Topham, P. (2007) Urine proteomics: the present and future of measuring urinary protein components in disease. *CMAJ.* 177(4), 361–368.
8. Thongboonkerd, V. (2005) Genomics, proteomics and integrative "omics" in hypertension research. *Curr. Opin. Nephrol. Hypertens.* **14**(2), 133–139.
9. Comte, B., et al. (2006) Detection of bile salt-dependent lipase, a 110 kDa pancreatic protein, in urines of healthy subjects. *Kidney Int.* **69**(6), 1048–1055.
10. Schaub, S., et al. (2004) Urine protein profiling with surface-enhanced laser-desorption/ionization time-of-flight mass spectrometry. *Kidney Int.* **65**(1), 323–332.
11. O'Farrell PH (1975) High resolution two-dimensional electrophoresis of proteins. *J. Biol. Chem.* **250**(10), 4007–4021.
12. Unlu, M., Morgan, M. E., and Minden, J. S. (1997) Difference gel electrophoresis: a single gel method for detecting changes in protein extracts. *Electrophoresis* **18**(11), 2071–2077.
13. Marouga, R., David, S., and Hawkins, E. (2005) The development of the DIGE system: 2D fluorescence difference gel analysis technology. *Anal. Bioanal. Chem.* **382**(3), 669–678.
14. Tonge, R., et al. (2001) Validation and development of fluorescence two-dimensional differential gel electrophoresis proteomics technology. *Proteomics* **1**(3), 377–396.
15. DeSouza, L., et al. (2005) Search for cancer markers from endometrial tissues using differentially labeled tags iTRAQ and cICAT with multidimensional liquid chromatography and tandem mass spectrometry. *J. Proteome Res.* **4**(2), 377–386.
16. Yin, H., Killeen, K., Brennen, R., Sobek, D., Werlich, M., and van de Goor, T. (2005) Microfluidic chip for peptide analysis with an integrated HPLC column, sample enrichment column, and nanoelectrospray tip. *Anal. Chem.* 77(2), 527–533.
17. Chan, H. L., et al. (2005) Proteomic analysis of redox- and ErbB2-dependent changes in mammary luminal epithelial cells using cysteine- and lysine-labelling two-dimensional difference gel electrophoresis. *Proteomics* **5**(11), 2908–2926.
18. Gharbi, S., et al. (2002) Evaluation of two-dimensional differential gel electrophoresis for proteomic expression analysis of a model breast cancer cell system. *Mol. Cell Proteomics* **1**(2), 91–98.
19. Weeks, M. E., et al. (2006) A parallel proteomic and metabolomic analysis of the hydrogen peroxide- and Sty1p-dependent stress response in Schizosaccharomyces pombe. *Proteomics* 6(9), 2772–2796.
20. Shaw, M. M. and Riederer, B. M. (2003) Sample preparation for two-dimensional gel electrophoresis. *Proteomics* 3(8), 1408–1417.

21. Jin, T., et al. (2007) Proteomic identification of potential protein markers in cerebrospinal fluid of GBS patients. *Eur. J. Neurol.* **14**(5), 563–568.
22. Kakisaka, T., et al. (2007) Plasma proteomics of pancreatic cancer patients by multi-dimensional liquid chromatography and two-dimensional difference gel electrophoresis (2D-DIGE): up-regulation of leucine-rich alpha-2-glycoprotein in pancreatic cancer. *J. Chromatogr. B Analyt. Technol. Biomed. Life Sci.* **852**(1–2), 257–267.
23. Katayama, M., et al. (2006) Protein pattern difference in the colon cancer cell lines examined by two-dimensional differential in-gel electrophoresis and mass spectrometry. *Surg. Today* **36**(12), 1085–1093.
24. Lee, I. N., et al. (2005) Identification of human hepatocellular carcinoma-related biomarkers by two-dimensional difference gel electrophoresis and mass spectrometry. *J. Proteome Res.* **4**(6), 2062–2069.
25. Orenes-Pinero, E., et al. (2007) Searching urinary tumor markers for bladder cancer using a two-dimensional differential gel electrophoresis (2D-DIGE) approach. *J. Proteome Res.* **6**(11), 4440–4448.
26. Ryu, O. H., Atkinson, J. C., Hoehn, G. T., Illei, G. G., and Hart, T. C. (2006) Identification of parotid salivary biomarkers in Sjogren's syndrome by surface-enhanced laser desorption/ionization time-of-flight mass spectrometry and two-dimensional difference gel electrophoresis. *Rheumatology (Oxford)* **45**(9), 1077–1086.
27. Thiede, B., et al. (2005) Peptide mass fingerprinting. *Methods* **35**(3), 237–247.
28. Aebersold, R. and Goodlett, D. R. (2001) Mass spectrometry in proteomics. *Chem. Rev.* **101**(2), 269–295.

Chapter 19

Analysis of Peptides in Biological Fluids by LC-MS/MS

Pedro R. Cutillas

Abstract

Urine contains large amounts of small peptides, which may represent a rich, yet largely unexplored, source of novel biomarkers for disease monitoring. This chapter describes detailed procedures for the analysis of urinary polypeptides by LC-MS/MS. Hundreds to thousands of small peptides (~700 to ~7000 Da) can be detected in urine with the described techniques. Extraction procedures, based on commercially available reagents, effectively remove interfering urinary organic and inorganic salts and neutral compounds, making this a robust and simple assay with the power to detect hundreds to thousands of polypeptides in urine. Analysis time is relatively short, making this protocol a valuable alternative to conventional proteomic techniques based on multidimensional separations. The methodology is therefore particularly useful when the aim is to analyse samples with sufficient depth and throughput so as to make it useful to compare large numbers of specimens. Procedures for enhancing quantitative and qualitative analysis of LC-MS/MS data are also detailed.

Key words: Urinary proteomics, peptidomics, urine analysis, quantification.

1. Introduction

Urine is an ideal source of biomarkers because it can be obtained non-invasively and because it is more stable and less complex than blood (1, 2). As a result, many research programs use urine as the biofluid of choice for biomarker discovery and monitoring, not only for kidney diseases (the obvious choice) but also for other diseases including cancer (3–7).

Contrary to the medical textbook view, urine is not protein free but it contains relatively large amounts of polypeptides (about 10–40 μg/ml in healthy individuals). Urinary proteins may originate from two different sources: they may appear in urine as a result of active protein secretion by epithelial cells that line

P.R. Cutillas, J.F. Timms (eds.), *LC-MS/MS in Proteomics*, Methods in Molecular Biology 658,
DOI 10.1007/978-1-60761-780-8_19, © Springer Science+Business Media, LLC 2010

the urogenital track or as a result of incomplete reabsorption of plasma proteins by kidney cells. Thus the urinary proteome contains a mixture of plasma and kidney proteins, whose ratios may be altered in disease (8, 9).

As with blood, urine contains proteases that actively digest proteins into small peptides, which also appear in urine in large quantities (10) (proteases are also present on luminal membranes of renal epithelial cells and are therefore in contact with the 'pre-urine'). These small peptides may not merely be waste products in their route to excretion. Indeed, it has been postulated that, since they are present in the tubular fluid (the 'pre-urine' in contact with tubular kidney cells), some of these peptides may have physiological roles by acting on luminal receptors of kidney tubular cells (11).

Protocols for urinary proteomics are in general designed to analyse large polypeptides (>10 kDa) and these are either based on size exclusion techniques (ultrafiltration, dialysis or gel filtration) or on protein precipitation with organic solvents or acids (12). These sample preparation techniques are needed because urine contains large amounts of organic and inorganic salts that interfere with downstream protein analysis by 2D gel electrophoresis or by direct LC-MS/MS. However, extraction techniques based on size exclusion or precipitation, in addition to removing salts, also remove small peptides, which are therefore not analysed in most urinary proteomics studies.

The present protocol is designed for enriching urinary peptides in a form that is compatible with downstream analysis by LC-MS and LC-MS/MS with electrospray as the ion source. These peptides have the same mass to charge ratio (m/z) and hydrophobicity as organic acids, which are present in large quantities in urine, and whose signals, therefore, mask the presence of peptides in LC-MS experiments of untreated urine. Removal of organic acids is therefore a prerequisite for the analysis of small urinary peptides by LC-MS (when electrospray is the ion source). In this protocol, this is accomplished by strong cation exchange (SCX) extraction. At low pH, peptides are positively charged and thus bind negatively charged sulfonic acid groups on SCX beads, while organic acids do not bind and can therefore be aspirated to waste. After extensive washing of beads to remove loosely bound molecules, peptides are then recovered from SCX beads using a solution of high ionic strength and relatively high pH. The extraction protocol makes use of volatile salts whenever possible in order to make it more compatible with LC-MS analysis. There is also a reverse-phase (RP) extraction step prior to SCX, which is needed to deplete the sample of inorganic salts. This desalting step reduces the ionic strength of the sample thus making possible the binding of peptides to the SCX material.

Desalted peptides are then analysed by LC-MS/MS. Although many types of mass spectrometers may be suitable, in this protocol, analysis is exemplified using a Q-TOF instrument equipped with a nano-electrospray ion source. Similarly, other separation techniques, in addition to LC, may also be suitable, with capillary electrophoresis showing impressive results (13). **Figure 19.1** illustrates the complexity of the urinary peptidome as analysed with this LC-MS workflow; hundreds (or even thousands) of peptides can be detected in a 30 min gradient LC-MS run from the analysis of 50 μl of urine (by extrapolation). Hundreds of MS/MS spectra can be collected and thousands of these small peptides may be present in urine (9, 11, 14), making the peptidome of this fluid an attractive, yet largely unexplored, source of potential biomarkers of disease. With minor modifications the utility of this protocol could also be extended to analyse larger polypeptides and proteins.

2. Materials

2.1. Reversed-Phase Solid-Phase Extraction

1. Peptide mixture obtained from the digestion of yeast enolase.
2. Reversed-phase (RP) solid-phase extraction (SPE) cartridges (HLB Oasis 94226, Waters, Milford, MA).
3. SPE vacuum manifold (optional).
4. RP conditioning solution: acetonitrile (ACN) (*see* **Note 1**).
5. RP equilibration solution: 0.1% (v/v) TFA/5% (v/v) ACN.
6. RP elution solution: 0.1% (v/v) TFA/60% (v/v) ACN.

2.2. Strong Cation Exchange Extraction

1. Strong cation exchange (SCX) magnetic beads (Dyna beads, Invitrogen).
2. Magnetic rack separator for 1.5 ml tubes (e.g. Invitrogen cat no. CS1500) and for 96-well plates.
3. SCX conditioning solution: 1 M NaCl/50 mM ammonium bicarbonate, pH 8.8 (ammonium solutions are volatile and should be prepared fresh).
4. SCX loading solution: 0.1% (v/v) TFA/20 (v/v)% ACN/0.01% Tween-20.
5. SCX wash solution: 0.1% (v/v) TFA/20 (v/v)% ACN.
6. SCX elution solution: 500 mM ammonium acetate in 20% (v/v) ACN.

2.3. LC-MS/MS (see Note 2)

1. Solvent A: 0.1% formic acid in LC-MS grade water.
2. Solvent B: 0.1% formic acid in LC-MS grade ACN.
3. Trap column (Symmetry C18, 180 μm × 20 mm, Waters) (*see* **Note 2**).

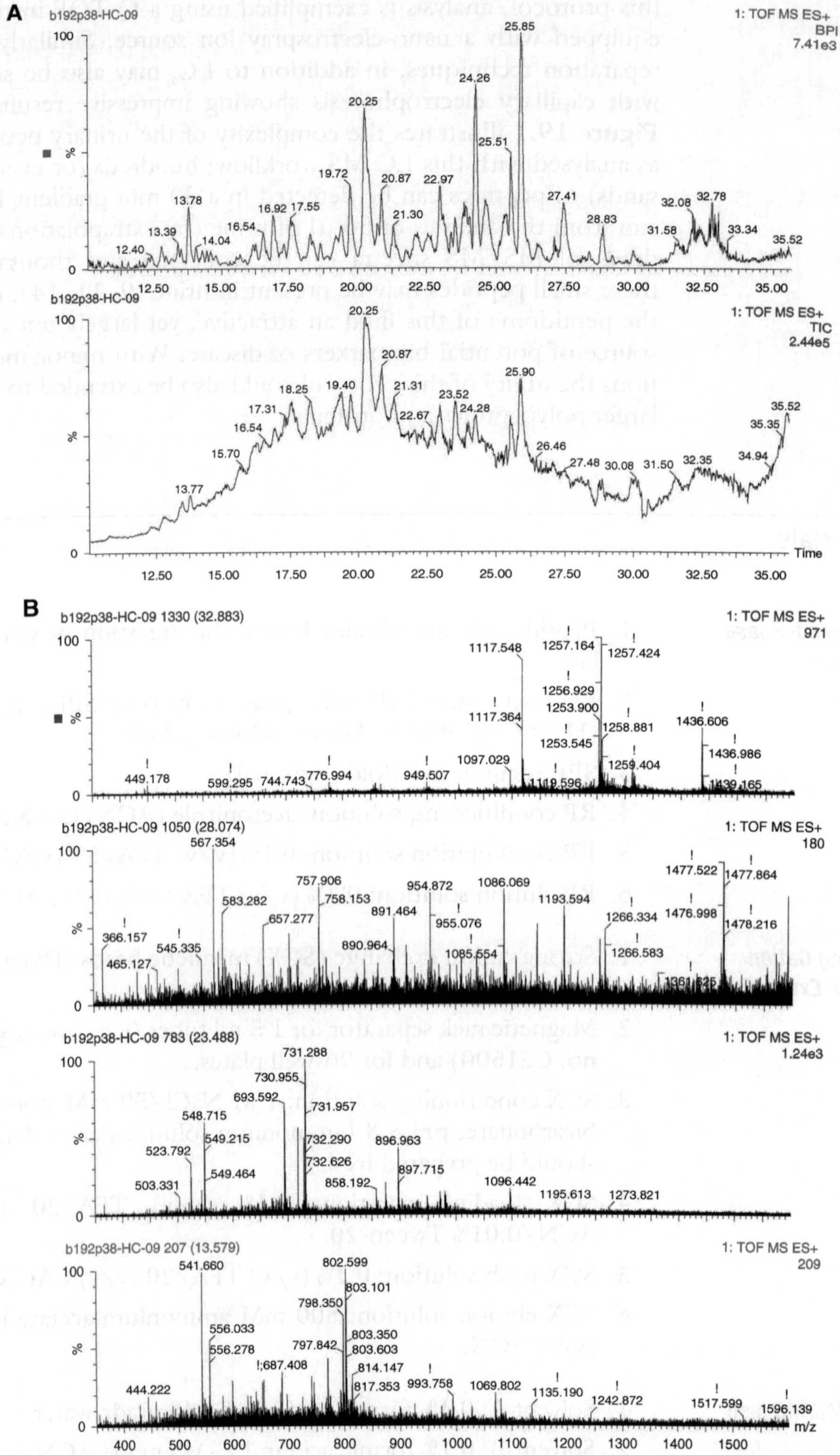

Fig. 19.1 Complexity of the urinary peptidome. (**A**) Base peak and total ion chromatograms of urine are shown in *top* and *bottom* panel respectively. (**B**) Four individual spectra (1 s acquisition) are shown at time points (from *top* to *bottom*) 32.9, 28.1, 23.5 and13.6 min.

4. Nanoflow analytical RP column (BEH130 C18, 100 μm × 100 mm, Waters).
5. LC-MS/MS system (this protocol will be exemplified with a nanoAcquity UPLC connected online with a Q-TOF Premier, Waters).

3. Methods

3.1. Urine Sample Preparation (see Note 3)

1. As soon as possible after collection, urine samples should be refrigerated to 0°C by placing them on ice.
2. Centrifuge specimens at 10,000×*g* for 10 min in a centrifuge refrigerated to 4°C. This step removes cells shedded by the urogenital track and other debris.
3. Specimens should be aliquoted into 1 ml aliquots and stored at −80°C until the time of analysis (optional).
4. On the day of analysis urine specimens should be defrosted at room temperature. Defrosted specimens should be stored at 0°–4°C until extraction is performed.
5. Add a known amount of peptide mixture to the samples. These will serve as internal standards for quantification. For human urine, 1 pmol of peptide derived from yeast enolase spiked into 1 ml urine can serve as a convenient internal standard peptide mixture.
6. Acidify specimens by adding 10 μl 10% TFA to 1 ml urine.
7. Centrifuge at 12,000×*g* for 10 min (*see* **Note 4**).

3.2. Extraction of Peptides by RP SPE (see Note 5)

1. Place RP SPE cartridges in a vacuum manifold (*see* **Note 6**).
2. Fill cartridges with RP conditioning solution. Apply vacuum to the manifold so that the flow rate of mobile phase through the cartridge is about 2 ml/min. The steps for SPE extraction are conditioning of the cartridge, equilibration, sample loading, washing and elution. In each case the solvent from the previous step should have completely passed through the column before adding the solvent of the next step (*see* **Note 7**).
3. Conditioning: apply a total of 10 ml of RP conditioning solution to the cartridge.
4. Equilibration: add 10 ml of RP equilibration solution.
5. Sample loading: add acidified urine specimens to the equilibrated SPE RP cartridges.
6. Washing: when specimens have passed through cartridges, wash the cartridge by adding 10 ml of RP equilibration

solution to the cartridges. This step removes hydrophilic salts loosely bound to the cartridge's packing material.

7. Elution: elute bound peptides by adding 1 ml of RP elution solution. Collect eluted peptides in a 1.5 ml tube.

3.3. SCX Extraction (see Note 8)

1. Use 5 μl of dynabeads® SCX per urine specimen (*see* **Note 9**).
2. Bead conditioning and washing can be done in bulk. For this purpose transfer the beads to a 1.5 ml protein low-bind Eppendorf tube. The magnetic beads are separated from the solvent using a magnetic rack for Eppendorf for 1 min, discarding the solvent afterwards (this applies to all step when separation is required).
3. Add 800 μl of SCX conditioning solution to the beads and incubate them for 10 min at room temperature (RT).
4. Wash the beads three times in 800 μl of SCX loading solution
5. Resuspend the beads in 1 ml of loading solution. At this point, the beads are ready to be used.
6. Transfer 10 μl of beads to the RP SPE sample from step above per well. Mix the contents thoroughly. Incubate for 45 min at RT shaking gently.
7. Apply to a magnetic rack until all SCX beads are collected on the wall of the tube (approximately 1–2 min) and discard the solvent by aspiration with a pipette or vacuum.
8. Wash the incubated beads twice with 100 μl SCX loading solution
9. Wash once with 100 μl SCX wash solution. This step removes residual Tween-20 from the solution in the last two washes (*see* **Note 10**).
10. Elute the beads with 20 μl of SCX elution solution.
11. Dry in a Speed-Vac to remove volatile components and store dry peptides at −20°C until the day of analysis by LC-MS/MS

3.4. LC-MS/MS

1. Prior to LC-MS/MS analysis, dried peptides from the SCX step should be resuspended in 20 μl 0.1% TFA.
2. Analyse 1 μl of extracted peptides by LC-MS and LC-MS/MS in data-dependent acquisition.
3. Separation of peptides can be accomplished by using gradient elution as follows:

Time	% solvent B (balance with solvent A)
0	1
30	35
31	80
40	80
40.1	1

4. Flow rate should be set at 400 nl/min (*see* **Note 11**).
5. The column should be equilibrated between runs at 1% B for at least 10 min.
6. Voltages at the tune page of the tandem mass spectrometer should be tuned for maximal sensitivity and resolution.
7. Data-dependent acquisition (DDA) experiments involve performing MS/MS on ions that meet predefined criteria. Typical settings are chosen to select multiply charged ions for MS/MS that produce at least 50 ion counts/s in a 0.5 s survey scan. This is then followed by three MS/MS scans of the three most intense peptide ions for 1 s each.

3.5. Data Analysis

Depending on the purpose of the study, qualitative or quantitative analyses (or both) may be performed. A qualitative analysis may just involve obtaining identities of as many peptides as possible using LC-MS/MS data; this is accomplished by matching MS/MS spectra to theoretical fragmentation of peptides in protein databases using search engines such as Mascot, Protein Prospector and Sequest. Relative quantification of peptides may be performed when the aim of the study is to compare polypeptide amounts across several samples. This workflow involves comparing peak areas or heights of extracted ion chromatograms of peptides across the samples to be evaluated. Absolute quantification of peptides in urine by LC-MS is also possible. This workflow ideally needs the construction of standard curves and isotopically labelled internal standards.

3.5.1. Qualitative Analysis (see Note 12)

1. Load LC-MS/MS raw data folders into Mascot Distiller in order to obtain peak lists of MS/MS spectra.
2. Load peak lists into Mascot. Settings should be chosen to match the performance of the mass spectrometer and the particulars of the experimental workflow and origin of the sample. Appropriate mass windows, database, modifications and enzyme should be chosen (*see* **Note 13**).

3. Submit searches and review data retuned by Mascot. A decoy database should be searched to estimate the false positive identification rate.

3.5.2. Quantitative Analysis (see Note 14)

Relative quantification may be performed by comparing peak areas or heights of extracted ion chromatograms of the peptide mass across samples (*see* also **Fig. 19.2**).

1. In the 'QuanLynx' function of MassLynx, create a new method by entering the m/z and retention time of the peptide(s) of interest as well as internal standard peptides (**Fig. 19.2A**).

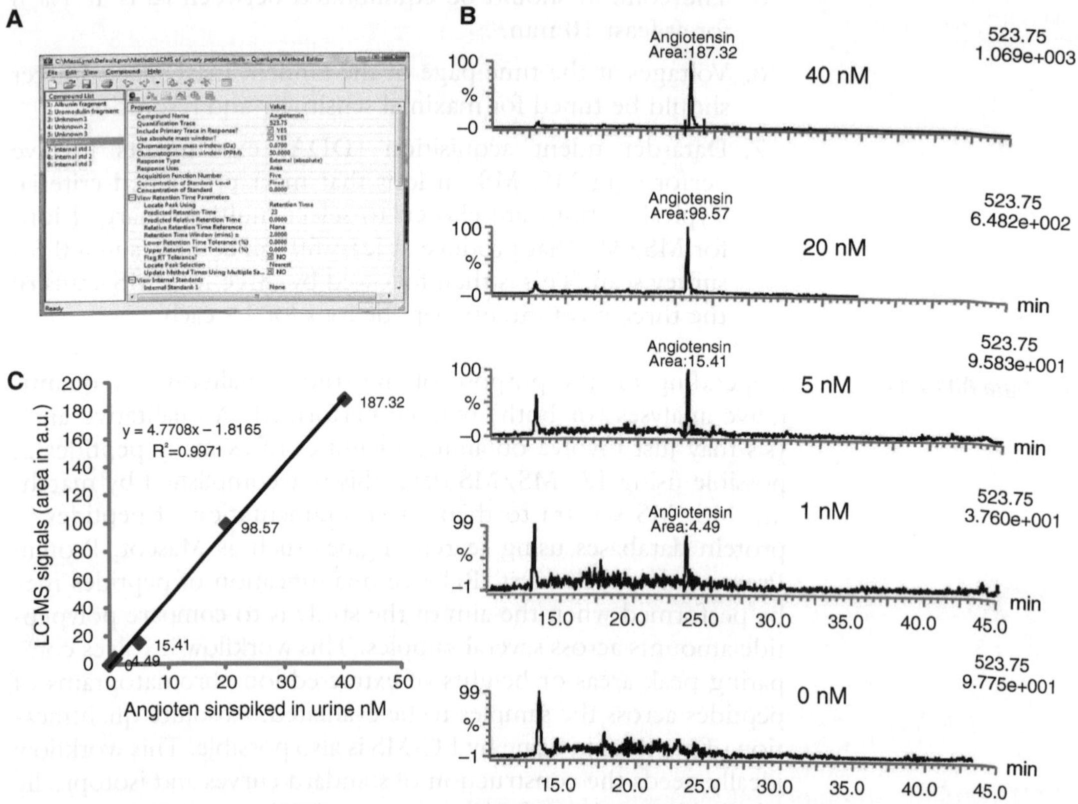

Fig. 19.2 Quantification of angiotensin in urine. To demonstrate quantification of a peptide in urine, control urine aliquots were spiked with increasing amounts of angiotensin ($m/z = 523.75$) and a standard curve constructed. Peptide m/z values were entered in QuanLynx **(A)** and used to obtain extracted ion chromatograms of the peptide ions of interest **(B)** whose intensities (areas or peak heights) can then be plotted to investigate linearity and dynamic range of quantification **(C)**.

2. Process samples by running this method across the samples to be compared.
3. Review XICs and correct peak integration as required (**Fig. 19.2B**).
4. Export area values of XICs into a spreadsheet file such as Excel.

5. Normalise these values to those of internal standard peptides (*see* **Note 15**).

4. Notes

1. All reagents should be prepared using LC-MS or HPLC grade solvents.
2. LC-MS/MS analysis is exemplified here using a Waters Q-TOF MS/MS instrument connected online with a nanoflow UPLC system (nanoAcquity). Any other instrumentation capable of performing MS/MS experiments and of LC separations at nanolitre per minute flow rates should also be adequate. Similarly, other C18 trap and analytical columns capable of accepting nanoflow per minute flow rates should also be adequate.
3. These sample preparation steps are designed to analyse small peptides. For the analysis of larger polypeptides and proteins, it is necessary to introduce a proteolytic step (e.g. using sequencing grade trypsin) prior to further sample preparation, so as to generate protein-derived peptides of a size compatible with the rest of the protocol. This optional proteolytic step may be performed between Steps 5 and 6 in **Section 3.1**.
4. This second centrifugation step may be omitted when samples are processed fresh without a freezing cycle.
5. The RP SPE step separates organic hydrophobic molecules, which includes peptides and proteins, from inorganic salts. This desalting step makes peptides ready for further extraction by SCX.
6. Manual application of mobile phases through SPE cartridges using syringes is an alternative to a vacuum manifold. However, by using a vacuum manifold the procedure is more efficient and several samples can be processed in parallel.
7. Cartridges should not be allowed to dry. Doing so may result in low recovery of peptides.
8. This SCX step separates peptides from hydrophobic organic acids, which dominate MS spectra of unprocessed urine and mask the presence of small peptides in urine.
9. This protocol uses magnetic SCX beads because of convenience of use. Other SCX material (i.e. silica based SCX beads) has been used in this laboratory with good results.
10. Tween-20 is a detergent that helps avoiding bead aggregation that may occur as a result of hydrophobic interactions

between protein-coated beads. Residual Tween-20 in samples may shorten the life of the column in the LC-MS/MS system. This final wash with a solvent that does not contain Tween-20 makes the assay more robust and more compatible with LC-MS.

11. The UPLC system is designed to operate at pressures of up to 10,000 psi and therefore flow rates higher than 400 nl/min are possible.

12. Here the workflow for the analysis of LC-MS/MS data is exemplified using the Mascot Search Engine (Matrix-Science) and other bioinformatics tools from Matrix-Science. Other search engines such as Sequest, Phenyx, Protein Prospector, and Protein Global Server should also be adequate.

13. The nature of the sample, the workflow employed and intrinsic instrument performance determine which search parameters one should use for the analysis of LC-MS/MS data. The database to be searched against should match the species from which urine was obtained. For human samples, the International Protein Index (IPI) Human database is recommended. The mass window tolerances should match typical mass accuracy achievable by the mass spectrometer. If using a Q-TOF instrument that has been internally calibrated, mass windows of 15 ppm and 50 mmu for parent and fragment ions are adequate. When analysing undigested urine, no enzyme should be specified. Variable modifications may include oxidised methionine and carbohydrate modifications, but no more than four variable modifications should be chosen in a single search. Several searches, each specifying a different set of modifications, are more advisable.

14. Several computer programs have been developed for quantifying by LC-MS using peak areas or heights. Some of them are based on alignment of ion chromatograms (15–17), while others directly quantify ions based on their elution profiles which are obtained by calculating the areas or peak heights of constructed extracted ion chromatograms (18, 19). Here, a quantification method is exemplified using the 'QuanLynx' function provided with the MassLynx software.

15. Peptide intensity levels may be further normalised to concentration of creatinine which serves as a surrogate measure of urine concentration.

References

1. Decramer, S., Gonzalez de Peredo, A., Breuil, B., Mischak, H., Monsarrat, B., Bascands, J. L., and Schanstra, J. P. (2008) Urine in clinical proteomics. *Mol. Cell. Proteomics.* 7, 1850–1862.
2. Thongboonkerd, V. (2008) Urinary proteomics: towards biomarker discovery, diagnostics and prognostics. *Mol. Biosyst.* **4**, 810–815.
3. Lam, T., and Nabi, G. (2007) Potential of urinary biomarkers in early bladder cancer diagnosis. *Expert. Rev. Anticancer Ther.* **7**, 1105–1115.
4. M'Koma, A. E., Blum, D. L., Norris, J. L., Koyama, T., Billheimer, D., Motley, S., Ghiassi, M., Ferdowsi, N., Bhowmick, I., Chang, S. S., Fowke, J. H., Caprioli, R. M., and Bhowmick, N. A. (2007) Detection of pre-neoplastic and neoplastic prostate disease by MALDI profiling of urine. *Biochem. Biophys. Res. Commun.* **353**, 829–834.
5. Perroud, B., Lee, J., Valkova, N., Dhirapong, A., Lin, P. Y., Fiehn, O., Kultz, D., and Weiss, R. H. (2006) Pathway analysis of kidney cancer using proteomics and metabolic profiling. *Mol. Cancer* **5**, 64.
6. Schiffer, E. (2007) Biomarkers for prostate cancer. *World J. Urol.* **25**, 557–562.
7. Celis, J. E., Wolf, H., and Ostergaard, M. (2000) Bladder squamous cell carcinoma biomarkers derived from proteomics. *Electrophoresis* **21**, 2115–2121.
8. Vilasi, A., Cutillas, P. R., Maher, A. D., Zirah, S. F., Capasso, G., Norden, A. W., Holmes, E., Nicholson, J. K., and Unwin, R. J. (2007) Combined proteomic and metabonomic studies in three genetic forms of the renal Fanconi syndrome. *Am. J. Physiol. Renal. Physiol.* **293**, F456–F467.
9. Cutillas, P. R., Chalkley, R. J., Hansen, K. C., Cramer, R., Norden, A. G., Waterfield, M. D., Burlingame, A. L., and Unwin, R. J. (2004) The urinary proteome in Fanconi syndrome implies specificity in the reabsorption of proteins by renal proximal tubule cells. *Am. J. Physiol. Renal. Physiol.* **287**, F353–F364.
10. Zimmerli, L. U., Schiffer, E., Zurbig, P., Good, D. M., Kellmann, M., Mouls, L., Pitt, A. R., Coon, J. J., Schmieder, R. E., Peter, K. H., Mischak, H., Kolch, W., Delles, C., and Dominiczak, A. F. (2008) Urinary proteomic biomarkers in coronary artery disease. *Mol. Cell. Proteomics.* **7**, 290–298.
11. Cutillas, P. R., Norden, A. G., Cramer, R., Burlingame, A. L., and Unwin, R. J. (2004) Urinary proteomics of renal Fanconi syndrome. *Contrib. Nephrol.* 141, 155–169.
12. Thongboonkerd, V., Chutipongtanate, S., and Kanlaya, R. (2006) Systematic evaluation of sample preparation methods for gel-based human urinary proteomics: quantity, quality, and variability. *J. Proteome Res.* 5, 183–191.
13. Mischak, H., Coon, J. J., Novak, J., Weissinger, E. M., Schanstra, J. P., and Dominiczak, A. F. (2008) Capillary electrophoresis-mass spectrometry as a powerful tool in biomarker discovery and clinical diagnosis: An update of recent developments. *Mass Spectrom. Rev.* **28**, 703–724.
14. Cutillas, P. R., Norden, A. G., Cramer, R., Burlingame, A. L., and Unwin, R. J. (2003) Detection and analysis of urinary peptides by on-line liquid chromatography and mass spectrometry: application to patients with renal Fanconi syndrome. *Clin. Sci. (Lond)* 104, 483–490.
15. America, A. H., and Cordewener, J. H. (2008) Comparative LC-MS: a landscape of peaks and valleys. *Proteomics* 8, 731–749.
16. America, A. H., Cordewener, J. H., van Geffen, M. H., Lommen, A., Vissers, J. P., Bino, R. J., and Hall, R. D. (2006) Alignment and statistical difference analysis of complex peptide data sets generated by multidimensional LC-MS. *Proteomics* 6, 641–653.
17. de Groot, J. C., Fiers, M. W., van Ham, R. C., and America, A. H. (2008) Post alignment clustering procedure for comparative quantitative proteomics LC-MS data. *Proteomics* 8, 32–36.
18. Cutillas, P. R., and Vanhaesebroeck, B. (2007) Quantitative profile of five murine core proteomes using label-free functional proteomics. *Mol. Cell. Proteomics.* 6, 1560–1573.
19. Park, S. K., Venable, J. D., Xu, T., and Yates, J. R., 3rd. (2008) A quantitative analysis software tool for mass spectrometry-based proteomics. *Nat. Methods* 5, 319–322.

Part VI

Novel Applications of LC-MS

Chapter 20

Quantification of Protein Kinase Activities by LC-MS

Maria P. Alcolea and Pedro R. Cutillas

Abstract

Measuring the enzymatic activity of protein kinases in cell and tissue extracts represents a difficult task owing to the complex regulation and dynamics of such enzymes. Here we describe a sensitive and specific approach for the quantitative analysis of PI3K-dependent protein kinase activity based on the mass spectrometry measurement of reaction products. The principle of this method can be applied to develop other kinase assays and thus should contribute to the understanding of processes controlled by protein kinases. Because of the enhanced sensitivity of this technique, it may be applied to the multiplex measurement of pathway activities when sample amounts are limiting.

Key words: PI3K signalling pathway, AKT/PKB, kinase enzymatic assay, strong cation exchange (SCX), kinase activity quantification, mass spectrometry, cancer.

1. Introduction

Protein kinases are enzymes of significant importance in biochemistry since they control the regulation of pathways involved in many different biological processes, including energy metabolism, cellular proliferation, differentiation, migration and cell death (1). All of these processes are known to be deregulated in cancer, making protein kinases drug targets of increasing interest in oncology (2).

Accurate and sensitive methods to quantify the activation of protein kinases are an essential requirement in preclinical research to evaluate the potential of developing drugs against specific kinases or pathways. These approaches are invaluable during the drug developmental process to monitor the extent to which lead compounds inhibit their target in cells and tissues (i.e.

P.R. Cutillas, J.F. Timms (eds.), *LC-MS/MS in Proteomics*, Methods in Molecular Biology 658, DOI 10.1007/978-1-60761-780-8_20, © Springer Science+Business Media, LLC 2010

to determine their pharmacodynamic properties). In addition, patient stratification is an important issue when designing clinical trials (3). Since the activity of signalling pathways in cancer is the result of genetic, epigenetic and proteomic alterations, a direct measure of oncogenic pathway activation may have a greater predictive power than proxy assays such as those based on DNA mutation analyses or on the concentration of the target in tumours, which may be poorly correlated with pathway activity (4).

Quantifying kinase activity is not an easy task: many kinases may be expressed in individual cells, each protein kinase has specificity for a particular subset of proteins, specific sites are phosphorylated by more than one kinase, and up to 30% of all cellular proteins may be phosphorylated at a certain point (5, 6).

Among the molecular pathways regulated by protein kinases, the phosphoinositide 3-kinase/Akt axis is one of the crucial nodes in growth factor and antigen signalling networks and is commonly found to be aberrantly activated in cancer and inflammation, thus constituting an established drug target in oncology and immunology (7, 8). The product of class I PI3K is phosphatidylinositol-3,4,5-trisphosphate (PIP3), a lipid second messenger that recruits several PH domain containing proteins to membranes, including protein kinase B (also known as Akt) and phosphoinositide-dependent kinase-1 (PDK1). PDK1 phosphorylates and activates Akt, which in turn phosphorylates numerous downstream effectors that control cell growth, proliferation and survival (9). The PI3K reaction is opposed by the tumour suppressor PTEN, one of the most frequently deactivated genes in cancer. Therefore the PI3K pathway can be altered at several levels in cancer cells, including mutations and amplifications of the PI3K catalytic subunit, or on receptor tyrosine kinase that lies upstream of PI3K, by loss of PTEN and by Akt overexpression or activating mutations (10). However, it is not yet clear whether or not this pathway is active in all cancer patients. Should this not be the case, identifying the patient subpopulation with overactive PI3K signalling may be of extreme importance for the success of therapies designed to target the PI3K/Akt axis.

Given the importance of monitoring PI3K activation, several methods for quantifying the activity of this pathway have been developed, the most common of them consisting of assessing the phosphorylation status of downstream effectors of PI3K by Western blotting and immunofluorescence using commercially available antibodies. Although useful for some experiments, these approaches are nonetheless limited by their low sensitivity, narrow dynamic range and their subjective and semiquantitative nature. In order to tackle this issue, our laboratory has developed a technique (Aktide assay) for the absolute quantitation of the PI3K/Akt signalling pathway activation based on the measurement of PI3K-dependent kinase enzymatic activity by mass spectrometry (11). This approach allows quantitative analysis of the PI3K/Akt axis with great sensitivity, precision and specificity, representing an advantage when compared with alternative methods based on radioactivity, immunochemistry or fluorescence (12, 13). In this chapter we will provide a thorough description of this new strategy, which allows for accurate measurement of the PI3K signalling pathway. Furthermore, this approach can in principle be applied to measure the activity of any kinase-driven signalling pathway for which specific peptide substrates are available, making this method suitable for multiplexing kinase measurements.

2. Materials

2.1. Cell Culture and Lysis

1. Human acute monocytic leukaemia cell line (P31/FUJ) as example of a cancer cell line showing PI3K/Akt signalling pathway activation. P31/FUJ cell line can be obtained from the Health Science Research Resources Bank, Japan (*see* **Note 1**).
2. Phosphate-buffered saline (PBS): Prepare 10X stock with 1.37 M NaCl, 27 mM KCl, 100 mM Na_2HPO_4, 18 mM KH_2PO_4 (adjust to pH 7.4 with HCl if necessary) and autoclave before storage at room temperature. Prepare working solution by dilution of one part with nine parts water (*see* **Note 2**).
3. Roswell Park Memorial Institute (RPMI) medium (Gibco/BRL, Bethesda, MD) supplemented with 10% foetal bovine serum (FBS, HyClone, Ogden, UT), 100 units/ml penicillin and 100 μg/ml streptomycin (Invitrogen, Carlsbad, CA).
4. Example of compound to be tested: Wortmannin (Calbiochem, San Diego, CA). Prepare a stock solution of 10 mM in DMSO and store small aliquots at −20°C (*see* **Note 3**).
5. Triton buffer for cell lysis: 50 mM Tris–HCl, pH 7.4, 150 mM NaCl, 1 mM EDTA, 1% (w/v) Triton X-100, 0.5 mM NaF, 1 mM Na_3VO_4, 0.05 TIU/ml aprotinin, 10 μM leupeptin, 0.7 μg/ml pepstatin A, 10 μg/ml TLCK, 1 mM DTT, 1 mM PMSF, 1 μM okadaic acid. Prepare the stock lysis buffer (50 mM Tris–HCl, pH 7.4, 150 mM NaCl, 1 mM EDTA, 1% (w/v) Triton X-100) and store in aliquots at −20°C. The day of the experiment supplement 1 ml of stock lysis buffer with 1 μl 0.5 N NaF, 10 μl 100 mM Na_3VO_4, 10 μl 5TIU/ml aprotinin, 1 μl 10 mM leupeptin, 1 μl 0.7 mg/ml pepstatin A, 1 μl 10 mg/ml TLCK, 1 μl

1 M DTT, 10 μl 100 mM PMSF, 2 μl 0.5 mM okadaic acid (*see* **Notes 4** and **5**).

6. Bio-Rad Protein Assay solution (Bio-Rad, Hercules, CA) (*see* **Note 6**).

2.2. Aktide Assay

1. 96-well plate (V-shaped bottom).
2. Reaction buffer: 20 mM Tris, pH 7.4, 1 mM Na_3VO_4, 25 mM β-glycerol-phosphate, 75 mM $MgCl_2$, 0.75 mM ATP.
3. Peptide used as substrate (Aktide) RPRAATF, stock solution 1 mM (*see* **Note 7**).
4. Internal standard (IS) peptide: RP*RAApTF, where P* is L-proline-$^{13}C_5$, ^{15}N, a 6 Da heavier version of L-proline and pT is phosphothreonine. Dissolve at 0.05 μM in 20% ACN, 0.1% trifluoroacetic acid.
5. Peptide to construct standard curves (phospho-Aktide) RPRAApTF (*see* **Note 8**).
6. Stopping solution: 20% ACN, 0.1% trifluoroacetic acid.

2.3. SCX Extraction

1. Dynabeads® SCX (Invitrogen, Carlsbad, CA).
2. 96-well PCR plates.
3. Magnetic racks for both Eppendorf and for 96-well plates (Invitrogen, Carlsbad, CA).
4. Conditioning solution: 1 M NaCl, 50 mM ammonium bicarbonate.
5. Loading solution: 25% ACN, 0.1% TFA, 0.01% Tween-20.
6. Washing solution: 25% ACN, 0.1% TFA.
7. Eluting solution: 150 mM ammonium bicarbonate, 5% ACN, 0.1% TFA.
8. Reconstituting solution: 0.1% trifluoroacetic acid.

2.4. MS Quantification of Enzymatic Reactions

1. When peptide separation is required, it may be performed by reverse-phase chromatography using a C18 column in a high-pressure liquid chromatography system. In our example we use a nanoflow ultrahigh-pressure liquid chromatograph (Acquity, Waters/Micromass) connected on line to a quadrupole time-of-flight (Q-TOF) mass spectrometer (Waters/Micromass UK Ltd, Manchester, UK). Example of chromatography column: BEH C18 100 μm × 100 mm column (Waters/Micromass UK Ltd, Manchester, UK).
2. LC-MS/MS mobile phases: solution A (0.1% FA in LC-MS grade water) and solution B (0.1% FA in LC-MS grade ACN).
3. Sample resuspension solution: 0.1% trifluoroacetic acid.

3. Methods

The current non-radioactive technique allows for the absolute quantification of the PI3K/Akt signalling pathway in cell lysates. This strategy uses a mass spectrometry-based approach to quantify the phosphorylation of RPRAATF peptide (Aktide) by a PI3K-dependent protein kinase activity. The Aktide represents a highly selective substrate of protein kinases downstream of PI3K including Akt/PKB and GSK. The concept of this MS application is to exploit enzymatic activity to amplify the signal of the enzyme itself.

It is important to highlight that in order to perform an accurate absolute measurement of the activity, the use of non-radioactive isotopes of the phospho-Aktide peptide is required. When only relative quantification is required, this isotope-labelled internal standard may be omitted. It is also imperative that the specificity of the assay is analysed by using inhibitors of the studied pathway, such as Wortmannin or LY294002 for the case example presented here.

3.1. Sample Preparation for Aktide Assay

1. Human acute myeloid cell line P31/FUJ is routinely cultured at 37°C in a humidified atmosphere of 5% CO_2 and grown in RPMI medium supplemented with 10% bovine serum, 100 units/ml penicillin and 100 μg/ml streptomycin. Cells are maintained from 50×10^4 to 100×10^4 cells/ml in 75 cm^2 flasks.
2. The experimental conditions require that the cells are seeded at 50×10^4 cells/ml 24 h before the enzymatic assay (*see* **Note 9**). For this purpose, cells are centrifugated at $200\times g$ for 5 min, washed twice with PBS and resuspended in fresh RPMI medium.
3. The day of the experiment we supplement the stock lysis buffer as indicated in Step 5 of **Section 2.1**. Please note that the lysis buffer must be prechilled at 4°C before using (*see* **Note 10**).
4. Cells are transferred to a 96-well plate where the cell lysis will take place. Each well must contain at least 1×10^4 cells in 200 μl of RPMI medium. It is important to plan a convenient plate layout according to the conditions to be tested, taking into account that they should ideally be in triplicate (*see* **Note 11**). Seed the plate accordingly and let it stand for 30 min in the cell culture incubator (37°C) (*see* **Note 12**).
5. Prepare serial dilutions of the compound to be tested, e.g. Wortmannin dissolved in DMSO at the following concentrations: 10 mM, 1 mM, 100 μM, 10 μM and 0 μM.

Then dilute them 10 times in cell culture medium to obtain the concentrations: 1 mM, 100 μM, 10 μM, 1 μM and 0 μM. Add 2 μl of these solutions to the wells containing 200 μl of cell suspension. The following final concentrations will be obtained: 10 μM, 1 μM, 100 and 10 nM and 0 nM (*see* **Note 13**). After treatment, cells must be incubated for at least 30 min with the inhibitor at 37°C.

6. Centrifuge the plate at 200×*g* for 5 min at 4°C. Discard the supernatant by aspiration and add 20 μl of lysis buffer. Mix thoroughly and leave the plate on ice for 30 min for the lysis to take place. Centrifuge the plate at 2500×*g* for 15 min at 4°C and transfer the lysate to a fresh plate leaving behind the pellet.
7. At this point the protein content of the cell lysates should be measured using standard proteins quantification procedures.
8. Keep the cell lysates on ice until performing the Aktide assay (*see* **Note 14**).

3.2. Aktide Assay

1. Prepare the reaction mix by adding 200 μl of reaction buffer to 150 μl of Aktide stock + 650 μl water in a test tube.
2. Distribute 5 μl of the reaction mix in the wells of a 96-well plate (V-shaped bottom) and start the assay by adding 2.5 μl of cell lysate (*see* **Note 15**). Mix the contents thoroughly by pipetting up and down five times. The final concentrations of each substrate are 100 μM ATP and 100 μM Aktide.
3. Incubate the reaction for 10 min at 37°C shaking gently. Stop the reaction by adding 100 μl of 0.05 μM IS peptide dissolved in stop solution (*see* **Note 16**).

3.3. Product Extraction by SCX in 96-Well Plate Format

1. Use 2.5 μl of Dynabeads® SCX (50% slurry) per reaction, for 100 reactions use 250 μl (*see* **Note 17**).
2. Conditioning and washing the beads can be done in bulk. For this purpose transfer the beads to a 1.5 ml protein low-bind Eppendorf tube. The magnetic beads are separated from the solvent using a magnetic rack for Eppendorf for 1 min, discarding the solvent afterwards (this applies to all step when separation is required).
3. Add 800 μl of conditioning solution (1 M NaCl, 50 mM ammonium bicarbonate, freshly prepared) to the beads and incubate them for 10 min at room temperature (RT) (*see* **Note 18**).
4. Wash the beads three times in 800 μl of loading solution (25% ACN, 0.1% TFA, 0.01% Tween-20) (*see* **Note 19**).

5. Resuspend the beads in 1 ml of loading solution. At this point, the beads are ready to be used.
6. Transfer the Aktide reactions to a 96-well PCR plate and dispense 10 μl of beads per well (*see* **Note 20**). Mix thoroughly the content of the wells to help the diffusion of the beads. Incubate 45 min at RT shaking gently.
7. Apply to a 96-well plate magnetic rack and discard the solvent.
8. Wash the beads twice with loading solution and once with washing solution (*see* **Note 21**).
9. Elute reaction products and IS from the beads by adding 20 μl of elution solution. Incubate for 10 min at room temperature shaking at 1000 rpm (*see* **Note 22**).
10. Transfer the eluents to a V-shaped 96-well plate. Dry samples in a speed vac for approximately 2 h or until dry and subsequently resuspended in 50 μl 0.1% TFA (LC-MS grade) (*see* **Notes 23 and 24**).

3.4. Quantification of the Enzymatic Reaction by Mass Spectrometry (MS)

1. Several different mass spectrometry techniques may be used to quantify the phospho-Aktide, product of the Aktide reaction. Multiple reaction monitoring (MRM) LC-MS/MS performed in triple quadruples should in principle provide the most specific and sensitive MS method for this analysis. MALDI-MS or ESI-MS (without chromatography or MS/MS) are not recommended because the presence of isobaric ions complicates the analysis based on mass only, but MALDI-MS/MS or ESI-MS/MS in which fragments are monitored after fragmentation by MS/MS may be suitable alternatives to LC-MS/MS. Here, we give an example of phospho-Aktide quantification by LC-MS/MS using a Q-TOF in pseudo-MRM mode. The method consists of monitoring the parent to daughter ion transition, i.e. m/z 449.7–400.7 for phospho-Aktide and m/z 453.7–403.7 for the IS phospho-Aktide (*see* **Note 25**).
2. Prior to LC-MS/MS analysis, the mass spectrometer should be tuned for maximal sensitivity. In order to tune the collision energy (CE) value, infuse synthetic phospho-Aktide and select the ion at m/z 449.7 (2+) for MS/MS. Gradually increase the CE potential and record the intensity of the fragment ion at m/z 400.7. Use the CE that gave a more intense fragment ion. In our case, the best CE is around 22 eV.
3. Perform the analysis of phospho-Aktide and IS by gradient elution at 600 nl/min flow rate using the following gradient (the balance solvent is A):

Time (min)	%B
0	2
0.1	2
5	25
5.1	80
8	80
8.1	2

The composition of mobile phases A and B is given in **Section 2.4**. These settings allow analysing about 100 samples/day/mass spectrometer.

4. Run a series of standards together with the samples; this will later be used to construct a standard curve in order to perform an absolute quantification of the product. For this purpose prepare serial dilutions of phospho-Aktide in 0.1% formic acid: 0.05, 0.025, 0.1, 0.5, 5, 50, 125 (μM) containing 0.1 μM of the IS peptide. Run these standards in parallel with the unknown samples (*see* **Note 26**).
5. Analyse 1 μl of unknown samples and the standard serial dilution samples by LC-MS/MS using the settings indicated above.
6. Phospho-Aktide quantification in the different samples is performed by measuring the area under the curve of the extracted ion chromatogram for the fragment ion at m/z 400.7 in both standards and unknown samples (**Fig. 20.1A**, *see* **Note 27**). The obtained areas must be normalised by the area of the phospho-Aktide internal standard fragment at m/z 403.7.
7. Results from the standards are used to generate a standard curve by plotting phospho-Aktide/IS signal in the y-axis and phospho-Aktide concentration in the x-axis. The resulting graph will allow to obtain a linear regression equation ($y = a\,x + b$, where a represents the slope of the curve, b is the y-intercept, y is the normalised area at m/z 400.7 and x is the micromolar concentration of phospho-Aktide) and the correlation coefficient (R^2) value provides a measure of the reliability of the linear relationship between the x and the y values (values close to 1 indicate excellent linear reliability).
8. By fitting the area values from unknown samples to the linear regression equation (**Fig. 20.1B**), the amount of phospho-Aktide generated by each individual reaction can be measured in absolute units. Finally, these results can be normalised by the different protein content of the cell lysate

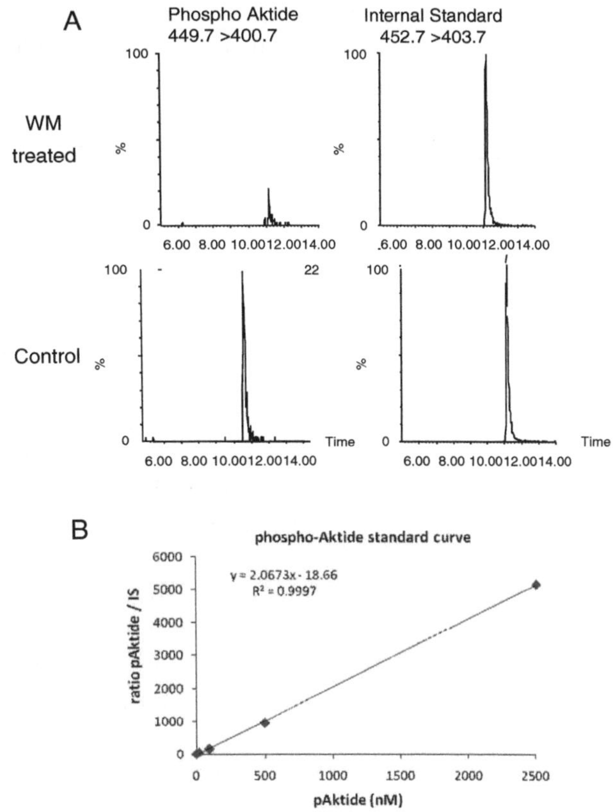

Fig. 20.1. Quantification of phospho-Aktide by LC-MS/MS. Products of the kinase reaction can be quantified by LC-MS/MS. This involves obtaining the signals of fragments at *m/z* 400.7 and 403.7 for the analyte and internal standard, respectively. (**A**) Examples of chromatograms of phospho-Aktide and internal standards obtained from the analysis of cells treated or not with Wortmannin (WM), a PI3-kinase inhibitor. (**B**) An example of standard curve constructed by plotting the intensities of increasing concentration of phospho-Aktide relative to a fixed amount of internal standard.

loaded in each well and incubation time. The following equation can be used to compute the Aktide kinase-specific activity in the studied samples.

$$\text{Specific Activity} = \frac{\text{Normalised Area} - \text{b}}{\text{a} \times 10^{6}} \times \frac{\text{Reaction volume}(\mu\text{l})}{\text{Reaction time (min)}} \times \frac{\text{Sample dilution}}{\text{Protein (mg/reaction)}} = \mu\text{mol/min/mg}$$

where normalised area is the area of m/z 400.7 divided by that at m/z 403.7, reaction time is the incubation time (10 min in this protocol), protein is the amount of protein in the reaction (protein concentration in mg/μl times the volume used in the reaction in μl) and sample dilution is any dilution applied to the sample (50 μl in this case). The activity is then reported in specific activity, i.e. enzyme units (μmol of phospho-Aktide produced per unit of time in minutes) per milligram of protein in the studied samples.

4. Notes

1. This assay can be performed with any cell line to be studied. Here we exemplify the approach by using the acute myeloid leukaemia cell line P31/FUJ.
2. Routine solutions should be prepared in water that has a resistivity of 18 MΩ-cm, unless otherwise stated. The resuspension solution used straight before running the samples in the LC-MS along with all the solvents used in the LC-MS system must be LC-MS grade. LC-MS grade solvents used here were provided by LGC Promochem (Middlesex, UK), but any other LC-MS grade solvents would be adequate.
3. Ideally one should test the specificity of the enzymatic assay; for this purpose a compound that inhibits directly or indirectly the kinase to be tested can be used. In our case, the use of Wortmannin, a specific covalent pan-inhibitor of phosphoinositide 3-kinases (class I, II and III PI3Ks) (14), inactivates the PI3K pathway and drastically reduces the amount of phospho-Aktide produced in the reaction. Wortmannin is a very potent PI3K inhibitor, even more so than LY294002, another commonly used PI3K pan-inhibitor (IC50 s 5 nM and 1.4 μM, respectively) (15, 16).
4. The lysis buffer used in this protocol is Triton X-100 based. Triton X-100 is a non-ionic detergent that intercalates in the lipid bilayers disrupting the cellular structure and acting as a lysing agent. This compound is very viscous at room temperature; hence it is easier to use after being gently warmed.
5. For this kind of experiment, in which we study protein kinase activities, it is essential that the lysis buffer contains different protease and phosphatase inhibitors such as Na_3VO_4 and NaF (tyrosine phosphatase inhibitors); aprotinin, PMSF and TLCK (serine protease inhibitors); leupeptin (thiol and serine proteases inhibitor); pepstatin A (acid proteases inhibitor) and okadaic acid (serine/threonine phosphatase inhibitor).
6. Any other reagent for protein quantification would be valid as well.
7. The peptide used in the assay RPRAATF peptide is a highly selective substrate of protein kinases downstream

PI3K (12, 17–19), with Akt/PKB being particularly active towards this peptide. However, other peptides whose specificity has been validated for a certain kinase could, in principle, be used for a similar kinase assay.

8. Peptides may be synthesised by any standard solid-phase method.
9. By seeding the cells at 50 × 10^4 cells/ml 24 h before the experiment, the cells will start proliferating actively switching on the PI3K/Akt axis, among other pathways. This fact certainly facilitates studying the differences in Akt kinase activity between the experimental conditions.
10. Unless otherwise indicated, everything should be done on ice in order to preserve the enzymatic activity to be studied and to reduce any phosphatase activity that may remain in the sample.
11. Plate layout according to the different conditions to be tested:

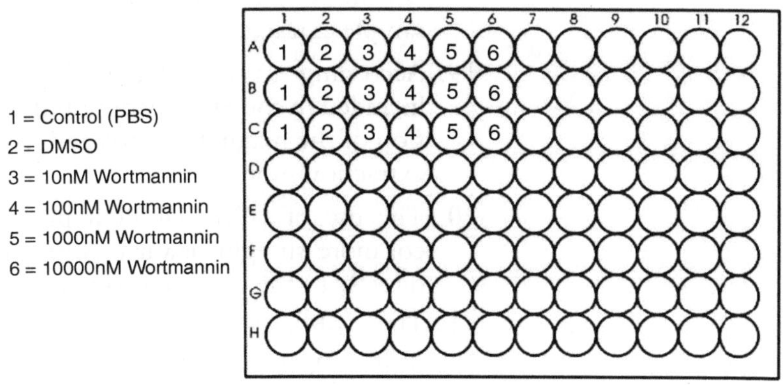

12. If the cell line to be studied does not grow in suspension, it is recommended to seed it directly in the 96-well plate 24 h before the experiment. This will allow the adhesion of the cells, which will afterwards be treated and lysed in the plate. In such cases, Steps 1–3 of Section 3.1 "Sample preparation for Aktide assay" section must be modified accordingly to the specific cell line requirements.
13. The use of DMSO cannot be avoided since it is required to solubilise the inhibitors used. However, because the DMSO is toxic for the cells, it is better to have it diluted at least 1000 times in the culture medium. This will be achieved by following the suggested strategy.
14. The impact of total cell lysate freezing/thawing cycles has not been fully investigated. For this reason we recommend to perform the enzymatic activity assay with fresh cell lysates.
15. The 96-well plate V-shaped bottom is more convenient for the Aktide assay due to the small volume in which the reac-

tion takes place (10 μl). In addition it will help to recover the whole content of the well.

16. After stopping the Aktide assay by acidifying the sample with 0.1% TFA, the plate can be snap frozen and kept at −80°C prior to desalting and LC-MS analysis.
17. In the Aktide assay an important excess of Aktide peptide is used; this causes overloading in the nano-HPLC column when the reactions are directly analysed by LC-MS. In order to avoid this problem, the phospho-Aktide is separated from the Aktide based on their different basicity (pKb). The different isoelectric point (pI) that these two compounds present (approximately 6.8 for phospho-Aktide and 12 for Aktide) makes it possible to effectively separate them by SCX. The SCX will also help to desalt the samples extending the life of the chromatographic column (20).
18. Conditioning the beads with a stringent solution liberates any unspecific binding and washes them before starting the sample extraction.
19. According to manufacturer's instructions Tween-20 is included in the SCX loading buffer to favour the resuspension of the magnetic beads and hence to make the product extraction more efficient.
20. The use of a 96-well PCR plate makes the SCX protocol more straightforward. A magnetic rack can be applied directly to the plate to easily wash the beads.
21. The washing solution does not contain Tween-20 to avoid its incompatibility with LC-MS.
22. Because ammonium bicarbonate solutions are volatile they must always be used fresh.
23. For LC-MS analysis peptides can be resuspended in 0.1% TFA, an ion pair reagent for reverse-phase liquid chromatography.
24. If required, cell lysates can be frozen and stored at −20°C before the LC-MS analysis.
25. The signal of the labelled internal standard (m/z 403.7 phospho-Aktide*) will be used both for normalising the normal experimental variability between the different samples and for the correct assignment of the phospho-Aktide peak in the LC-MS spectrum. The latter is possible because the mass to charge difference between the labelled and the unlabelled phospho-Aktide is $\Delta m/z$=3 (charge = +2).
26. Please note that the standard range of concentrations may need to be adjusted according to the signal from specific samples.

27. For quantification, the m/z for phospho-Aktide must be extracted from the total chromatogram. The result will be the extracted ion chromatogram (XIC), the elution profile of which will provide us with the area proportional to the studied peptide.

References

1. Manning, G., Plowman, G. D., Hunter, T., and Sudarsanam, S. (2002) Evolution of protein kinase signaling from yeast to man. *Trends Biochem Sci* **27**, 514–520.
2. Cohen, P. (2002) Protein kinases–the major drug targets of the twenty-first century?, *Nat Rev Drug Discov* **1**, 309–315.
3. Sawyers, C. L. (2008) The cancer biomarker problem. *Nature* **452**, 548–552.
4. Haber, D. A., and Settleman, J. (2007) Cancer: drivers and passengers. *Nature* **446**, 145–146.
5. Johnson, S. A., and Hunter, T. (2005) Kinomics: methods for deciphering the kinome. *Nat Methods* **2**, 17–25.
6. Turk, B. E. (2008) Understanding and exploiting substrate recognition by protein kinases. *Curr Opin Chem Biol* **12**, 4–10.
7. Martelli, A. M., Tazzari, P. L., Evangelisti, C., Chiarini, F., Blalock, W. L., Billi, A. M., Manzoli, L., McCubrey, J. A., and Cocco, L. (2007) Targeting the phosphatidylinositol 3-kinase/Akt/mammalian target of rapamycin module for acute myelogenous leukemia therapy: from bench to bedside. *Curr Med Chem* **14**, 2009–2023.
8. Fasolo, A., and Sessa, C. (2008) mTOR inhibitors in the treatment of cancer. *Expert Opin. Investig. Drugs* **17**, 1717–1734.
9. Vanhaesebroeck, B., Leevers, S. J., Ahmadi, K., Timms, J., Katso, R., Driscoll, P. C., Woscholski, R., Parker, P. J., and Waterfield, M. D. (2001) Synthesis and function of 3-phosphorylated inositol lipids. *Annu. Rev. Biochem.* **70**, 535–602.
10. Yuan, T. L., and Cantley, L. C. (2008) PI3K pathway alterations in cancer: variations on a theme. *Oncogene* **27**, 5497–5510.
11. Cutillas, P. R., Khwaja, A., Graupera, M., Pearce, W., Gharbi, S., Waterfield, M., and Vanhaesebroeck, B. (2006) Ultrasensitive and absolute quantification of the phosphoinositide 3-kinase/Akt signal transduction pathway by mass spectrometry. *Proc. Natl. Acad. Sci. USA* **103**, 8959–8964.
12. Bozinovski, S., Cristiano, B. E., Marmy-Conus, N., and Pearson, R. B. (2002) The synthetic peptide RPRAATF allows specific assay of Akt activity in cell lysates. *Anal. Biochem.* **305**, 32–39.
13. Shults, M. D., Janes, K. A., Lauffenburger, D. A., and Imperiali, B. (2005) A multiplexed homogeneous fluorescence-based assay for protein kinase activity in cell lysates. *Nat. Methods* **2**, 277–283.
14. Hazeki, O., Hazeki, K., Katada, T., and Ui, M. (1996) Inhibitory effect of wortmannin on phosphatidylinositol 3-kinase-mediated cellular events. *J. Lipid Mediat. Cell Signal* **14**, 259–261.
15. Vlahos, C. J., Matter, W. F., Hui, K. Y., and Brown, R. F. (1994) A specific inhibitor of phosphatidylinositol 3-kinase, 2-(4-morpholinyl)-8-phenyl-4H-1-benzopyran-4-one (LY294002). *J. Biol. Chem.* **269**, 5241–5248.
16. Powis, G., Bonjouklian, R., Berggren, M. M., Gallegos, A., Abraham, R., Ashendel, C., Zalkow, L., Matter, W. F., Dodge, J., Grindey, G., and et al. (1994) Wortmannin, a potent and selective inhibitor of phosphatidylinositol-3-kinase. *Cancer Res.* **54**, 2419–2423.
17. Alessi, D. R., Caudwell, F. B., Andjelkovic, M., Hemmings, B. A., and Cohen, P. (1996) Molecular basis for the substrate specificity of protein kinase B; comparison with MAPKAP kinase-1 and p70 S6 kinase. *FEBS Lett.* **399**, 333–338.
18. Kobayashi, T., Deak, M., Morrice, N., and Cohen, P. (1999) Characterization of the structure and regulation of two novel isoforms of serum- and glucocorticoid-induced protein kinase. *Biochem. J.* **344** *Pt* 1, 189–197.
19. Park, J., Leong, M. L., Buse, P., Maiyar, A. C., Firestone, G. L., and Hemmings, B. A. (1999) Serum and glucocorticoid-inducible kinase (SGK) is a target of the PI 3-kinase-stimulated signaling pathway. *Embo. J.* **18**, 3024–3033.
20. Villen, J., and Gygi, S. P. (2008) The SCX/IMAC enrichment approach for global phosphorylation analysis by mass spectrometry. *Nat. Protoc.* **3**, 1630–1638.

Chapter 21

A Protocol for Top-Down Proteomics Using HPLC and ETD/PTR-MS

Sarah R. Hart

Abstract

Analysis of intact proteins by tandem mass spectrometry has mostly been confined to high-end mass spectrometry platforms. This protocol describes the application of routine HPLC to separate proteins, MALDI-ToF mass spectrometry to interrogate intact protein species and electron transfer dissociation/proton transfer reaction within a quadrupole ion trap to perform tandem mass spectrometry.

Key words: Mass spectrometry, intact protein, electron transfer dissociation, proton transfer reaction high-performance liquid chromatography.

1. Introduction

Mass spectrometric characterisation of proteomic samples has matured over the last decade to the point where identification of protein components within biological samples of moderate complexity is now routine (1). The overwhelming majority of proteomic studies rely on a few basic principles, reviewed extensively elsewhere (2–4). Briefly, protein-containing samples are subjected to either one- or two-dimensional polyacrylamide gel electrophoresis and stained proteins excised and proteolysed (5). Alternatively in 'shotgun' proteomics, crude extracts are proteolysed and the peptides separated, often by ion exchange chromatography (IEX) (6). Both gel and shotgun approaches converge upon reversed-phase liquid chromatography of peptides,

P.R. Cutillas, J.F. Timms (eds.), *LC-MS/MS in Proteomics*, Methods in Molecular Biology 658,
DOI 10.1007/978-1-60761-780-8_21, © Springer Science+Business Media, LLC 2010

coupled online to an electrospray tandem mass spectrometer (7), or spotting on the target plate of a MALDI-ToF/ToF instrument (8). Whilst such approaches have clear advantages for high-throughput, quantitative, automated analyses, and in terms of the suitability of polypeptide analytes for collision-induced dissociation (CID) to generate meaningful product ion spectra, this strategy has some disadvantages. First, distinction of isoforms and closely related proteins is severely impeded by peptide-centric strategies, where the inherent limitations of observed sequence coverage (size of tryptic peptides being the most significant) confound distinction of highly homologous or polymorphic proteins (9). A further confounding possibility is the presence of post-translationally modified protein variants (10). Whilst targeted methods have been extremely effective in the generation of residue-specific information pertaining to the sites of particular modifications (11–13), the proteolysis step, necessary to generate (mostly unmodified) peptides, results in loss of connectivity between specific protein molecules and modification status. Given that post-translational modifications such as phosphorylation are known to strongly influence other modifications to the same protein molecule, quantitative understanding of every modification present upon multiply modified proteins is important in elucidating the modification-induced dynamic changes to structure and function. Some classes of proteins may also prove difficult to analyse using conventional shotgun proteomics; for example small, highly basic proteins such as histones are notoriously difficult to analyse, even without considering the additional complication of multiple post-translational modifications (14).

Mass spectrometric analyses of intact proteins were detailed within the earliest reports of the 'soft' ionisation techniques (15, 16), although the popularity of such methods increased dramatically with the commercial availability of high-resolution FTMS instrumentation in the late 1990s (10). Development of electron capture dissociation (ECD) enabled ready characterisation of large analytes with many degrees of freedom (17) which are poorly characterised by CID and other ion heating methods due to vibrational dissipation of collision energy (18). Most commercial ECD-enabled instruments are FT-ICR platforms, which are extremely costly. More recently, electron transfer dissociation (ETD), where the ion/electron reaction of ECD is substituted for an ion/ion reaction between radical anions, usually derived from fluoranthene, and peptide cations, has enabled the generation of ECD-like products, primarily c and z ions, from highly charged polypeptide precursors on a variety of mass spectrometry platforms (19–22). ETD of large poly-protonated species generates highly protonated product ions, the charge state and hence identity of which can be difficult to determine on the low-resolution quadrupole ion trap instrumentation to which this fragmentation technology is commonly coupled. A secondary ion/ion reaction of the products of ETD fragmentation, which can be achieved using basic fluoranthene anions, proton transfer reaction (PTR), is used to reduce the charge state of these product ions (20, 23) and thus improve their identification.

This chapter does not aim to provide a one-fits-all solution for top-down protein analysis, since the analysis of intact proteins remains difficult. Optimal results therefore require case-by-case optimisation of the analytical conditions. Moreover, the range of equipment which may be used for such analyses is now of sufficient breadth to merit a review in itself (10). Instead, this protocol chapter attempts to exemplify some methods which can be adapted for use with a range of biological systems, using the low-to-medium resolution mass spectrometry instrumentation of the type available within many laboratories. An awareness of the limitations associated with low-resolution mass spectrometry and the caveats therefore associated with data generated via such methods is essential with such an approach.

2. Materials

1. Proteins were purchased from Sigma-Aldrich (Dorset, UK).
2. Plasma samples were obtained with consent from healthy adult volunteers according to a standard operating protocol, collected into heparin-coated tubes, centrifuged at 4°C to remove cellular material and stored at −80°C until use.
3. *HPLC Columns*: Ion exchange columns: ProSwift WCX-1S 1 mm × 50 mm, ProSwift SAX 1S 1 × 50 mm (Dionex, Camberley, Surrey, UK); *see* **Note 2**. Reversed-phase column: ProSwift RP 4 mm × 2 mm × 250 mm (Dionex).
4. *Equipment*: Chromatographic separation was performed using an UltiMate 3000 capillary pump with UV detector (Dionex), equipped with an external switching valve (Rheodyne MX Series II, Presearch, Hitchin, Hertfordshire), operated from Chromeleon software (Dionex) via a serial connection. A Probot liquid handling system (Dionex) was used as a fraction collector. Vacuum centrifugation was performed using an Ez-2Plus (GeneVac, Ipswich, Suffolk, UK). Microcentrifuge tubes (500 μl and 1.5 ml) were supplied by Starlab (Milton Keynes, Buckinghamshire, UK) and 96-well plates from Abgene (Epsom, Surrey, UK).
5. *Mass spectrometry* was performed using an Ultraflex II MALDI-ToF (Bruker Daltonics, Coventry, UK), a QTof Ultima Global (Waters Micromass MS Technologies,

Wythenshawe, Manchester, UK) and an HCT PTM Discovery quadrupole ion trap (Bruker Daltonics) coupled to a NanoMate Triversa nanoelectrospray (nESI) robot (Advion Biosystems, Ithaca, NY, USA).

3. Methods

3.1. Two-Dimensional Protein Separation by High-Performance Liquid Chromatography

The inherent variation in the chemical properties of proteins, e.g. charge, polarity and hydrophobicity, provides ready means to fractionate the proteome. Separation of protein components is thus readily achieved using high-performance liquid chromatography methods. Some biological samples have additional considerations, for instance the investigation of serum and plasma samples is complicated by the presence of a relatively small number of very high-concentration protein components, e.g. albumin and immunoglobulins (24). Such components can significantly confound the observation of protein species present at lower concentrations. Hence an initial step to remove these abundant components, frequently by the use of immobilised antibodies, is common (25).

Ion exchange (IEX) methods are used to separate proteins on the basis of their solution-phase charge (26). Here, we employ anion and cation exchange columns together to achieve separation of proteins, both bearing net basic and acidic charge. This is followed by monolithic reversed-phase separation of IEX fractions.

1. Proteins to be separated are dissolved in a suitable volume (200–500 μl, depending upon solubility) of an appropriate low-salt solvent (e.g. Buffer A: 100 mM KH_2PO_4 in HPLC grade water; *see* **Note 1**). Biological fluids are best prepared via buffer exchange into Buffer A using a molecular weight cut-off filter within a chilled centrifuge (10–30 kDa nominal MWCO, e.g. VivaSpin 15 (Generon, Maidenhead, Berkshire)).
2. Sample is injected onto the ion exchanger using an appropriate volume loop on an HPLC switching valve.
3. Gradient elution is used to displace the proteins from the column bed at a typical flow rate of 100 μl/min, over a gradient from 100% buffer A (low salt) to 45% B (same components as A but supplemented with 1 M KCl) over 20–25 min, monitoring UV absorbance at 214 and 280 nm. A typical UV chromatograph of a plasma protein mixture separated using mixed-bed ion exchange chromatography is illustrated in **Fig 21.1A**.

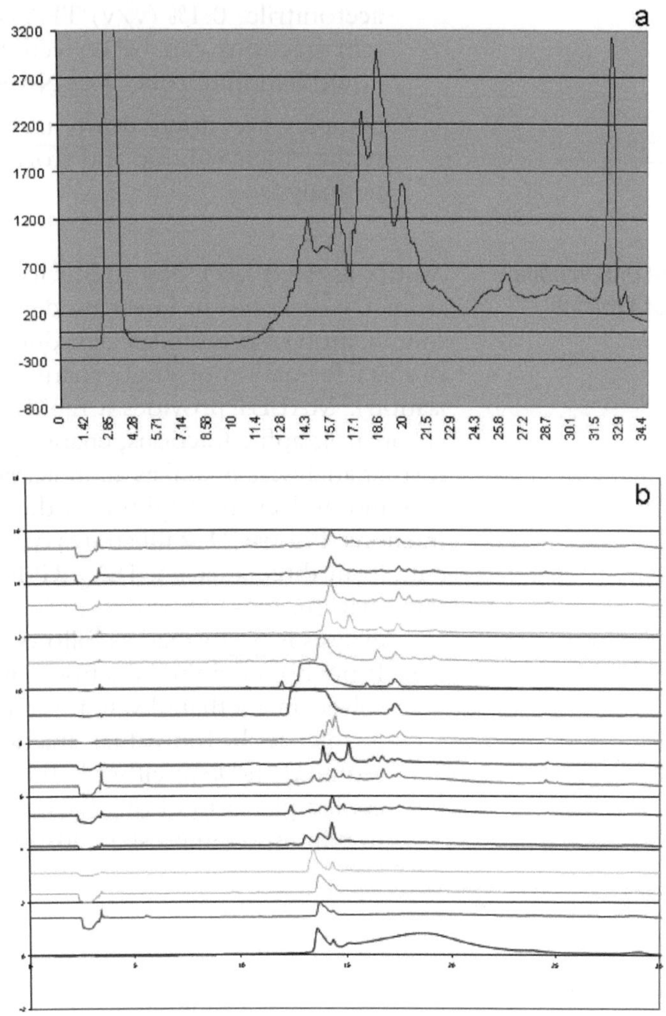

Fig. 21.1. Ion exchange/reversed-phase chromatographs of complex mixtures. Proteins from an immunodepleted plasma sample were separated by two dimensions of chromatography; IEX (**A**) separates proteins on the basis of charge. Fractions from the first separation were further subjected to RP separation on the basis of hydrophobicity (**B**; chromatographs normalised to a scale of 1 and stacked).

4. Eluted proteins are collected into clean microcentrifuge tubes. Depending upon peak profile, between 8 and 20 first-dimension samples of 30–90 s can be collected. Fractions from the first dimension can be injected directly onto the second, reversed-phase column.
5. IEX fractions are injected onto the reversed-phase column using a switching valve. The column is washed for 5 min with low organic solvent (2% (v/v) acetonitrile, 0.1% (v/v) TFA or formic acid in water) to remove salt and urea.
6. Gradient elution of proteins is performed at 100 μl/min using a linear gradient from 20 to 100% B (90% (v/v)

acetonitrile, 0.1% (v/v) TFA or formic acid in water). Protein fractions can be collected into 96-well plates using a liquid handling robot (*see* **Note 3**).

7. Samples are dried down using vacuum centrifugation to reduce their volume and stored at −20°C if necessary prior to analysis.

3.2. MALDI Screening of Intact Proteins

Matrix-assisted laser desorption ionisation (MALDI) mass spectrometry operated in linear mode is an effective method for the examination of protein species, due to its high sensitivity and the primary formation of singly charged species (vs. electrospray ionisation). MALDI provides a rapid screening tool for analysis of chromatographic fractions, enabling prioritisation of fractions for further analyses, as well as an indication of the complexity of samples and molecular weight(s) of the protein(s) present within the fractions. **Figure 21.2** illustrates typical linear mode MALDI-ToF data from three adjacent IEX/RP fractions from a complex protein mixture.

Analysis is performed as follows:

1. Protein-containing fractions in the 96-well plate are dissolved in methanol:water 1:1 (v/v) containing 0.1% (v/v) formic acid. Individual fractions can be resuspended as required and kept either in the refrigerator or on the Peltier-cooled platform of the Triversa maintained at 8°C, in order to minimise unnecessary adsorptive protein losses.

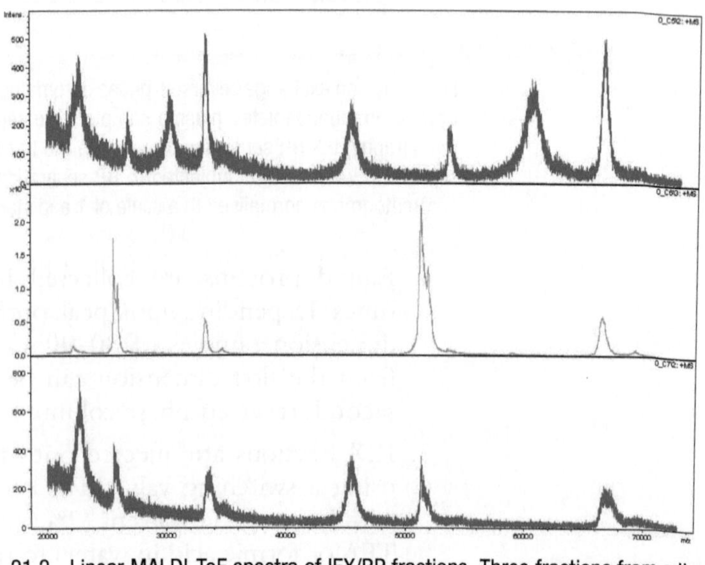

Fig. 21.2. Linear MALDI-ToF spectra of IEX/RP fractions. Three fractions from adjacent RP fractions were subjected to MALDI-ToF; strong differences in the spectra indicate the effective resolution of protein components.

2. Samples are mixed on-target 1:1 (v/v) with a saturated solution of sinapinic acid (50 g/l in 50% (v/v) acetonitrile, 0.1 (v/v)% TFA in water) and allowed to dry prior to analysis.
3. The Ultraflex II MALDI-ToF/ToF instrument is operated in linear mode to improve sensitivity for intact protein analysis. Instrument calibration and tuning were achieved in the spectra presented (**Fig. 21.2**) using Bruker Protein Calibration Standard 2, which enables calibration over the range 22–66 kDa, accumulating 2000–3000 laser shots into each summed spectrum, depending upon signal intensity.

3.3. Electrospray Analysis of Proteins and Electron Transfer Dissociation (ETD)/Proton Transfer Reaction (PTR) of Polypeptide Species

Electrospray ionisation of intact proteins typically yields highly charged ion species, with the distribution between different charge states providing an indication of condensed-phase protein charge (27).

Data presented in **Figs. 21.3** and **21.4** were generated as follows:

1. Preliminary screening of protein molecular weight and observed charge states is performed by offline nanoelectrospray (nESI) via static infusion from silica tips, using the QTof Global, using MaxEnt deconvolution to calculate intact masses.
2. The 96-well plate is placed upon the Peltier-cooled stage of the Triversa Nanomate, 10 μl sample picked up and introduced to a 400 tip nESI array chip. Typical operating voltages of 1.5–1.8 kV are used, with nESI current being monitored using ChipSoft (Advion).
3. nESI was collected using the HCT ion trap. MS was performed upon a suitable protein standard to establish appropriate sensitivity and basic tune parameters, for experimental samples fine tuning of settings is usually necessary. Optimal trap drive, skimmer and octopole voltages vary substantially with analyte.
4. ETD/PTR product ion spectra are generated: ion charge control (ICC) values for precursor ion (usually 200,000), ETD ions (80–100,000) and PTR ions (100–120,000) will also need to be adjusted per sample. Reaction time for ETD is also varied with substrate (typical values 10–25 ms). Regular monitoring of the spray current and intensity of product ions is necessary, particularly if samples are highly concentrated. Extended acquisition times (~15 min) can aid in improving the signal-to-noise of the product ion peaks, particularly for larger precursors, the low flow rate of the Triversa nESI robot aids in maintaining extended spray times with limited sample volumes.
5. Proteins can be subjected to reduction immediately prior to this step and either diluted or subjected to RP-SPE. In

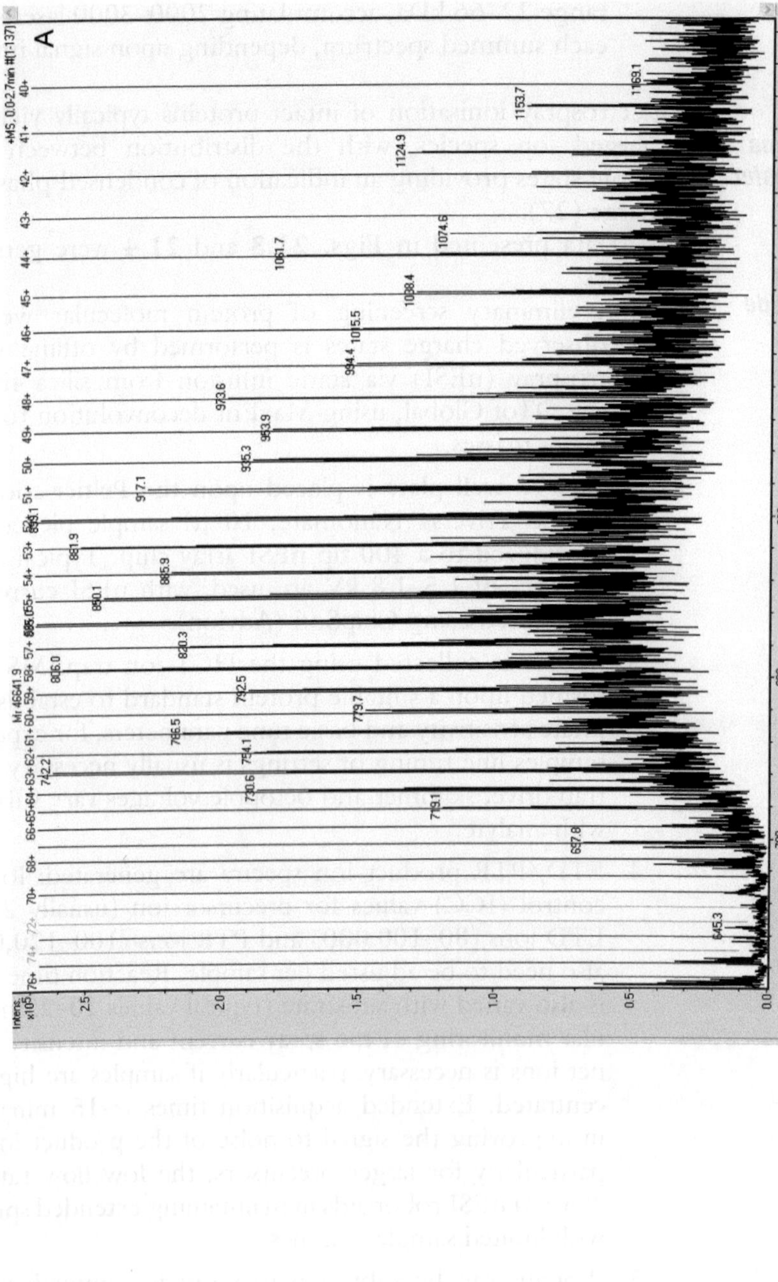

Fig. 21.3. ETD/PTR with database searching to infer identity. ETD/PTR data generated from intact enolase (*Saccharomyces cerevisae*) **(A)** intact protein with simple charge state deconvolution;

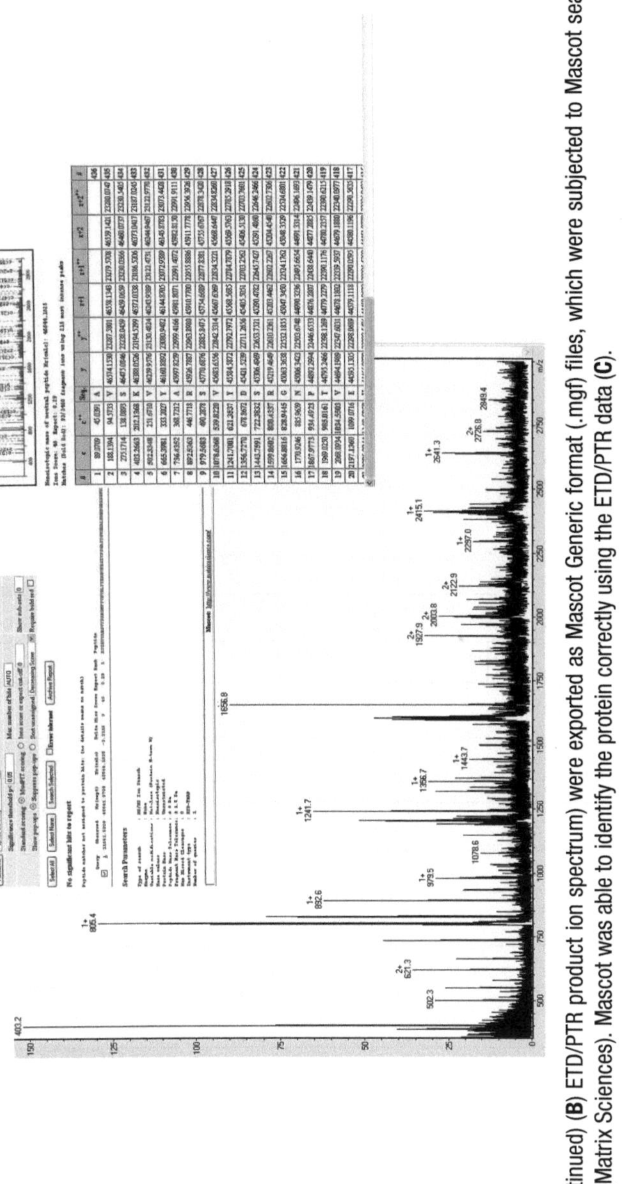

Fig. 21.3. (continued) (**B**) ETD/PTR product ion spectrum) were exported as Mascot Generic format (.mgf) files, which were subjected to Mascot searching using a six-processor Mascot cluster (Matrix Sciences). Mascot was able to identify the protein correctly using the ETD/PTR data (**C**).

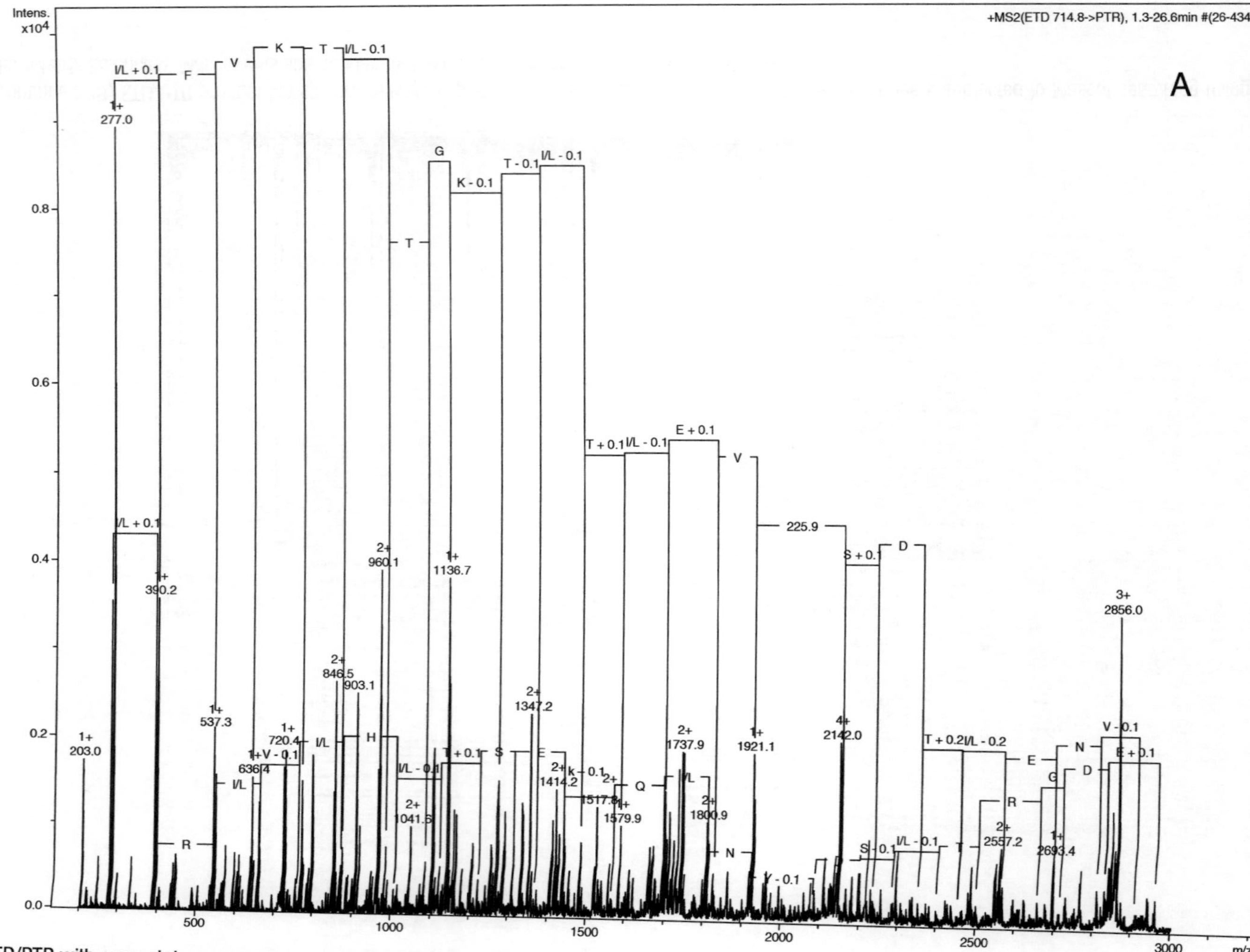

Fig. 21.4. ETD/PTR with manual de novo sequence interpretation. ETD/PTR data generated from intact bovine ubiquitin (**A**) intact mass data.

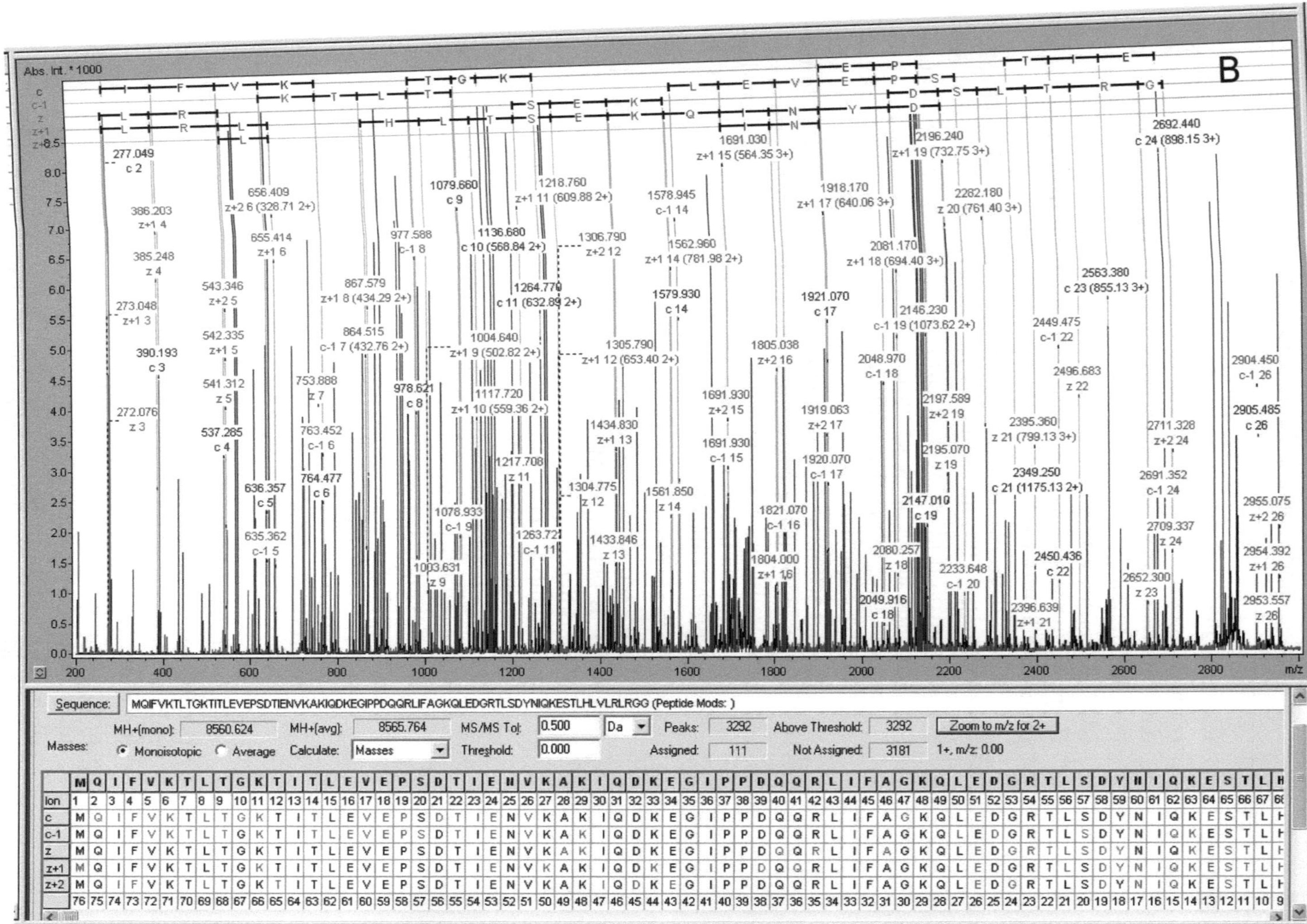

Fig. 21.4. (continued) (**B**) ETD/PTR product ion spectrum) were examined for differences between adjacent peaks using the Annotate tool in Data Analysis (Bruker). Two strong sets of masses corresponding to amino acid residue masses were observed, and sequence tags thus derived were subjected to manual BLAST searching using the NBCI BLAST server. Sequences identified as having high sequence identity were then downloaded and compared to the observed product ion spectrum. Seventy-eight percent top-down sequence coverage was observed for this 8.6 kDa protein.

ECD, disulphide bonds are known to strongly influence fragmentation (28); therefore spectra taken before and after such treatment can provide confirmation of the presence of cystines and sequence localisation (29).

3.4. Data Analysis

Rapid analysis of data generated using bottom-up methods is readily achieved using automated algorithms (30, 31). Top-down data analysis tools are currently rather more nascent and low-throughput, with ProSight PTM being the foremost method (32). Recently, a version of Mascot lacking an upper mass limit for polypeptide size has been released, enabling preliminary indication of identity for low-resolution data.

1. ETD data are processed using Data Analysis (Bruker Daltonics). Product ion data are exported as a .mgf file. Full deconvolution of low-resolution ETD spectra remains cumbersome at the time of writing. Mascot searching provides an effective method of rapidly matching product ion spectra against sequence data where they exist (*see* **Fig. 21.3**). The correct sequence or a very close homologue must be in the database to enable correct matching, and information about post-translational modifications (cleavage of signal peptides, covalent modifications) must also be known and included as variable modifications in the database search for this strategy to be effective.
2. Manual sequence tags can be generated from contiguous ion series as per standard de novo sequencing methods (33). Mass differences between product ion peaks are compared with amino acid residue masses to suggest peptide sequence. This information can then be fed into a BLAST search (34), allowing for isobaric amino acids (leucine/isoleucine) and masses which cannot be distinguished on low-resolution instruments (lysine/glutamine). Theoretical product ion spectra from matched protein sequence hits can then be compared to experimental mass spectrometric data to enable additional sequence ions to be verified, for instance using BioTools (*see* **Fig. 21.4**).

3.5. Conclusions

Top-down protein sequencing is by no means routine in comparison to current best practise in bottom-up methods. However, recent advances in ion activation methods and database searching routines are now beginning to catch up with existing separation methods, to the extent that low-throughput analysis of proteins from mixtures on inexpensive instrumentation is now feasible. The impact of higher-resolution instrumentation, such as hybrid orbitrap and QqToF instruments enabled with ETD capability, and hence not requiring proton transfer to reduce product ion charge state will undoubtedly prove extremely useful in top-down analyses.

Although methods for top-down proteomics continue to be developed, comparatively few studies feature analysis of large intact proteins, partly due to difficulties in gaining adequate precursor and product ion intensity from large, thermally labile analytes with polydisperse signals. Very few reports exist in the literature for proteins larger than 60 kDa, although promising recent work with heated inlets (termed prefolding dissociation) indicates that this limitation can be surpassed (35).

4. Notes

1. Urea (up to 2 M) can be included in the ion exchange mobile phases to enable dissociation of protein complexes if this is desirable. If this is the case, ensure that samples are not heated prior to the reversed-phase separation step as this could result in irreversible protein carbamylation.
2. An advantage of using two ion exchange columns (a strong anion exchanger and weak cation exchanger) in tandem is that highly acidic proteins which would not be retained on anion exchange material will interact with the cation exchanger, and a common mobile phase will effect protein elution from both columns.
3. At this stage a portion of the sample (10% should suffice) can readily be removed for conventional reduction, alkylation, proteolysis and LC-MS/MS analysis for additional identification confirmation purposes.

Acknowledgements

I would like to thank current and former members of the Michael Barber Centre, University of Manchester for constructive comments during this work. John Cottrell (Matrix Sciences) provided demonstration licensing of top-down Mascot. Ken Cook (Dionex) provided demonstration of HPLC columns and assisted in setting up gradients. Carsten Baessmann, Andrea Kiehne, Markus Lubeck, Andrea Schneider and Julia Smith (Bruker Daltonics) have provided essential guidance both with ETD/PTR experiments and with data processing.

References

1. Cravatt, B. F., Simon, G. M., and Yates, J. R., 3rd (2007) The biological impact of mass-spectrometry-based proteomics. *Nature* **450**, 991–1000.
2. Hart, S. R., and Gaskell, S. J. (2005) LC-tandem MS in proteome characterization. *TrAC Trends Anal. Chem.* **24**, 566–75.
3. Hart, S. R., and Gaskell, S. J. (2008) Methods of proteome analysis: challenges and opportunities. *SEB Exp. Biol. Ser.* **61**, 37–64.
4. Aebersold, R., and Mann, M. (2003) Mass spectrometry-based proteomics. *Nature* **422**, 198–207.
5. Li, G., Waltham, M., Anderson, N. L., Unsworth, E., Treston, A., and Weinstein, J. N. (1997) Rapid mass spectrometric identification of proteins from two-dimensional polyacrylamide gels after in gel proteolytic digestion. *Electrophoresis* **18**, 391–402.
6. Washburn, M. P., Wolters, D., and Yates, J. R., 3rd (2001) Large-scale analysis of the yeast proteome by multidimensional protein identification technology. *Nat. Biotechnol.* **19**, 242–247.
7. Yates, J. R., 3rd, Carmack, E., Hays, L., Link, A. J., and Eng, J. K. (1999) Automated protein identification using microcolumn liquid chromatography-tandem mass spectrometry. *Methods Mol. Biol.* **112**, 553–569.
8. Zhen, Y., Xu, N., Richardson, B., Becklin, R., Savage, J. R., Blake, K., et al. (2004) Development of an LC-MALDI method for the analysis of protein complexes. *J. Am. Soc. Mass Spectrom.* **15**, 803–822.
9. Siuti, N., and Kelleher, N. L. (2007) Decoding protein modifications using top-down mass spectrometry. *Nat. Methods* **4**, 817–21.
10. Kelleher, N. L. (2004) Top-down proteomics. *Anal. Chem.* **76**, 197A-203A.
11. Garcia, B. A., Shabanowitz, J., and Hunt, D. F. (2005) Analysis of protein phosphorylation by mass spectrometry. *Methods* **35**, 256–264.
12. Wiesner, J., Premsler, T., and Sickmann, A. (2008) Application of electron transfer dissociation (ETD) for the analysis of posttranslational modifications. *Proteomics* **8**, 4466–4483.
13. Thingholm, T. E., Jorgensen, T. J., Jensen, O. N., and Larsen, M. R. (2006) Highly selective enrichment of phosphorylated peptides using titanium dioxide. *Nat. Protoc.* **1**, 1929–1935.
14. Burlingame, A. L., Zhang, X., and Chalkley, R. J. (2005) Mass spectrometric analysis of histone posttranslational modifications. *Methods* **36**, 383–394.
15. Tanaka, K., Waki, H., Ido, Y., Akita, S., Yoshida, Y., Yoshida, T., Matsuo, T. (1988) Protein and polymer analyses up to *m/z* 100 000 by laser ionization time-of-flight mass spectrometry. *Rapid Commun. Mass Spectrom.* **2**, 151–153.
16. Fenn, J. B., Mann, M., Meng, C. K., Wong, S. F., and Whitehouse, C. M. (1989) Electrospray ionization for mass spectrometry of large biomolecules. *Science* **246**, 64–71.
17. Zubarev, R. A., Kelleher, N. L., and McLafferty, F. W. (1998) Electron capture dissociation of multiply charged protein cations. A nonergodic process. *J. Am. Chem. Soc.* **120**, 3265–3266.
18. Lifshitz, C. (2006) Intramolecular vibrational energy redistribution and ergodicity of biomolecular dissociation. In "Principles of mass spectrometry applied to biomolecules," Laskin, J., Lifshitz, Chava (Eds.). Wiley, Hoboken, New Jersey, pp. 239–75.
19. Syka, J. E., Coon, J. J., Schroeder, M. J., Shabanowitz, J., and Hunt, D. F. (2004) Peptide and protein sequence analysis by electron transfer dissociation mass spectrometry. *Proc. Natl. Acad. Sci. USA* **101**, 9528–33.
20. Coon, J. J., Ueberheide, B., Syka, J. E., Dryhurst, D. D., Ausio, J., Shabanowitz, J., et al. (2005) Protein identification using sequential ion/ion reactions and tandem mass spectrometry. *Proc. Natl. Acad. Sci. USA* **102**, 9463–9468.
21. Kaplan, D. A., Hartmer, R., Speir, J. P., Stoermer, C., Gumerov, D., Easterling, M. L., et al. (2008) Electron transfer dissociation in the hexapole collision cell of a hybrid quadrupole-hexapole Fourier transform ion cyclotron resonance mass spectrometer. *Rapid Commun. Mass Spectrom.* **22**, 271–278.
22. McAlister, G. C., Berggren, W. T., Griep-Raming, J., Horning, S., Makarov, A., Phanstiel, D., et al. (2008) A proteomics grade electron transfer dissociation-enabled hybrid linear ion trap-orbitrap mass spectrometer. *J. Proteome Res.* **7**, 3127–3136.
23. Hartmer, R., Lubeck, M., Bäßmann, C., Brekenfeld, A. (2007) New setup for top-down characterization of proteins via consecutive ion/ion reactions. *J. Biomol. Tech.* **18**, 34–35.
24. Anderson, N. L., and Anderson, N. G. (2002) The human plasma proteome: his-

tory, character, and diagnostic prospects. *Mol. Cell Proteomics* **1**, 845–867.

25. Liu, T., Qian, W. J., Gritsenko, M. A., Xiao, W., Moldawer, L. L., Kaushal, A., et al. (2006) High dynamic range characterization of the trauma patient plasma proteome. *Mol. Cell Proteomics* **5**, 1899–1913.
26. Westerlund, B. (2004) Ion–exchange chromatography. In "Purifying proteins for proteomics: a laboratory manual," Simpson, R. J., (Ed.). Cold Spring Harbor Laboratory Press, Cold Spring Harbor, NY, pp. 121–146
27. Gaskell, S. J. (1997) Electrospray: principles and practice. *J Mass Spectrom* **32**, 677–688.
28. McLafferty, F. W., Horn, D. M., Breuker, K., Ge, Y., Lewis, M. A., Cerda, B., Zubarev, R. A., and Carpenter, B. K. (2001) Electron capture dissociation of gaseous multiply charged ions by Fourier-transform ion cyclotron resonance. *J. Am. Soc. Mass Spectrom.* **12**, 245–249.
29. Samgina, T. Y., Artemenko, K. A., Gorshkov, V. A., Lebedev, A. T., Nielsen, M. L., Savitski, M. L., et al. (2007) Electrospray ionization tandem mass spectrometry sequencing of novel skin peptides from Ranid frogs containing disulfide bridges. *Eur. J. Mass Spectrom. (Chichester, Eng)* **13**, 155–163.
30. Lundgren, D. H., Han, D. K., and Eng, J. K. (2005) Protein identification using TurboSEQUEST. *Curr Protoc Bioinformatics.* **Chapter 13**, Unit 13 3.
31. Perkins, D. N., Pappin, D. J., Creasy, D. M., and Cottrell, J. S. (1999) Probability-based protein identification by searching sequence databases using mass spectrometry data. *Electrophoresis* **20**, 3551–3567.
32. Leduc, R. D., and Kelleher, N. L. (2007) Using ProSight PTM and related tools for targeted protein identification and characterization with high mass accuracy tandem MS data. *Curr. Protoc. Bioinformatics* **Chapter 13**, Unit 13 6.
33. Taylor, J. A., and Johnson, R. S. (1997) Sequence database searches via de novo peptide sequencing by tandem mass spectrometry. *Rapid Commun. Mass Spectrom.* **11**, 1067–1075.
34. Johnson, M., Zaretskaya, I., Raytselis, Y., Merezhuk, Y., McGinnis, S., and Madden, T. L. (2008) NCBI BLAST: a better web interface. *Nucleic Acids Res.* **36**, W5–W9.
35. Han, X., Jin, M., Breuker, K., and McLafferty, F. W. (2006) Extending top-down mass spectrometry to proteins with masses greater than 200 kilodaltons. *Science* **314**, 109–112.

Subject Index

P.R. Cutillas, J.F. Timms (eds.), *LC-MS/MS in Proteomics*, Methods in Molecular Biology 658,
DOI 10.1007/978-1-60761-780-8,

MIX
Papier aus verantwortungsvollen Quellen
Paper from responsible sources
FSC® C105338
FSC
www.fsc.org

If you have any concerns about our products,
you can contact us on
ProductSafety@springernature.com

In case Publisher is established outside the EU,
the EU authorized representative is:
Springer Nature Customer Service Center GmbH
Europaplatz 3, 69115 Heidelberg, Germany

Printed by Libri Plureos GmbH
in Hamburg, Germany